Mark S. Udevitz

FRESHWATER WETLANDS
Ecological Processes and Management Potential

ACADEMIC PRESS RAPID MANUSCRIPT REPRODUCTION

Proceedings of the Symposium
Freshwater Marshes: Present Status, Future Needs,
held in February 1977 at Rutgers University,
New Brunswick, New Jersey with support from the
U.S. Environmental Protection Agency,
Rutgers—The State University,
and Rider College

FRESHWATER WETLANDS
Ecological Processes and Management Potential

Edited by

Ralph E. Good

*Department of Biology, Rutgers University
Camden, New Jersey*

Dennis F. Whigham

*Chesapeake Bay Center for Environmental Studies
Smithsonian Institution
Edgewater, Maryland*

Robert L. Simpson

*Biology Department, Rider College
Lawrenceville, New Jersey*

Technical Editor
Crawford G. Jackson, Jr.
San Diego Natural History Museum
San Diego, California

ACADEMIC PRESS   New York   San Francisco   London   1978

A Subsidiary of Harcourt Brace Jovanovich, Publishers

COPYRIGHT © 1978, BY ACADEMIC PRESS, INC.
ALL RIGHTS RESERVED.
NO PART OF THIS PUBLICATION MAY BE REPRODUCED OR
TRANSMITTED IN ANY FORM OR BY ANY MEANS, ELECTRONIC
OR MECHANICAL, INCLUDING PHOTOCOPY, RECORDING, OR ANY
INFORMATION STORAGE AND RETRIEVAL SYSTEM, WITHOUT
PERMISSION IN WRITING FROM THE PUBLISHER.

ACADEMIC PRESS, INC.
111 Fifth Avenue, New York, New York 10003

United Kingdom Edition published by
ACADEMIC PRESS, INC. (LONDON) LTD.
24/28 Oval Road, London NW1 7DX

Library of Congress Cataloging in Publication Data

Main entry under title:

Freshwater wetlands.

    Proceedings of a conference held at Rutgers Univer-
sity, New Brunswick, N. J., Feb. 13-16, 1977.
    Includes bibliographical references and index.
    1. Wetland ecology—Congresses.   2.   Wetland
conservation—Congresses.   I.   Good, Ralph E.
II.   Whigham, Dennis F.   III.   Simpson, Robert Lee,
Date
QH541.5.M3F74   574.5'2632   78-2836
ISBN 0-12-290150-9

PRINTED IN THE UNITED STATES OF AMERICA

CONTENTS

List of Contributors   vii
Preface   xi
Foreword and Introduction   xiii

I PRIMARY PRODUCTION PROCESSES

Biomass and Primary Production in Freshwater Tidal
Wetlands of the Middle Atlantic Coast   3
Dennis F. Whigham, Jack McCormick, Ralph E. Good, and
Robert L. Simpson

Primary Production of Prairie Glacial Marshes   21
A. G. van der Valk and C. B. Davis

Life History Aspects of Primary Production in Sedge
Wetlands   39
John M. Bernard and Eville Gorham

Primary Production in Northern Bog Marshes   53
R. J. Reader

The Role of Hydrology in Freshwater Wetland
Ecosystems   63
J. G. Gosselink and R. E. Turner

Primary Production Processes: Summary and
Recommendations   79
Armando A. de la Cruz

II DECOMPOSITION PROCESSES

Decomposition of Intertidal Freshwater Marsh Plants   89
William E. Odum and Mary A. Heywood

Litter Decomposition in Prairie Glacial Marshes   99
C. B. Davis and A. G. van der Valk

Decomposition in Northern Wetlands   115
Jim P. M. Chamie and Curtis J. Richardson

Decomposition in the Littoral Zone of Lakes   131
Gordon L. Godshalk and Robert G. Wetzel

Decomposition Processes: Summary and
Recommendations   145
John L. Gallagher

III NUTRIENT DYNAMICS

Chemical Composition of Wetland Plants   155
  Claude E. Boyd

Nutrient Movements in Lakeshore Marshes   169
  R. T. Prentki, T. D. Gustafson, and M. S. Adams

**Nutrient Dynamics of Freshwater Riverine Marshes
and the Role of Emergent Macrophytes**   195
  Jeffrey M. Klopatek

Nutrient Dynamics of Northern Wetland Ecosystems   217
  Curtis J. Richardson, Donald L. Tilton, John A. Kadlec, Jim P. M.
  Chamie, and W. Alan Wentz

**Seasonal Patterns of Nutrient Movement in a Freshwater
Tidal Marsh**   243
  Robert L. Simpson, Dennis F. Whigham, and Raymond Walker

Nutrient Dynamics: Summary and Recommendations   259
  Ivan Valiela and John M. Teal

IV MANAGEMENT POTENTIAL

Management of Freshwater Marshes for Wildlife   267
  Milton W. Weller

**Tidal Freshwater Marsh Establishment in Upper
Chesapeake Bay:** *Pontederia cordata* **and** *Peltandra virginica* 285
  E. W. Garbisch, Jr. and L. B. Coleman

**Effects of Canals on Freshwater Marshes in Coastal
Louisiana and Implications for Management**   299
  James H. Stone, Leonard M. Bahr, Jr., and John W. Day, Jr.

**Management of Freshwater Wetlands for Nutrient
Assimilation**   321
  William E. Sloey, Frederic L. Spangler, and C. W. Fetter, Jr.

Ecology and the Regulation of Freshwater Wetlands   341
  Jack McCormick

Management Potential: Summary and Recommendations   357
  Forest Stearns

Index   *365*

LIST OF CONTRIBUTORS

Numbers in parentheses indicate the pages on which authors' contributions begin.

M. S. Adams (169), Institute for Environmental Studies and Botany Department, University of Wisconsin–Madison, Madison, Wisconsin 53706

Leonard M. Bahr, Jr. (299), Department of Marine Sciences, Center for Wetland Resources, Louisiana State University, Baton Rouge, Louisiana 70803

John M. Bernard (39), Department of Biology, Ithaca College, Ithaca, New York 14850

Claude E. Boyd (155), Department of Fisheries and Allied Aquacultures, Auburn University Agricultural Experiment Station, Auburn, Alabama 36830

Jim P. M. Chamie (115, 217), Botany–Biology Department, South Dakota State University, Brookings, South Dakota 57007

L. B. Coleman (285), Environmental Concern Inc., P.O. Box P, St. Michaels, Maryland 21663

Armando A. de la Cruz (79), P.O. Drawer Z, Mississippi State University, Mississippi State, Mississippi 38762

C. B. Davis (21, 99), Department of Botany and Plant Pathology, Iowa State University, Ames, Iowa 50011

John W. Day, Jr. (299), Department of Marine Sciences, Center for Wetland Resources, Louisiana State University, Baton Rouge, Louisiana 70803

C. W. Fetter, Jr. (321), Geology Department, University of Wisconsin–Oshkosh, Oshkosh, Wisconsin 54901

John L. Gallagher (145),[1] Marine Institute, The University of Georgia, Sapelo Island, Georgia 31327

[1] Present address: U.S. Environmental Protection Agency, Corvallis Environmental Research Laboratory, 200 S.W. 35th Street, Corvallis, Oregon 97330.

E. W. Garbisch, Jr. (285), Environmental Concern Inc., P.O. Box P, St. Michaels, Maryland 21663

Gordon L. Godshalk (131), W. K. Kellogg Biological Station, Michigan State University, Hickory Corners, Michigan 49060

Ralph E. Good (3), Department of Biology, Rutgers University, Camden, New Jersey 08102

Eville Gorham (39), Department of Ecology and Behavioral Biology, University of Minnesota, Minneapolis, Minnesota 55455

J. G. Gosselink (63), Center for Wetland Resources, Louisiana State University, Baton Rouge, Louisiana 70803

T. D. Gustafson (169),[2] Institute for Environmental Studies and Botany Department, University of Wisconsin—Madison, Madison, Wisconsin 53706

Mary A. Heywood (89), Department of Environmental Sciences, University of Virginia, Charlottesville, Virginia 22903

John A. Kadlec (217), Department of Wildlife Sciences, Utah State University, Logan, Utah 84321

Jeffrey M. Klopatek (195), Environmental Sciences Division, Oak Ridge National Laboratory, Oak Ridge, Tennessee 37820

Jack McCormick (3, 341), WAPORA, Inc., 6900 Wisconsin Ave., N.W., Washington, D.C. 20015

William E. Odum (89), Department of Environmental Sciences, University of Virginia, Charlottesville, Virginia 22903

R. T. Prentki (169), Institute for Environmental Studies and Botany Department, University of Wisconsin–Madison, Madison, Wisconsin 53706

R. J. Reader (53), Department of Botany and Genetics, University of Guelph, Guelph, Ontario, Canada N1G 2W1

Curtis J. Richardson (115, 217),[3] School of Natural Resources, The University of Michigan, Ann Arbor, Michigan 48109

Robert L. Simpson (3, 243), Biology Department, Rider College, Lawrenceville, New Jersey 08648

William E. Sloey (321), Biology Department, University of Wisconsin–Oshkosh, Oshkosh, Wisconsin 54901

Frederic L. Spangler (321), Biology Department, University of Wisconsin–Oshkosh, Oshkosh, Wisconsin 54901

Forest Stearns (357), Department of Botany, University of Wisconsin–Milwaukee, Milwaukee, Wisconsin 53201

James H. Stone (299), Department of Marine Sciences, Center for Wetland Resources, Louisiana State University, Baton Rouge, Louisiana 70803

[2] Present address: Department of Biology, Juniata College, Huntingdon, Pennsylvania 16652.

[3] Present address: School of Forestry and Environmental Studies, Duke University, Durham, North Carolina 27706.

John M. Teal (259), Woods Hole Oceanographic Institution, Woods Hole, Massachusetts 02543

Donald L. Tilton (217), Rockefeller Fellow in Environmental Affairs, The University of Michigan, Ann Arbor, Michigan 48109

R. E. Turner (63), Center for Wetland Resources, Louisiana State University, Baton Rouge, Louisiana 70803

Ivan Valiela (259), Boston University Marine Program, Marine Biological Laboratory, Woods Hole, Massachusetts 02543

A. G. van der Valk (21, 99), Department of Botany and Plant Pathology, Iowa State University, Ames, Iowa 50011

Raymond Walker (243), Biology Department, Rutgers University, Camden, New Jersey 08102

Milton W. Weller (267), Department of Entomology, Fisheries, and Wildlife, University of Minnesota, St. Paul, Minnesota 55108

W. Alan Wentz (217), Department of Wildlife Sciences, South Dakota State University, Brookings, South Dakota 57007

Robert G. Wetzel (xiii, 131), W. K. Kellogg Biological Station, Michigan State University, Hickory Corners, Michigan 49060

Dennis F. Whigham (3, 243),[4] Biology Department, Rider College, Lawrenceville, New Jersey 08648

[4] Present address: Chesapeake Bay Center for Environmental Studies, Smithsonian Institution, Route 4, Box 622, Edgewater, Maryland 21037.

PREFACE

This volume resulted from a conference entitled "Freshwater Marshes: Present Status, Future Needs" held 13–16 February 1977 at Rutgers—The State University of New Jersey, New Brunswick, New Jersey.

The idea for the New Brunswick conference originated during an Ecological Society of America (Aquatic Ecology Section) sponsored symposium entitled "Production Ecology of Freshwater Tidal and Nontidal Marshes" held during the 1976 American Institute of Biological Sciences annual meeting in New Orleans. Participants at that meeting believed that additional ecological aspects of freshwater wetlands should be considered and encouraged the editors of this volume to seek funding for an expanded conference. With that charge, the editors prepared a proposal to hold a major conference focusing on ecological processes and management potential of freshwater wetlands. The need for a compilation of data on freshwater wetlands was recognized by the Environmental Protection Agency, resulting in funding from their Monitoring Technology Division to Rutgers University (Grant R805090-01-0). Additional financial assistance for the conference and volume was provided by Rider College and Rutgers University. The New Brunswick conference brought together 40 wetland ecologists from the United States and Canada to discuss the current state of knowledge and the future needs of research on freshwater wetlands dominated by herbaceous vegetation.

The conference was divided into sessions on primary production, decomposition, mineral cycling, and management potential with a moderator in charge of each session. The sessions contained papers on freshwater tidal wetlands, littoral wetlands of rivers, streams, lakes, and ponds, prairie glacial pothole wetlands, and northern bog wetlands. The final day of the conference was devoted to summary papers presented by the session moderators (Armando de la Cruz, John Gallagher, Ivan Valiela, John Teal, and Forest Stearns) with an overview of the conference by Robert Wetzel.

This volume similarly organized attempts to summarize our present knowledge of the ecology of freshwater wetlands. Whereas we recognize that none of the papers present the final answers, we believe that they will provide for a more adequate framework for wetland management and will serve to stimulate much needed research on freshwater wetlands.

We express our sincere appreciation to all who have made the volume possible. Special thanks go to the many individuals who served as peer reviewers and to Crawford G. Jackson, Jr., our Technical Editor. We are also indebted to Mr. Vernon Laurie, Conference Coordinator, Monitoring Technology Division of the U.S. Environmental Protection Agency, who is environmental liason to the Department of State UNESCO Man in the Biosphere Program (MAB). We also thank the staff of the Continuing Education Center—Rutgers University for its splendid handling of the conference. Appreciation is due Mr. John Cooney of the President's Office and Dr. Norbert Psuty (Director, Center for Coastal and Environmental Studies) for their financial support of this undertaking. Also Dr. Paul Pearson, Dr. Benjamin Stout, and Dr. David Pramer, Rutgers University administrators, were kind enough to serve as conference hosts. The editors thank Raymond Walker and Herbert Grover for their technical support during the symposium. The efforts of Academic Press personnel are gratefully acknowledged. Support from the Office of Research and Sponsored Programs at Rutgers University allowed for typesetting the volume in cooperation with Allen Press, Inc. Finally, we extend special thanks to our wives Norma Good, Janice Whigham, and Penelope Simpson for their patience and support during this endeavor, which started as an idea in the spring of 1975.

FOREWORD AND INTRODUCTION

Freshwater wetlands mean different things to many persons. To some, the vitality and beauty of marshes can only be described in words, as Errington has done so poignantly.[1] At the opposing extreme, wetlands represent to some a wasteland of unused space that can be exploited via drainage or as a partial panacea as the possible receptacle of human waste products.

To the limnologist, freshwater wetlands are complex systems and generally represent an extremely important resource. Definitions vary markedly from marshes of glacial origin, to prairie wetlands, to freshwater marshes of estuarine areas that experience severe fluctuations in water movement under tidal influences. In glaciated regions, marshes are often remnant wetlands of shallow lake systems in which macrovegetation, largely emergent, extends over the entire water surface. Technically a swamp contains persistent standing water among the vegetation, whereas marshes contain water-saturated sediments with no or little standing water among the vegetation.[2] These definitions, however, are commonly useful only in detailed analyses of successional changes of wetland conditions and biota. As will be seen in the ensuing compendium, the conditions of wetlands change markedly and rapidly in response to fluctuations in climate and precipitation. Many responses are cyclical in long-term periodicity.

As one analyzes the following reviews and contributions to our understanding of the operation of freshwater wetlands, areas of ignorance emerge everywhere. Primary productivity is probably the best understood of all parameters in wetland systems. Yet estimates of growth of the major root–rhizome systems are inadequately evaluated. The most critical determination of annual turnover times of foliar and root biomass is essential, regardless of the tedium and difficulties. Both problems are surmountable and accurate assessments of productivity can be made.

The mechanisms and couplings of metabolism of detrital dissolved and

---

[1] Errington, P. L. (1957). "Of Men and Marshes." Iowa State Univ. Press, Ames.
[2] Wetzel, R. G. (1975). "Limnology." W. B. Saunders, Philadelphia.

particulate organic matter provide a fundamental flexibility and stability to the dynamics of wetland metabolism and productivity.[2] Particulate and dissolved organic pools have different metabolic mobilities. A small pool of labile, readily available organic matter is strongly coupled to active biotic metabolism, especially photosynthesis. A much larger pool, some 90% or more of the total, consists of much more refractory compounds, largely of allochthonous and wetland macrophytic plant sources. The rates of decomposition are governed by molecular structure, concentration of substrates, and enzyme availability and competition, as well as physical factors such as temperature and O_2. The relatively slow rates of utilization of the more resistant detrital reservoir give stability to the system, tying it over during periods of low detrital inputs and recharging the reserve during excessive photosynthetic inputs. It is essential that these control mechanisms be better elucidated in marsh systems.

The data on nutrient cycling presented in the following papers are an improvement on what was available previously for freshwater wetlands. Emphasis is on measurements of chemical mass and change in these nutrient concentrations with time from one system component (leaves, rhizomes, etc.) to another. In other words, the chemical pools are being identified. Kinetic studies of nutrient flux rates are essential in all future studies, particularly employing modern techniques such as radioisotope tracer methods. Much greater coupling is needed of nutrient dynamics to the physiology of the marsh flora and rates of microbial metabolism in the rhizosphere. These physiological relationships cannot be neglected for a comprehensive understanding of wetland metabolism and responses of these systems to perturbations of environmental change.

Productivity measurements must be made with sufficient accuracy to ensure that error is less than variations in photosynthetic responses resulting from environmental conditions. The highly variable biotic and environmental gradients within marsh systems impose a formidable heterogeneity that makes accurate measurements difficult. One must, however, reject the implications that accurate evaluations of *in situ* rates of photosynthesis, respiration, and photorespiration are impractical. The technological problems are quite solvable with application of necessary experimental resources. We have reached the point where these questions must be addressed through modern technology and expertise.

By far the weakest component in understanding how a wetland operates focuses on a knowledge of decomposition. In very general terms, we know enough in this area to evaluate the more likely important qualitative pathways among decomposition and nutrient components of marsh systems. A consideration of the quantitative dynamics and controlling mechanisms of decomposition, however, reveals a nearly total ignorance.

The primitive understanding of our knowledge of the decompositional processes has several important consequences that must be recognized.

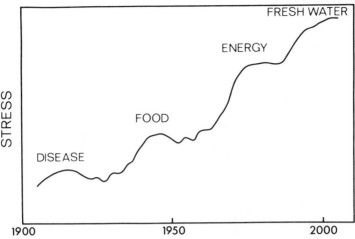

Fig. 1.   Generalized peaks and partial stabilization of human stress experienced in response to limiting resources in the twentieth century (see text).

Analysis of loss of organic matter and nutrients from litterbags concerns only the particulate fractions in an initial approximation of gross degradation processes and ignores the major microflora decomposition of soluble organic compounds. Similarly, gross chemical analyses, such as nitrogen or fiber content, indicate certain general trends but are of limited value. Measurements of community respiratory metabolism can indicate some responses of the decompositional system to perturbations but yield little insight into regulatory mechanisms involved. It is apparent that detailed physiological and biochemical approaches are needed in relation to environmental controls, e.g., toxicity of metabolic end products and their direct effects on nutrient mobility and assimilation rates by the plants.

It is easy to address obvious voids in our understanding of marsh systems. One must, however, view with great concern the ease with which engineers and others view wetlands as potential sites for effluent treatment and other exploitation. Hopefully this view is not shared by biologists who realize our ignorance of the operation of natural wetland systems, let alone those undergoing artificially induced stress. Relatively little attention of the symposium is directed toward the value of freshwater wetland hydrology and groundwater recharge on a regional basis. Little is known of the effects of, for example, long-term manipulation of nutrient influxes on permanent changes in plant species composition, primary productivity, and detrital metabolism.

We are experiencing a destructive loss of freshwater marshes in the United States of from 0.5 to 1% per year through drainage and other forms of exploitation (J. McCormick, *personal communication*). This loss is indeed cause for major alarm. Since the beginning of the twentieth century, we have experienced a number of acute stresses in response to resource exploitation with expanding human population. Soon after the turn of the century, human disease prevailed, was gradually confronted, and the stress stabilized (Fig. 1). Subsequently crises associated with food limitations were eased with massive deployment of pesticides and improved agricultural practices. We are now in a crisis of severe stress from energy limitations. By necessity alone, this stress should ease to tolerable levels within the next few decades through means of strict energy conservation and use of alternatively available resources.

Toward the end of the present century we shall probably encounter an acute stress for usable freshwater. Contemporary needs for freshwater resources exceed availability in a number of large regions: the Northeast, Florida, and the Southwest. The reasons for the stress and impending increasingly acute demands are many: (1) excessive use, particularly for industrial purposes, (2) unabated degradation of water quality and management, and especially (3) the distribution of population density and demand in areas very distant from Midwestern and Northwestern concentrations of freshwater.[2,3] Methods of counteracting demands in relation to available freshwater sources by reasonable means of acquisition will require a significant alteration of perspective in the future. While population increase is a variable factor, it is a physiological fact that the population, and associated freshwater demands, will more than double within the next half-century before probable stabilization. Geographical distribution to regions where water is abundant will certainly occur to a greater extent than has been the case in the past.

Dependence upon groundwater reservoirs will also increase in all areas. To be utilized effectively within fluctuating climatic replenishment, these reservoirs must be refilled in a systematic manner from surface catchment sources (cf., for example, Ambroggi[4]). Wetlands play an important role in such replenishment. Decisions to perturb violently or to destroy wetlands, as with nutrients, organic matter, or heavy metals, should be made by biologists backed with extensive research knowledge on the systems and not by exploitative interest groups to satisfy short-term needs. We have this obligation to future generations.

Communication is fundamental in recognition of the importance of freshwater wetland systems. Reciprocal understanding is needed among biologists, engineers, and hydrologists, in addition to officials that control finan-

---

[3] Vallentyne, J. R. (1972). Freshwater supplies and pollution: Effects on the demophoric explosion on water and man, *In* "The Environmental Future" (N. Polunin, ed.), pp. 181–211. Macmillan, London.

[4] Ambroggi, R. P. (1977). Underground reservoirs to control the water cycle. *Sci. Am.* 236, 21–27.

cial support of urgently required research. The ensuing proceedings represent a small step in that direction——hopefully an embryonic seed that will flourish.

Robert G. Wetzel
Michigan State University

PART I

Primary Production Processes

BIOMASS AND PRIMARY PRODUCTION IN FRESHWATER TIDAL WETLANDS OF THE MIDDLE ATLANTIC COAST

Dennis F. Whigham[1]

Biology Department, Rider College
Lawrenceville, New Jersey 08648

Jack McCormick

WAPORA, Inc., 6900 Wisconsin Ave., N.W.
Washington, D.C. 20015

Ralph E. Good

Department of Biology, Rutgers University
Camden, New Jersey 08102

and

Robert L. Simpson

Department of Biology, Rider College
Lawrenceville, New Jersey 08648

Abstract   Although there are many measurements of peak standing crop and several estimates of aboveground annual net production, there are few accurate estimates of total net primary production for Middle Atlantic coastal freshwater tidal wetlands. Estimates of biomass and production vary widely both within and between vegetation types. The variability appears to be due to sampling techniques and/or the heterogeneous nature of freshwater tidal wetland

[1] Present address: Chesapeake Bay Center for Environmental Studies, Smithsonian Institution, Route 4, Box 622, Edgewater, Maryland 21037.

vegetation. Undoubtedly, all estimates of primary production are low because they do not include data on growing season mortality of plants and plant parts, herbivore consumption, and belowground production. Most estimates of primary production are also low because freshwater tidal wetland communities usually undergo a series of physiognomic changes during the growing season and the measurements of primary production do not account for those seasonal patterns. Comparatively, freshwater macrophyte production in tidal wetlands appears to be at least equal to that of brackish water wetlands at the same geographic latitude.

Key words   *Biomass, freshwater, Middle Atlantic coast, productivity, tidal, wetland.*

INTRODUCTION

Three major types of tidal wetlands form a continuum within the Middle Atlantic coastal region. At one extreme of the continuum are saline wetlands that are characterized by the presence of extensive stands of salt-marsh cordgrass (*Spartina alterniflora*). At the other extreme, freshwater wetlands occur near the inland limit of the tide in areas that are never exposed to water of elevated salinity. Species diversity is greatest at the freshwater end of the continuum with wild rice (*Zizania aquatica*) and spatterdock (*Nuphar advena*) being characteristic species. Brackish wetlands occupy the areas where fresh and salt water mix and are characteristically dominated by several species. Salt-meadow cordgrass (*Spartina patens*) and Olney three-square (*Scirpus olneyi*) frequently form the bulk of plant cover. Such species as cattail (*Typha* sp.) and marsh mallow (*Hibiscus* sp.) may be prominent either in brackish or freshwater wetlands but distinctions between them are often made based on the presence of big cordgrass (*Spartina cynosuroides*) which may be conspicuous along tidal channels of brackish wetlands.

The primary production of tidal wetlands has been purported to be very high (Odum, 1971; Westlake, 1963) and Kirby and Gosselink (1976) have recently demonstrated that saline and brackish wetland primary production may be even higher than previously suspected. With the exception of a summary by Keefe (1972) and a brief review in Whigham and Simpson (1976*a*), there has been no systematic attempt to compile and analyze data on primary production of freshwater tidal wetlands. This paper summarizes data on biomass and primary production of freshwater tidal wetlands and makes comparisons with production estimates for saline and brackish wetlands within the Middle Atlantic coastal zone.

Kirby and Gosselink (1976) have recently shown that standing crop measurements may be significant underestimates of actual annual production of Gulf Coast saline wetlands. We herein present data showing that most published estimates of production in freshwater tidal wetlands are underestimates of the actual annual net production and that the differences may be very large. Most investigations underestimate actual net production for several reasons. For example, the single harvest method of estimating annual net production has been frequently used in the past, but is not suitable for most freshwater tidal wetlands. Most sites are dominated by several species during one growing season and the single harvest not only misses senesced vegetation but it does not include production of species that may dominate later. Our recent estimates of belowground production and leaf mortality show that these components can also account for significant amounts of the total annual net production. Therefore we provide estimates of total annual net production (for 1 vegetation type), that were derived from peak biomass and multiple harvest techniques plus measurements of leaf mortality and belowground production.

ESTIMATES OF BIOMASS AND PRODUCTION

Published and unpublished data on biomass and primary production were available for freshwater tidal wetlands in Maryland, Virginia, New Jersey and Pennsylvania (Table 1). For comparison, we have also incorporated data on brackish wetlands into Table 1.

The standing crop and/or production estimates were obtained primarily from Delaware River, Delaware Bay or Chesapeake Bay freshwater tidal wetlands, mostly since 1970. All data were obtained by using 2 techniques. The single harvest method entails measuring peak biomass during the period when standing crops are thought to be maximum. There is an inherent variability and potential error in the method because there are no corrections for: (1) plant tissues that developed and died prior to the sampling time, (2) herbivory, and (3) tissues that will develop after the harvest. The multiple harvest technique, involving data collected at intervals of several days to several weeks throughout the growing season, yields better estimates of total annual net production; however, most investigations utilizing this technique have been in brackish and saline wetlands (Gosselink et al., 1977; Kirby and Gosselink, 1976; Smalley, 1959; Stroud and Cooper, 1968; Williams and Murdoch, 1969). Gosselink et al. (1977) measured production of a Gulf Coast freshwater tidal wetland species using multiple harvest techniques which also included estimates of mortality and leaf turnover.

Estimates of average peak aboveground biomass in freshwater tidal wetlands range from 432 g/m² in *Sagittaria latifolia* vegetation to 2,311 g/m² in *Spartina cynosuroides* vegetation (Table 1). Peak biomass is almost always greatest in wetlands dominated by grasses such as *S. cynosuroides* and *Phragmites* (1,850 g/m²), or else vegetation dominated by semi-shrubs such as *Hibiscus* (1,714 g/m²) and *Lythrum* (1,616 g/m²). Average peak aboveground biomass is slightly less in wetlands dominated by *Typha* (1,215 g/m²), *Zizania* (1,218 g/m²), *Ambrosia* (1,205 g/m²), *Bidens* (1,017 g/m²), *Acnida* (940 g/m²), and *Acorus* (857 g/m²). Average peak standing crop is least in stands dominated by typical emergent macrophytes such as *Pontederia/Peltandra* (682 g/m²), *Nuphar* (627 g/m²), *Scirpus* (606 g/m²), and *Sagittaria*

Table 1. Peak Standing Crop and Net Annual Production Estimates for Freshwater (Part 1) and Brackish (Part 2) Tidal Wetlands Within the Middle Atlantic Coastal Region[a]

Vegetation type[b]	Peak standing crop[c]			Annual production[c]	Source
	Above-ground	Below-ground	Dead		
Part 1—Freshwater tidal wetlands					
Polygonum/ Leersia	2,142 [2]	...	...	...	J. McCormick (*personal observations*)
	1,547	...	...	...	Wass and Wright (1969)
	769	507	...	...	Good and Good (1975)
	523	...	...	...	McCormick (1970)
	1,425				
Nuphar advena	513 [2]	...	...	...	J. McCormick (*personal observations*)
	245	...	...	...	Wass and Wright (1969)
	743 [2]	...	...	863 [2]	J. McCormick (*personal observations*)
	516	...	...	...	McCormick and Ashbaugh (1972)
	605	1,146	...	...	Good and Good (1975)
	529 [6]	...	...	...	Whigham and Simpson (1975)
	...	4,799	...	780	Whigham and Simpson (1975)
	1,175 [2]	...	...	...	McCormick (1970)
	627				
Pontederia/ Peltandra	648 [2]	...	...	...	J. McCormick (*personal observations*)
	988	...	132	...	Flemer et al. (*In press*)
	1,286	2,463	...	...	Good et al. (1975)
	594 [2]	...	...	...	McCormick and Ashbaugh (1972)
	486 [2]	...	...	...	Whigham and Simpson (1975)
	553	...	...	...	Good and Good (1975)
	657 [3]	...	...	650	Whigham and Simpson (1975)
	677	...	...	1,126	J. McCormick (*personal observations*)
	686			**888**	
Acorus calamus	1,174 [2]	...	...	...	J. McCormick (*personal observations*)
	605	...	...	...	McCormick and Ashbaugh (1972)
	819 [4]	...	...	...	Whigham and Simpson (1975)
	623	...	...	1,071	J. McCormick (*personal observations*)
	857				

Table 1. Continued

| Vegetation type[b] | Peak standing crop[c] | | | Annual produc-tion[c] | Source |
	Above-ground	Below-ground	Dead		
Typha sp.	2,338	...	167	...	Heinle et al. (1974)
	1,190 [2]	...	330 [2]	...	Flemer et al. (*In press*)
	966	...	...	1,868	Johnson (1970)
	...	...	...	956[d]	Stevenson et al. (1976)
	987	...	...	...	McCormick and Ashbaugh (1972)
	850	1,800	...	...	Good and Good (1975)
	894	1,371	...	...	Good et al. (1975)
	1,297 [3]	...	...	1,320	Whigham and Simpson (1975)
	1,199	...	...	1,534	J. McCormick (*personal observations*)
	1,310 [3]	...	...	...	McCormick (1970)
	804	5,053	...	...	Walker and Good (1976)
	1,215			**1,420**	
Hibiscus palustris	1,714 [2]	...	...	...	J. McCormick (*personal observations*)
	...	...	...	489[d]	Stevenson et al. (1976)
	1,714				
Zizania aquatica	2,091 [2]	...	...	...	J. McCormick (*personal observations*)
	1,178 [4]	...	135 [3]	...	Flemer et al. (*In press*)
	560	...	...	...	Wass and Wright (1969)
	1,390	...	...	...	McCormick and Ashbaugh (1972)
	1,600	721	...	...	Good and Good (1975)
	866 [4]	...	...	1,589 [5]	Whigham and Simpson (1975)
	1,346	...	...	1,520	J. McCormick (*personal observations*)
	1,117 [2]	...	...	...	McCormick (1970)
	1,218			**1,578**	
Spartina cynosuroides	3,543[e] [2]	...	...	...	J. McCormick (*personal observations*)
	951	...	241	...	Flemer et al. (*In press*)
	1,207	...	...	1,572	Johnson (1970)
	2,311				
Phragmites communis	3,999 [2]	...	...	...	J. McCormick (*personal observations*)
	1,367 [3]	...	347 [3]	...	Flemer et al. (*In press*)
	1,451	...	...	1,678	Johnson (1970)
	1,727	...	...	...	McCormick and Ashbaugh (1972)
	1,493	...	...	2,066	J. McCormick (*personal observations*)

Table 1. Continued

Vegetation type[b]	Peak standing crop[c]			Annual produc- tion[c]	Source
	Above- ground	Below- ground	Dead		
	654	...	...	...	McCormick (1970)
	1,074	7,180	...	...	Walker and Good (1976)
	1,850			**1,872**	
Acnida cannabina	1,112	...	...	1,547	J. McCormick (*personal observations*)
	768	560	...	...	Good et al. (1975)
	940				
Bidens sp.	1,026 [3]	...	...	910	Whigham and Simpson (1975)
	1,109	...	...	1,771	J. McCormick (*personal observations*)
	900	...	...	...	McCormick (1970)
	1,017			**1,340**	
Phalaris arundinacea	566	...	...	...	Whigham and Simpson (1975)
Lythrum salicaria	2,104	...	...	2,100	Whigham and Simpson (1975)
	1,373 [2]	...	...	...	McCormick (1970)
	1,616				
Ambrosia trifida	1,160	...	...	1,160	Whigham and Simpson (1975)
	1,227 [2]	...	...	...	McCormick (1970)
	1,205				
Sagittaria latifolia	649	...	...	1,071	J. McCormick (*personal observations*)
	214	...	...	...	Good et al. (1975)
	432				
Part 2—Brackish tidal wetlands					
Spartina patens/ Distichlis spicata	1,123 [3]	...	...	...	J. McCormick (*personal observations*)
	449 [3]	...	...	...	Heinle (1972)
	680	...	1,209	...	Flemer et al. (*In press*)
	1,525 [2]	...	...	...	Jack McCormick and Associates, Inc. (1973)
	480	...	...	572	Mendelssohn and Marcellus (1976)
	445	...	...	...	Drake (1976)
	1,145 [2]	...	...	...	de la Cruz (1973)
	897				
Iva/Baccharis/ Spartina patens	1,075 [2]	...	...	...	J. McCormick (*personal observations*)
	534	...	...	...	Drake (1976)
	895				

Table 1. Continued

Vegetation type[b]	Peak standing crop[c]			Annual production[c]	Source
	Above-ground	Below-ground	Dead		
Juncus roemerianus	1,602 [2]	...	...	...	J. McCormick (*personal observations*)
	1,082 [3]	...	...	...	Heinle (1972)
	1,290				
Typha sp.	1,453 [2]	...	...	...	J. McCormick (*personal observations*)
	1,668 [3]	...	698 [3]	...	Flemer et al. (*In press*)
	1,435	...	...	...	Jack McCormick and Associates, Inc. (1973)
	919	...	...	...	Wass and Wright (1969)
	626	...	...	...	Drake (1976)
	1,361				
Hibiscus palustris	**1,354** [2]	...	...	...	J. McCormick (*personal observations*)
Panicum virgatum	4,029 [2]	...	...	...	J. McCormick (*personal observations*)
	652	...	...	...	Anderson et al. (1968)
	326	...	...	...	Drake (1976)
	2,259				
Scirpus sp.	802 [2]	...	...	...	J. McCormick (*personal observations*)
	366 [2]	...	...	...	Anderson et al. (1968)
	833 [3]	...	295 [3]	...	Flemer et al. (*In press*)
	561	...	...	...	Wass and Wright (1969)
	472	...	...	...	Drake (1976)
	193	...	...	...	Good (1965)
	606				
Spartina cynosuroides	968 [2]	...	...	...	Anderson et al. (1968)
	1,900 [2]	...	685 [2]	...	Flemer et al. (*In press*)
	1,401 [3]	...	...	...	Wass and Wright (1969)
	560	...	...	...	Mendelssohn and Marcellus (1976)
	672	...	...	563	Drake (1976)
	826 [4]	...	349 [4]	1,175 [4]	Odum and Fanning (1973)
	1,113			**1,053**	
Spartina alterniflora	1,003 [2]	...	...	...	J. McCormick (*personal observations*)
	1,020 [2]	...	...	...	Jack McCormick and Associates, Inc. (1973)
	2,410	...	...	...	Wass and Wright (1969)
	725 [4]	6,353 [4]	...	3,500	R. E. Good and R. Walker (*personal observations*)
	587	...	...	...	Drake (1976)

Table 1. Continued

Vegetation type[b]	Peak standing crop[c]			Annual production[c]	Source
	Above-ground	Below-ground	Dead		
	1,184	...	...	...	G. T. Potera and E. E. MacNamara (*personal observations*)
	563 [2]	...	...	...	Jack McCormick and Associates, Inc. (1974)
	943				

[a] The number of estimates used to calculate the entries is shown in brackets. All estimates were used to calculate the averages shown in boldface.

[b] The classification system used in construction of this table was provided by Jack McCormick and is available from him.

[c] All data are expressed as grams per square metre dry weight.

[d] Stevenson et al. (1976) data were converted to grams per square metre using 18.83 kJ/g as a conversion factor.

[e] These are adjusted estimates; the weights of all harvested materials in the 2 samples were 5,415 and 5,875 g/m².

(432 g/m²). This sequence must be viewed with caution because of the large amount of variability within and between vegetation types (Table 1). Some variation is undoubtedly due to sampling techniques, but much variation may be an expression of the highly variable spatial and species composition of freshwater tidal wetland vegetation. Assuming that biomass and production are related, data in Table 1 fall toward the low range of expected values of net production in freshwater wetlands (Lieth and Whittaker, 1976) but correspond to primary production and standing crop measurements made by Jervis (1969) in a New Jersey nontidal freshwater wetland.

Estimates of annual net production are higher than biomass estimates for the same vegetation types but there have been few estimates of annual production and the available data suggest that there is much variation within and between vegetation types (Table 1). The lowest reported estimates are for *Pontederia/Peltandra*-dominated vegetation (650 g/m²) and the highest reported values are for *Zizania* stands in New Jersey ($2,321$ g/m²). The latter estimate included root production. The range of average peak aboveground standing crop estimates in freshwater tidal wetlands (566–$2,311$ g/m²) is as high as that measured in brackish wetlands (216–$2,270$ g/m²) and, in most cases, biomass was greatest in those portions of brackish wetlands where freshwater tidal wetland species occur (Table 1). Comparison of data presented in Table 1 with data on biomass and production of saline wetlands (Keefe, 1972; Turner, 1976) reveals that aboveground biomass in freshwater tidal wetlands is greater than that measured in saline wetlands at the same geographic latitudes. *Juncus roemerianus* and *Spartina alterniflora* (tall form) wetlands had the highest average peak standing crops (1,160 and 958 g/m², respectively) while *Spartina patens/Distichlis spicata* and *Spartina alterniflora* (short form) wetlands had the lowest average peak standing crops (341–216 g/m², respectively). Estimated annual net primary production ranged from 1,036 g/m² in tall form *Spartina alterniflora* wetlands to 365 g/m² in short form *S. alterniflora* wetlands. The range of values reported is very wide and there are still not enough reliable data to document clearly the differences between net annual production of the 3 types of tidal wetlands.

It would also be interesting to know the comparative relationship between macrophyte production in tidal and nontidal freshwater wetlands because Turner (1976) has recently suggested that tidal subsidy may not be an important factor in determining production of saline wetlands. Based on macrophyte production data in other chapters of this volume and in Keefe (1972) and Whigham and Simpson (1976a), it is impossible to make reliable comparisons between analogous communities in tidal and nontidal freshwater wetlands because the production estimates for both types of wetlands are quite variable. It is also difficult to make any definite conclusions when comparisons are made of production estimates of individual species. For example, it appears that Delaware River tidal populations of *Zizania aquatica* var. *aquatica* are more productive than Minnesota nontidal populations but the pattern does not always hold because there are no differences between tidal and nontidal populations in New Jersey (Whigham and Simpson, 1977). Similarly, *Typha* sp. production has been measured many times but there are no apparent patterns when comparisons are made between tidal and nontidal populations (Whigham and Simpson, 1976a). Only further research will provide an answer to the question of whether or not factors associated with tidal activity cause the net community production of freshwater tidal wetlands to be greater than their nontidal counterparts.

FACTORS RESPONSIBLE FOR
UNDERESTIMATION OF PRODUCTIVITY

Jervis (1969) and Auclair et al. (1976) have documented the existence of pronounced seasonal changes in dominance in nontidal freshwater wetlands. Data from Whigham and Simpson (1975, 1976a, 1976b) in the Hamilton Marshes clearly demonstrate that there are seasonal patterns of accumulation of biomass in freshwater tidal wetlands. Seasonal patterns are also reported in McCormick (1970), McCormick and Ashbaugh (1972) and Good and Good (1975). Data from Whigham and Simpson are used to demonstrate 4 patterns of biomass accumulation (Fig. 1). Distinct seasonal variations in the distribution of aboveground biomass within most vegetation types are also illustrated (Fig. 2). In general, physiognomic changes are associated with early season dominance by perennials, senescence of the perennials

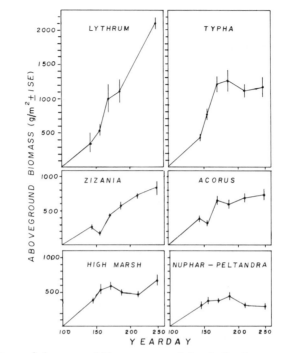

Fig. 1. Patterns of aboveground biomass accumulation in the 6 most common vegetation types in the Hamilton Marshes. Data were collected during the 1974 and 1975 growing season using 0.25 m² quadrats. Methods are detailed in the Whigham and Simpson (1975, 1976a).

and their replacement by a succession of annuals and a few perennials that reach peak biomass later in the season.

In vegetation dominated by the annual, *Zizania*, and a semi-shrub perennial, *Lythrum*, aboveground biomass increased linearly throughout the growing season and reached average peak biomass of 800 g/m² and 2,100 g/m², respectively (Fig. 1). A second pattern of biomass accumulation occurred in vegetation along stream banks and in pond-like areas. Those habitats were dominated by *Nuphar* and *Peltandra*, both perennials, early in the growing season and then by annuals (*Acnida* and *Polygonum punctatum*) and perennials (*Pontederia* and *Sagittaria*) following senescence of *Nuphar* and *Peltandra* leaves. Peak aboveground biomass (450 g/m²) occurred prior to the senescence of *Nuphar* and *Peltandra* leaves (Fig. 1). A third pattern was observed in vegetation dominated by *Typha* and *Acorus* (Fig. 1). Aboveground biomass peaked by June and remained rather constant for the remainder of the growing season. This type of growth of *Typha* has been well documented by Boyd (1970a, 1970b, and 1971), and Dykyjová et al. (1972) reported a similar pattern for individual clones of *Acorus* grown in greenhouse cultures. In both species, initial high growth rates are due to the

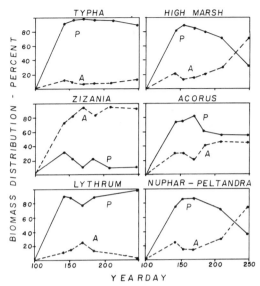

Fig. 2. Data from Fig. 1 were subdivided into perennials (P) and annuals (A) to show the changes in dominance patterns within the 6 vegetation types sampled.

mobilization of energy stored belowground from the previous growing season. Cattail vegetation was dominated by *Typha* throughout the growing season (Fig. 2) but sweet flag dominated vegetation went through a series of physiognomic changes because most *Acorus calamus* had senesced by mid-summer. *Acorus* sites were then commonly dominated by a perennial (*Sagittaria*) and several annuals (*Zizania*, *Polygonum arifolium*, and *Bidens*). Peak biomass was ≈700 g/m² in *Acorus*-dominated sites and 1,200 g/m² in *Typha*-dominated vegetation (Fig. 1). A common pattern on high marsh sites (Whigham and Simpson, 1976a) was characterized by early growing season dominance of *Peltandra* followed by succession of other species (Fig. 2). Similar to *Acorus*- and *Typha*-dominated vegetation, biomass increased rapidly after the onset of the growing season (Fig. 1) and most of the biomass was due to the dominance of *Peltandra*. By midsummer, almost all *Peltandra* leaves had senesced and the stands usually were dominated by 2 annuals (*Zizania and Impatiens*). Following senescence of *Zizania* and *Impatiens*, *Bidens* and *Polygonum arifolium* dominated the sites when peak standing crop (680 g/m²) occurred.

There have been few measurements of belowground biomass and production in saline wetlands (Turner, 1976) and freshwater tidal wetlands (Table 1). The paucity of data reflects the inherent difficulty in sampling wetland substrates and also the difficulty in measuring root production of perennial species. Belowground production measurements must, however, be made to estimate reliably net community production. It is quite likely that below-ground production is very high in some species as suggested by the belowground data for *Peltandra* and *Typha* presented in Table 1. This section details D. F. Whigham and R. L. Simpson's *(personal observations)* attempt to develop methods to estimate belowground production of perennials and annuals in a freshwater tidal wetland.

Belowground production of 2 perennials (*Acorus* and *Sagittaria*, examples of clonal and nonclonal species respectively) was estimated by measuring biomass changes that occurred during a 55 day observation period. Belowground production of annuals was estimated by measuring root biomass changes that occurred within the upper 10 cm of substrate where $\approx$90–95% of the annual roots are located.

Entire clones of *Acorus* and individuals of *Sagittaria* were hand extracted and gently washed in a nearby stream, returned to the laboratory, washed a second time, and then divided into leaves, roots, and stems/rhizome components before they were dried to constant weight at 105°C. Samples were collected in June and August and root and stem/rhizome production estimated by calculating the rate of biomass increment during the observation period. Daily production rates (grams per individual per day) were multiplied by the average plant density (individuals per square metre) and growing season (days) to estimate yearly production (grams per square metre per day). As stated, root production (primarily roots of annuals) within the upper 10 cm of marsh substrate was estimated by measuring the increase in root material in cores. Ten cores were collected in June and August in each of 8 study areas using a 5 cm diameter corer. The cores were washed in a series of nested soil sieves to remove inorganic material and the remaining materials were washed into white enamel trays. Roots were then handpicked from the trays and dried to determine the biomass in each core. Dry weights (grams per core) were converted to grams per square metre and daily production rates calculated (Table 2).

Acorus belowground biomass increased at a rate of 1.49 $g \cdot m^{-2} \cdot day^{-1}$ with the increase occurring in both root and rhizome components. *Sagittaria* subterranean biomass increased at a rate of 0.02 $g \cdot m^{-2} \cdot day^{-1}$ while root biomass in the upper 10 cm of marsh substrate increased at a rate of 0.4 $g \cdot m^{-2} \cdot day^{-1}$. Estimated yearly production of belowground biomass was 223.5 g/m^2 for *Acorus*, 4.0 g/m^2 for *Sagittaria* and 160 g/m^2 for all roots in the upper 10 cm of substrate. This method of estimating annual root production is probably valid for most species that form a dense root mat near the surface.

Large amounts of biomass are omitted from production and biomass measurements because of senescence of leaves and other plant parts. There are, however, few data on the magnitude of those losses within freshwater tidal wetlands. Authors Whigham and Simpson measured leaf mortality for 4 dominant species during the 1976 growing season in the Hamilton Marshes. Aluminum tags were attached to petioles of *Bidens*, *Acorus*, *Peltandra* and *Sagittaria* leaves and their disappearance rate observed over a 55-

Table 4. Peak Aboveground Biomass Measurements and an Estimate of Net Annual Aboveground Production for a High Marsh Site in the Hamilton Marshes near Trenton, New Jersey, During 1976

Taxa	Peak biomass (g/m²)	Date
Acorus calamus	73.6	11 Jun
Peltandra virginica	385.4	11 Jun
Scirpus validus	21.9	11 Jun
Leersia oryzoides	2.3	11 Jun
Impatiens capensis	119.3	19 Jul
Cicuta maculata	0.7	19 Jul
Cuscuta sp.	1.4	19 Jul
Zizania aquatica	84.7	11 Aug
Boehmeria cylindrica	2.6	11 Aug
Sagittaria latifolia	12.3	11 Aug
Bidens laevis	282.2	11 Aug
Polygonum arifolium	199.9	11 Aug
Typha latifolia	87.2	11 Aug
Acnida cannabina	7.2	11 Aug
Polygonum punctatum	5.8	11 Aug
	Total = 1,286.5[a]	

[a] Annual net aboveground production: estimate determined by summing peak aboveground biomass for all species.

species independent of when the samples were collected. Accounting for estimates of leaf mortality (600 g/m²) and belowground production (160 g/m² for annuals and 140 g/m² for 2 perennials), total estimated annual net production was 2,346.5 g/m². Leaf mortality and belowground production estimates are the same as those described previously. The latter is still an underestimate of total production because it does not include data on belowground production for several perennial species (especially *Peltandra*) and there is no estimate for herbivore consumption.

Clearly, the measurement of peak aboveground standing crop was not a good estimator of annual production. In addition to yielding significantly low estimates of production, peak aboveground standing crop may also have given incorrect relative estimates of species productivity due to differences in belowground production, leaf mortality and consumption (which may be expected to vary substantially between the wide range of growth habits and life histories of annual and perennial marsh species). Peak aboveground standing crop must be viewed as only a rough estimate of a single parameter, aboveground production.

CONCLUSIONS

Based on our literature review, it appears that freshwater tidal wetlands may be more productive than saline wetlands at the same latitude and that typical freshwater tidal wetland species are usually the most productive species when they occur in brackish wetlands. Due to the extensive variability in available data, it is impossible to make any valid comparisons between production of tidal and nontidal freshwater wetlands. Production estimates for freshwater tidal wetlands are low because they do not account for belowground production, leaf mortality during the growing season, herbivory and plant turnover during the growing season. Our studies in a New Jersey tidal wetland demonstrate that those factors may account for at least 50% of the total net community production. If this value is accurate we estimate that production of freshwater tidal wetlands within the Middle Atlantic coastal region ranges from $\approx 1,000$ g/m² to 3,500 g/m² and that some wetlands can produce >4,000 g/m². Because of the pronounced physiognomic changes that occur in community composition, we suggest that investigators use multiple harvest techniques when measuring primary production within freshwater tidal wetlands. The need for more adequate data on the dynamics of belowground biomass, leaf mortality and herbivory is obvious, and future production studies should include estimates of those parameters.

ACKNOWLEDGMENTS

Studies of the aboveground production, belowground production and leaf mortality of Hamilton Marsh vegetation was supported by a grant from the United States Department of Interior, Office of Water Research and Technology (Grant No. B-060-NJ), Hamilton Township Environmental Commission and Hamilton Township Department of Water Pollution Control. We thank Gene Turner and an anonymous reviewer for their valuable comments.

LITERATURE CITED

Anderson, R. R., Brown, R. G., and Rappleye, R. D. (1968). Water quality and plant distribution along the upper Patuxent River, Maryland. *Chesapeake Sci.* 9, 145–156.

Auclair, A. N. D., Bouchard, A., and Pajaczkowski, J. (1976). Plant standing crop and productivity relations in a *Scirpus-Equisetum* wetland. *Ecology* 57, 941–952.

Boyd, C. E. (1970a). Production, mineral accumulation and pigment concentrations in *Typha latifolia* and *Alternanthera philoxeroides*. *Arch. Hydrobiol.* 66, 139–160.

Boyd, C. E. (1970b). Production, mineral accumulation, and pigment concentrations in *Typha latifolia* and *Scirpus americanus*. *Ecology* 51, 285–290.

Boyd, C. E. (1971). The dynamics of dry matter and chemical substances in a *Juncus effusus* population. *Am. Midl. Nat.* 86, 28–45.

de la Cruz, A. A. (1973). The role of tidal marshes in the productivity of coastal waters. *ASB Bull.* 20, 147–156.

Drake, B. G. (1976). Seasonal changes in reflectance of standing crop biomass in three salt marsh communities. *Plant Physiol.* 58, 696–699.

Dykyjová, D., Ondok, P. J., and Hradecká, H. (1972). Growth rate and development of the root/shoot ratio in reedswamp macrophytes grown in winter hydroponic cultures. *Folia. Geobot. Phytotax.* 7, 259–268.

Flemer, D. A., Heinle, D. R., Keefe, C. W., Hamilton, D. H., and Johnson, M. Standing crops of marsh vegetation of two tributaries of Chesapeake Bay. *Chesapeake Sci.* 19, (*In press*).

Good, R. E. (1965). Salt marsh vegetation, Cape May, New Jersey. *Bull. N. J. Acad. Sci.* **10**, 1–11.

Good, R. E., and Good, N. F. (1975). Vegetation and production of the Woodbury Creek–Hessian Run freshwater tidal marshes. *Bartonia* **43**, 38–45.

Good, R. E., Hastings, R. W., and Denmark, R. (1975). "An environmental assessment of wetlands: A case study of Woodbury Creek and associated marshes." Rutgers University, Marine Sciences Center Technical Report 75-2. New Brunswick, New Jersey. 49 pp.

Gosselink, J. G., Hopkinson, C. S., and Parrondo, R. T. (1977). "Minor marsh plant species. Vol. I. Prodicion of marsh vegetation." Final report to dredged material research program. U.S.A.E.C., Waterways Experiment Station, Vicksburg, Mississippi.

Heinle, D. R. (1972). Estimate of standing crop (dry weight) of marsh vegetation on two Eastern Shore sites (Somerset County). *In*: "Program Planning and Evaluation." Water Resources Administration, Chesapeake Biological Laboratory, Natural Resources Institute, Solomons, Maryland.

Heinle, D. R., Flemer, D. A., Ustach, J. F., Murtagh, R. A., and Harris, R. P. (1974). "The role of organic debris and associated microorganisms in pelagic estuarine food chains." University of Maryland, Water Resources Research Center, Technical Report 22, College Park, Maryland. 54 pp.

Jack McCormick and Associates, Inc. (1973). "An environmental inventory of the Queen Anne's harbor tract, Anne Arundel County, Maryland." Jack McCormick and Associates, Inc., Devon, Pennsylvania.

Jack McCormick and Associates, Inc. (1974). "Standing crop vegetation analysis of SPA-1. (Hackensack Meadowlands Development Commission, specially planned area 1, Secaucus, Hudson County, New Jersey.)" Correspondence between James Schmid, Jack McCormick and Associates, Inc. and Hartz Mountain Industries. Jack McCormick and Associates, Inc., Devon, Pennsylvania.

Jervis, R. A. (1969). Primary production in the freshwater marsh ecosystem of Troy Meadows, New Jersey. *Bull. Torrey Bot. Club* **96**, 209–231.

Johnson, M. (1970). "Preliminary report on species composition, chemical composition, biomass, and production of marsh vegetation in the upper Patuxent Estuary, Maryland." Chesapeake Biological Laboratory. Reference No. 70-130. Solomons, Maryland.

Keefe, C. (1972). Marsh production: a summary of the literature. *Contrib. Mar. Sci.* **16**, 163–181.

Kirby, C. J., and Gosselink, J. G. (1976). Primary production in a Louisiana gulf coast *Spartina alterniflora* marsh. *Ecology* **57**, 1052–1059.

Lieth, H., and Whittaker, R. H. (1976). "Primary productivity of the biosphere." Springer-Verlag. New York.

McCormick, J. (1970). The natural features of Tinicum marsh, with particular emphasis on the vegetation. *In*: "Two studies of Tinicum marsh, Delaware and Philadelphia Counties, Pa." (J. McCormick, R. R. Grant, Jr., and R. Patrick, eds.), pp. 1–123. The Conservation Foundation, Washington, D.C.

McCormick, J., and Ashbaugh, T. (1972). Vegetation of a section of Oldmans Creek tidal marsh and related areas in Salem and Gloucester Counties, New Jersey. *Bull. N. J. Acad. Sci.* **17**, 31–37.

Mendelssohn, I. A., and Marcellus, K. L. (1976). Angiosperm production of three Virginia marshes in various salinity and soil nutrient regimes. *Chesapeake Sci.* **17**, 15–23.

Odum, E. P. (1971). "Fundamentals of Ecology" 3rd ed. W. B. Saunders Co., Philadelphia, Pa.

Odum, E. P., and Fanning, M. (1973). Comparison of the productivity of *Spartina alterniflora* and *Spartina cynosuroides* in Georgia Coastal marshes. *Georgia Acad. Sci.* **31**, 1–12.

Smalley, A. E. (1959). "The role of two invertebrate populations, *Littorina errorata* and *Orchelium fidicinum* in the energy flow of a salt marsh ecosystem." Ph.D. Thesis, University of Georgia, Athens.

Stevenson, J. C., Cahoon, D. R., and Seaton, A. (1976). Energy flow in freshwater zones of a brackish Chesapeake Bay marsh ecosystem. *Aquatic Ecology Newsletter* 9, 24 (abstract).

Stroud, L. M., and Cooper, A. W. (1968). "Color-infrared aerial photographic interpretation and net primary productivity of a regularly flooded North Carolina salt marsh." Report No. 14. Water Resources Institute, University of North Carolina, Chapel Hill.

Turner, R. E. (1976). Geographic variation in salt marsh macrophyte production: a review. *Contrib. Mar. Sci.* 20, 48–68.

Walker, R., and Good, R. E. (1976). Vegetation and production for some Mullica River–Great Bay tidal marshes. *Bull. N. J. Acad. Sci.* 21, 20 (abstract).

Wass, M., and Wright, T. (1969). "Coastal wetlands of Virginia. Interim report to the Governor and General Assembly," Virginia. Institute of Marine Sci., Spec. Rept. in Appl. Mar. Sci. and Oceanogr. Engr. No. 10.

Westlake, D. F. (1963). Comparisons of plant productivity. *Biol. Rev.* 38, 385–429.

Whigham, D. F., and Simpson, R. L. (1975). "Ecological studies of the Hamilton Marshes Progress report for the period June 1974–January 1975." Rider College, Biology Dept., Lawrenceville, New Jersey.

Whigham, D. F., and Simpson, R. L. (1976a). The potential use of freshwater tidal marshes in the management of water quality in the Delaware River. *In* "Biological Control of Water Pollution" (J. Tourbier, and R. W. Pierson, Jr., eds.), pp. 173–186. University of Pennsylvania Press, Philadelphia, Pa.

Whigham, D. F., and Simpson, R. L. (1976b). Sewage spray irrigation in a Delaware River freshwater tidal marsh. *In* "Freshwater Wetlands and Sewage Effluent Disposal." (D. L. Tilton, R. H. Kadlec, and C. J. Richardson, eds.), pp. 119–144. The University of Michigan, Ann Arbor, Michigan.

Whigham, D., and Simpson, R. L. (1977). Growth, mortality and biomass partitioning in freshwater tidal wetland populations of wild rice (*Zizania aquatica* var. *aquatica*). *Bull. Torrey Bot. Club* 104, 347–351.

Williams, R. B., and Murdoch, M. B. (1969). The potential importance of *Spartina alterniflora* in conveying zinc, manganese, and iron into estuarine food chains. *In* "Proc. 2nd Natl. Symposium on Radioecology." (D. J. Nelson, and F. C. Evans, eds.), pp. 431–439. U. S. Atomic Energy Commission, Washington, D. C.

PRIMARY PRODUCTION OF PRAIRIE GLACIAL MARSHES[1]

A. G. van der Valk and C. B. Davis

Department of Botany and Plant Pathology
Iowa State University
Ames, Iowa 50011

Abstract   On the basis of clip plot studies, the average maximum shoot standing crop of emergent species in Iowa prairie marshes ranges from 330 g/m^2 to 1,160 g/m^2 depending on the species. Submersed communities have a maximum standing crop of 190 g/m^2 and mud flat communities of 600 g/m^2. Belowground maximum standing crops range from 360 g/m^2 to 1,945 g/m^2 among emergent species and annual belowground production of emergents ranges from 508 g/m^2 to 640 g/m^2. Annual shoot net primary production is underestimated by maximum standing crop values by 2–22% for 5 emergent species. For these same species, net primary productivity varies from 4–23 g$\cdot$m$^{-2}\cdot$day^{-1}. Because of the cyclical changes that occur in the vegetation of Iowa prairie marshes over a period of 5 to 20 years, total annual shoot production in a marsh can change 18-fold among different stages in the cycle.

Key words   *Biomass*, Carex, *density, Iowa, marsh; prairie, primary production;* Scirpus, *standing crop*, Typha.

INTRODUCTION

Prairie glacial marshes have been studied extensively by animal ecologists interested in waterfowl and fur-bearing mammals. Unfortunately plant ecologists have not shown a comparable interest in these economically important wetlands. As a result, relatively little is known about the vegeta-

---

[1] Journal Paper No. J-8754 of the Iowa Agriculture and Home Economics Experiment Station, Ames, Iowa. Project 2071.

tion of prairie marshes and the biological and physical factors that control their composition and primary production.

In this paper, we review the scant published and unpublished information on the primary production of prairie glacial marshes in Iowa. We were unable to find published data on prairie marshes outside Iowa. Specifically, we summarize the available data on above- and belowground standing crops, look at the relationship between maximum standing crop and net primary production of shoots, and examine the reasons for any discrepancies between maximum standing crop and net primary production estimates. We also examine the significance of changes in the ratio of flowering to vegetative shoots on the primary production of emergent marsh species.

The Vegetation of Prairie Glacial Marshes

Iowa is on the southeastern boundary of the prairie marsh or pothole country. Most of the prairie marsh habitat occurs in the Dakotas, southwestern Minnesota, and the southern half of the Canadian prairie provinces of Manitoba, Saskatchewan, and Alberta (Shaw and Fredine, 1956). Stewart and Kantrud (1971, 1972) give a general description of prairie marshes and their vegetation. Additional descriptions of the vegetation of these marshes and their hydrology can be found in Walker and Coupland (1968) and Millar (1973) and in the references they contain.

The vegetation of Iowa's prairie marshes has been described by Hayden (1943) and Clambey (1975). In the Stewart and Kantrud (1971) classification of prairie marshes, all the marshes for which primary production data are available are Class IV A or B marshes; i.e., fresh to slightly brackish semipermanent ponds and lakes whose deepest zones are covered with marsh emergents and (or) submersed plants. There have been detailed vegetation studies of the vegetation of all the marshes where primary production measurements have been made. The Big Kettle was studied by van der Valk and Davis (1976a); Big Wall Lake by Van Dyke (1972); Eagle Lake by Clambey (1975) and Currier et al. (1977); Goose Lake by Weller and Spatcher (1965); and Silver Lake Fen by van der Valk (1975, 1976).

One feature of prairie marshes that has a marked influence on their production from year to year is the vegetation change that occurs in cycles of 5 to 20 years. This vegetation cycle can be divided into 4 recognizable stages: a dry marsh stage, a regenerating marsh stage, a degenerating marsh stage, and a lake stage. During drought years, Iowa marshes lose their standing water (i.e., have a natural drawdown), exposing the substrate and allowing the germination of the buried seeds of the perennial marsh emergents (*Typha* spp., *Scirpus validus, Scirpus fluviatilis, Sparganium eurycarpum, Sagittaria latifolia*, etc.) and annual mud flat species (*Bidens cernua, Polygonum* spp., *Cyperus* spp., *Rumex* spp., etc.). This is the dry marsh stage when emergent seedlings and mud flat species are the dominant vegetation. When rainfall returns to normal, standing water returns, and the mud flat species disappear, leaving only the emergents. Submerged species also quickly reappear since their seeds germinate with the return of standing water. The most common submersed macrophytes are *Potamogeton* spp., *Najas flexilis, Ceratophyllum demersum, Myriophyllum exalbescens, Chara* spp., and *Nitella* spp. For a year or so after the return of standing water, the emergent populations increase in shoot density as a result of vegetative growth. This is the regenerating marsh stage when expanding emergent populations are the dominant vegetation in the marsh. After a number of years, the emergent populations begin to decline. The reasons for this decline are poorly understood. It seems to be caused in part by the failure of some emergent species (e.g., *Scirpus validus, Sagittaria latifolia*) to continue to reproduce vegetatively; i.e., the populations become senescent and eventually die out. Muskrats, which can destroy large areas of emergents as a result of their feeding and lodge-building activities, also usually play a significant role in reducing emergent (particularly *Typha*) population levels during the degenerating marsh stage of the vegetation cycle. In the final stage, the lake stage, the marsh is almost devoid of emergent vegetation and resembles a pond or shallow lake. The dominant vegetation consists primarily of free-floating and submersed plants. This stage continues until a drought again exposes the marsh bottom allowing emergent and mud flat seeds to germinate. This cycle was first described by Weller and Spatcher (1965). Additional information on the cycle can be found in Weller and Fredrickson (1974) and van der Valk and Davis (1976*b*, 1978).

METHODS

Eagle Lake General Survey (1975 and 1976)   Eagle Lake was sampled using a stratified random sampling scheme in both 1975 and 1976. All the samples were harvested in both years in late July and early August. At each sampling point, a 1×1 m quadrat was clipped at ground level and sorted by species. These samples were ovendried at 80°C as were all other production samples. In 1975, 350 quadrats were harvested and in 1976, 322 quadrats.

Eagle Lake Selected Samples (1976)   Ten 1×1 m quadrats were harvested periodically during the summer in representative monodominant stands of 4 of the most abundant emergent species (*Typha glauca, Scirpus validus, Sparganium eurycarpum,* and *Carex* spp.).

Belowground biomass samples were collected in conjunction with the selected emergent shoot standing crop samples in June and late August–early September 1976. These belowground biomass samples were collected by using 50×50 cm quadrats. All the soil in the root and rhizome zone was removed from the quadrats. In the laboratory, the samples were washed free of soil, and dead and live material was separated on the basis of appearance. These samples were stored at 4°C between the time they were collected and the time they were washed.

Eagle Lake Turnover Study In May 1976, 3 permanent 1 × 1 m quadrats were established in monodominant stands of *Typha glauca, Sparganium eurycarpum, Scirpus validus, Scirpus fluviatilis,* and *Carex atherodes*. The *Carex* quadrats were located in areas cleared of all litter and vegetation in 1975. The corners of each quadrat were marked with wooden stakes, and a wire was strung around the periphery of each quadrat to mark its boundary. At weekly intervals during the summer and less frequently during the spring and fall, the number of vegetative, flowering or fruiting, and dead shoots in each quadrat was counted. At the time of each inventory, 50 to 60 shoots of each species were collected in a randomly selected area adjacent to the permanent quadrats. If the species were in flower or fruit, half of the shoots collected were flowering or fruiting shoots. The average weight of these shoots and the shoot density measurements were used to calculate the standing crop of the permanent quadrats during 1976.

Data From Other Marshes Below is an outline of the methods used to obtain standing crop data in the published and unpublished studies from other marshes in Iowa that are cited in this paper. Only aboveground samples were collected unless otherwise indicated.

   i). Van Dyke (1972) used systematic sampling to collect six 40 × 50 cm above- and belowground samples in monodominant stands of emergents at Big Wall Lake.
  ii). van der Valk and Davis (1976a) sampled systematically along a transect using 50 × 50 cm quadrats in late July.
 iii). D. Roosa (*personal communication*) and A. G. van der Valk both used 1 × 1 m quadrats laid out systematically along transects at Goose Lake.
  iv). van der Valk (1976) used 20 × 50 cm quadrats placed systematically along a transect.

Nomenclature follows Gleason and Cronquist (1963).

RESULTS AND DISCUSSION
Maximum Shoot Standing Crops

Average maximum shoot standing crops are summarized in Table 1. Too few standing crop samples of submersed species have been taken to give any reliable estimates of the maximum standing crops of particular species. The average maximum standing crop of submersed communities in prairie marshes is 192 g/m². This is in the lower end of the range of submersed standing crops given in the review by Spence et al. (1971).

If one ignores the data from the Eagle Lake turnover study (which will be discussed in the next paragraph) the average maximum standing crops of the dominant emergent species in prairie marshes are 1,156 g/m² for *Typha glauca*, 934 g/m² for *Phragmites communis*, 851 g/m² for *Scirpus acutus*, 667

Table 1. A Summary of the Maximum Standing Crops (g/m²) of Emergent and Submersed Species in Iowa Prairie Glacial Marshes

Species	Big Wall Lake[a]		Eagle Lake								Other marshes	
			1975[b]		1976[b]		1976 selected[c]		Turnover 1976			
	N	x̄	N	x̄	N	x̄	N	x̄	N	x̄	N	x̄
Bidens cernua	...	...	...	...	...	...	...	...	...	...	4	598[d]
Carex spp.[j]	6	545	36	850	24	523	10	885	3	2,231	10	530[e]
Phragmites communis	6	1,110	...	...	...	...	...	...	...	...	3	777[f]
Sagittaria latifolia	...	...	...	...	...	...	...	...	...	...	23	460[e]
Scirpus acutus	6	951	...	...	2	751	...	...	...	...	...	...
Scirpus fluviatilis	6	450	4	465	10	483	...	...	3	791	...	...
Scirpus validus	...	...	60	243	56	360	10	387	3	602	...	...
Sparganium eurycarpum	6	770	50	489	73	474	10	819	3	1,054	...	...
Typha glauca	6	1,549	68	758	91	1,011	10	1,281	3	2,106	19	1,180[g]
Emergents[h]	...	...	297	503	304	637	...	...	...	...	...	...
Submersed[i]	...	...	53	91	18	260	...	...	...	...	14	222[e]

[a] Van Dyke (1972).
[b] General survey.
[c] Samples collected in monodominant stands.
[d] Goose Lake, 1976 (A. G. van der Valk, *personal observation*).
[e] Big Kettle (van der Valk, 1976 and Davis, 1976a).
[f] Silver Lake Fen (van der Valk, 1976).
[g] Goose Lake 1974 and 1975 (D. Roosa, *personal observation*).
[h] All quadrats in which emergent species are dominants: all emergents listed in Table 1 plus species of *Phalaris*, *Spartina*, *Polygonum*, *Impatiens*, *Phragmites*.
[i] All quadrats in which submersed species are dominants: *Potamogeton pectinatus*, *Ceratophyllum demersum*, *Najas flexilis* and *Nitella* sp.
[j] *Carex rostrata*, *C. atherodes* and *C. lacustris*.

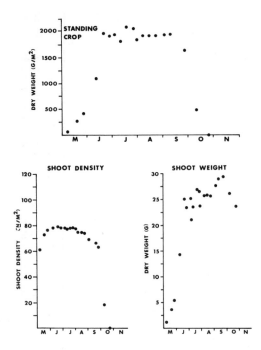

Fig. 1. Standing crops (g/m²), shoot densities (shoots/m²) and average shoot weights (g) for *Typha glauca* during the 1976 growing season at Eagle Lake.

g/m² for *Carex* spp., 638 g/m² for *Sparganium eurycarpum*, 466 g/m² for *Scirpus fluviatilis*, 460 g/m² for *Sagittaria latifolia*, and 330 g/m² for *Scirpus validus*. Not all emergents are equally productive. Eagle Lake was in the regenerating stage of the marsh cycle in 1975 and 1976. This is reflected in the increase in the average maximum standing crops of *Typha glauca* and *Scirpus validus* in the 1976 general survey over the 1975 general survey (Table 1).

Maximum standing crop values calculated from data collected in the turn-over study are, in all instances, higher than the clip plot values (Table 1). A study by Rieley and Jasnowski (1972) comparing clip plot and single plant procedures (similar to those used in the turnover study) for estimating stand-ing crop found that there was little difference (<2%) in the standing crops of the same area obtained from using these 2 methods. Therefore, it is unlikely that differences in sampling techniques are responsible for the observed differences in weight between the turnover and clip plot studies (Table 1). The main reason for the higher turnover values is that the permanent quad-rats were established in areas where these 5 dominant species seemed to be

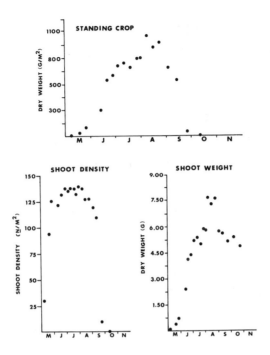

Fig. 2.   Standing crops (g/m²), shoot densities (shoots/m²), and average shoot weights (g) for *Sparganium eurycarpum* during the 1976 growing season at Eagle Lake.

growing vigorously. The turnover sample standing crops probably approach the highest standing crop values found for these species in Eagle Lake. This idea is supported by the fact that the highest clip-plot standing crops measured at Eagle Lake in 1975 and 1976 are very close to the turnover values except those for *Carex*. The highest standing crop recorded for *Typha glauca* in the general surveys was 2,118 g/m² (compared with 2,106 g/m² in the turnover study); for *Scirpus validus* was 627 g/m² (602 g/m²); for *Sparganium eurycarpum* 950 g/m² (1,054 g/m²); and for *Scirpus fluviatilis* 826 g/m² (791 g/m²). On the other hand, the highest value for *Carex* in the general surveys was only 1,336 g/m² compared with 2,231 g/m² in the turnover study. We examine the reasons for abnormally high standing crop of *Carex* in the section on NET PRIMARY PRODUCTION. Except for *Carex* in the turnover study, the maximum standing crop values of the marsh emergents are within the ranges previously observed for various emergent species (van der Valk and Bliss, 1971; Keefe, 1972; Gorham, 1974; Dykyjova and Pribil, 1975; Auclair et al., 1976).

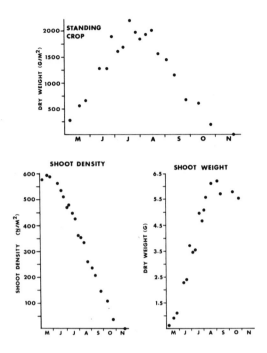

Fig. 3.   Standing crops (g/m²), shoot densities (shoots/m²), and average shoot weights (g) for *Carex atherodes* during the 1976 growing season at Eagle Lake.

Annual Standing Crop Pattern

Figures 1 through 5 give the calculated standing crops of the 5 emergent species in the turnover study from May to November 1976. Changes in shoot density and average shoot weight are also shown during the same period. The shoot weights given in these figures are frequency-weighted averages of the vegetative and flowering shoots.

The standing crop of all 5 species increased steadily until late July or August and then began to decline. *Typha glauca* (Fig. 1) reached its maximum standing crop on 17 July 1976. During most of July and August, however, standing crop values were very close to maximum. The maximum standing crop of *Sparganium eurycarpum* (Fig. 2) occurred in mid-August when the fruits reached their maximum size (see also Fig. 6). *Carex atherodes* reached its maximum standing crop by 17 July 1976; it remained close to this maximum until early August. Bernard (1974) in Minnesota found *C. rostrata* at or near its maximum standing crop from mid-July to late August. *Scirpus validus* reached and maintained maximum standing crop for only a very brief period around the beginning of August. *Scirpus fluviatilis*

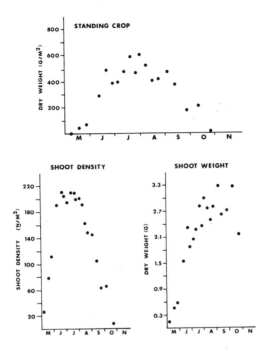

Fig. 4. Standing crops (g/m²), shoot densities (shoots/m²), and average shoot weight (g) for *Scirpus validus* during the 1976 growing season at Eagle Lake.

reached or was near its maximum standing crop from 17 July to 21 August 1976. Auclair et al. (1976) working in Quebec reported that this species did not reach maximum standing crop until September. It is possible that the maximum standing crops of Eagle Lake species occurred earlier in 1976 than usual because most of the marsh went dry in late June. In general, it would seem that prairie marsh emergents reach their maximum standing crops in late July and early August, and this would be the best time to conduct primary production studies using clip plots.

Belowground Biomass

There have been only 2 studies of the underground biomass of prairie marsh emergents, and the results of these studies are shown in Table 2. No underground biomass data are available for submersed species. Only 1 preliminary study has been done on mud flat species (at Goose Lake in September 1976). In areas dominated by *Bidens cernua* at Goose Lake, the underground biomass was ≈130 g/m². This yielded a root:shoot ratio of 0.22.

At Big Wall Lake, Van Dyke (1972) found that the underground biomass was highest in the summer at the time of peak shoot standing crop for

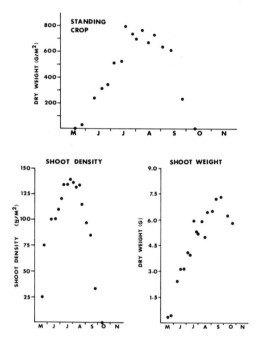

Fig. 5. Standing crops (g/m²), shoot densities (shoots/m²), and average shoot weights (g) for *Scirpus fluviatilis* during the 1976 growing season at Eagle Lake.

Phragmites communis (1,565 g/m²) and *Sparganium eurycarpum* (1,945 g/m²). But it was highest in the fall for *Carex lacustris* (1,172 g/m²), *Scirpus acutus* (1,870 g/m²), *Scirpus fluviatilis* (1,410 g/m²), and *Typha glauca* (1,450 g/m²). Over the winter months, 3 species lost no weight (*P. communis, T. glauca, S. fluviatilis*). On the other hand, 3 other species (*C. lacustris, S. acutus,* and *S. eurycarpum*) lost ≈450, 325, and 385 g/m² during the winter (Table 2). The spring underground biomass samples from Big Wall Lake were collected in April before shoot elongation had occurred. As a result, these data cannot be used to estimate annual underground production because the values do not represent the minimum for the year (see Jervis, 1969).

At Eagle Lake in 1976 (Table 2), underground biomass samples were collected in June and late August–early September. The June samples were taken after shoot initiation in the spring had reduced stored carbohydrates from the previous year (see Bernard, 1974). These June, Eagle Lake biomass values are considerably lower than those for the same species in April in Big Wall Lake, while the August–September Eagle Lake values are very similar to the summer and (or) fall maximum values at Big Wall Lake. If we presume

Table 2. A Summary of the Underground Biomass of Emergent Species in Iowa Prairie Glacial Marshes

Species	N	Underground biomass (g/m²)			
		Spring 1971[a]	Summer 1971	Fall 1971	Spring 1972
		Big Wall Lake (Van Dyke, 1972)			
Carex lacustris	6	907	825	1,172	721
Phragmites communis	6	1,121	1,565	1,148	1,221
Scirpus acutus	6	1,208	1,450	1,870	1,544
Scirpus fluviatilis	6	1,254	1,383	1,410	1,424
Sparganium eurycarpum	6	1,335	1,945	1,638	1,252
Typha glauca	6	1,167	1,302	1,450	1,442

Species	N	Eagle Lake			
		June 1976		August–September 1976	
		Dead	Live	Dead	Live
Carex spp.[b]	10	4	394	15	939
Scirpus fluviatilis	10	40	1,720	. . .	. . .
Scirpus validus	10	0	614	0	361
Sparganium eurycarpum	10	132	803	144	1,311
Typha glauca	10	161	791	244	1,431

[a] Spring—April before shoots produced; summer—at peak standing crop; fall—after shoots had died.

[b] *Carex atherodes* and *Carex rostrata*.

that the Eagle Lake June samples are a reasonable estimate of the minimum belowground biomass for 1976 and the August–September of the maximum, then we can estimate the annual belowground production for 4 of the dominant emergents at Eagle Lake. Our estimate of belowground production at Eagle Lake for 1976, then, is 640 g/m² for *Typha glauca*, 548 g/m² for *Carex atherodes*, and 508 g/m² for *Sparganium eurycarpum*. Previous estimates of the belowground annual production of these 3 species or similar species by Bernard and Bernard (1973), Bernard and MacDonald (1974), and Bernard (1974) are lower (180 to 400 g/m²) than those found at Eagle Lake. Jervis (1969) and Weller (1975), however, report higher annual belowground production of *Typha* than those found at Eagle Lake. Weller (1975) working at Round Lake in northwestern Iowa reported an increase of 4,800 g/m² of live rhizomes in a *Typha glauca* stand from 1963 to 1966.

Scirpus validus belowground biomass decreased by 41% during the summer of 1976 (Table 2), from 614 g/m² to 361 g/m². This reduction does not seem to be a sampling error since shoot standing crops harvested from the same quadrats are higher than average. There are too few data on which to base any firm conclusions, but the data available do suggest that the *Scirpus*

Table 3.   Total Shoot Biomass of Emergent, Submersed, and Mud Flat Species produced at Rush Lake Annually from 1964 to 1970[a]

Vegetation type	Year						
	1964	1965	1966	1967	1968	1969	1970
Mud flat	300	2	5	1	2	0	0
Emergent	428	544	640	561	453	97	16
Submersed	22	129	317	108	141	171	37
TOTAL	750	675	962	670	596	268	53

[a] Calculated from maximum standing crop data in Table 1 and frequency data in Weller and Frederickson (1973). Data are in kilograms $\times$ 10^3.

validus population may be on the decline at Eagle Lake. The decline of *Scirpus validus* several years after a drawdown has been previously documented by Weller and Fredrickson (1974). They reported that at Rush Lake the frequency of *Scirpus validus* in their samples dropped from 23% to 0% from 1964 to 1967 (see also Weller and Spatcher, 1965). The *Scirpus validus* population at Eagle Lake was infested with larvae of an unidentified dipteran living inside the shoots. Nearly all of the mature shoots showed some external evidence of larval damage. The energy drain caused by the larvae may be responsible for the decline in *Scirpus validus* belowground biomass. If this decline in belowground biomass is reflected in a decline in *Scirpus validus* shoot production in 1977, it would indicate Eagle Lake may soon be entering the degenerating marsh stage.

Primary Production and the Vegetation Cycle

To examine the effect of the vegetation cycle on the annual total primary production of a prairie marsh, we calculated the estimated total annual production of Rush Lake, a 160-ha marsh in northwestern Iowa, by using the general survey maximum standing crop data in Table 1 for Eagle Lake (except for the *Bidens cernua* data from Goose Lake used to estimate mud flat production) and the detailed data on species frequencies found in Weller and Fredrickson (1974). Weller and Fredrickson (1974) sampled Rush Lake at $\approx$ 6,900 sampling points annually from 1964 to 1970. During this period the marsh went through all 4 stages of the vegetation cycle. The results of these calculations are given in Table 3. These estimates are low because maximum standing crop data, not net primary production data, were used in the calculations; and also because floating plants (*Lemna* spp., *Spirodela polyrhiza*, *Riccia fluitans*) were not included in the calculations as no estimates of their standing crops are available. Neither of these factors should have any significant effect on the overall pattern of changes in annual production seen in Table 3.

Table 4. A Comparison of the Average Maximum Weights of Vegetative and Flowering Shoots of Emergent Species and of the Maximum Standing Crop (MSC) and Net Primary Production (NPP) of these Species in the Eagle Lake Turnover Study

Species	Max wt vege- tative shoot (g)	Date of max wt (1976)	Flowering shoot %	Max wt (g)	Date of max wt (1976)	In- flo- res- cence[a] (%)	Max shoot den- sity (N/m^2)	MSC (g/m^2)	NPP[b] (g/m^2)	$\frac{MSC}{NPP} \times 100$ (%)
C. atherodes	6.21	2 Sep	0	4.05	10 Jul	11	596	2,231	2,858	78
Sc. fluviatilis	8.28	25 Sep	24	6.32	24 Jul	27	140	791	943	83
Sc. validus	. . .	. . .	100	3.26	9 Oct	17	210	602	713	84
Sp. eurycarpum	6.96	7 Sep	13	13.81	21 Aug	41	139	1,054	1,066	98
T. glauca	28.8	25 Sep	55	32.47	17 Jul	15	79	2,106	2,297	91

[a] Percent of the flowering shoot maximum weight contributed by inflorescence.
[b] NPP—net primary production calculated as outlined in Mason and Bryant (1975).

During the dry marsh stage, total annual primary production was around 750,000 kg and it dropped to 675,000 kg in 1965 when the marsh was re-flooded, mostly because the submersed species were much less productive than the mud flat species. Total production in 1966 rose to 962,000 kg, an increase of 287,000 kg over the previous year. This increase was a result of an increase in the submersed and emergent communities that is a characteristic of the regenerating marsh stage. From 1967 to 1969, annual production dropped from 670, 000 kg to 268,000 kg. Most of this was a result of the decline of emergent production from 561,000 kg to only 97,000 kg during this period. This was the degenerating marsh stage in the cycle. Finally in 1970, Rush Lake had entered the lake stage. Total macrophyte production was only 53,000 kg, and most of that (37,000 kg) was submersed production. From a high in 1966, annual production dropped 18-fold in only 5 years; at a rate of about 180,000 kg per year. The impact of the vegetation cycle on the total annual production of a marsh from year to year is, as far as we know, unparallelled in any other wetland ecosystem.

Net Primary Production

From the data given in Figs. 1 through 5, it is possible to calculate an estimate of the net primary production of the 5 emergent species in the Eagle Lake turnover study. We have used the procedure outlined in Mason and Bryant (1975) to calculate the net primary production of these species. The results are shown in Table 4. The maximum standing crop values underestimate the net primary production values by only 2% for *Sparganium eurycarpum*, by 9% for *Typha glauca*, by 16% for *Scirpus validus*, by 17% for *Scirpus fluviatilis* and 22% for *Carex atherodes*.

Maximum standing crop can equal net primary production only if the maximum shoot density and average maximum shoot weight occurred simultaneously. If they are out of synchrony, then the maximum standing crop will always underestimate net primary production. Maximum shoot density in *Sparganium eurycarpum* occurred on 31 July 1976 and maximum shoot weight on 7 August 1976 (Fig. 2), hence, the small difference between maximum standing crop and net primary production. For *Carex atherodes*, maximum shoot density occurred on 18 May 1976 and maximum shoot weight did not occur until 2 September 1976. As a result, maximum standing crop underestimates net primary production by 22%. Shoot density had fallen 255 shoots/m² from May to September. Maximum shoot weight in *Scirpus fluviatilis* (Fig. 5) was reached on 25 September 1976, and shoot density reached a peak on 24 June 1976. Shoot density had dropped from 140 shoots/m² to only 32 shoots/m² at the time of maximum shoot weight. The pattern is very similar for *Scirpus validus* (Fig. 4). Both species have standing crops, as a result, that underestimate net primary production by more than 15%. The *Typha glauca* (Table 4) maximum standing crop estimate of net primary production is much better than the long period between maximum shoot density (19 June 1976) and maximum shoot weight (25 September 1976) would suggest it should be. This is because there is very little change in shoot density during the summer (Fig. 1).

Auclair et al. (1976) report that, for a marsh in Quebec, maximum standing crops underestimate net primary production by only 8% on the average. For *Typha angustifolia* Mason and Bryant (1975) and Kvĕt (1971) report that net primary production was underestimated by 23–28% and 7%, respectively. Maximum standing crops of *Phragmites communis* according to Mason and Bryant (1975) are 5–15% lower than net primary production. And in *Carex* the difference is 34% according to Bernard and MacDonald (1974). The available information suggests that maximum standing crop is often not as good an estimator of net primary production as Westlake (1963) and other earlier authors seemed to think. More emphasis should be placed on measuring actual net primary production in marshes.

As was noted earlier in this paper (MAXIMUM SHOOT STANDING CROPS section), the Eagle Lake turnover estimate of *Carex* maximum standing crop is considerably higher than the estimates based on clip plot studies. It seems to be the highest standing crop ever recorded for a *Carex* stand (Gorham, 1974; Bernard, 1974). The density of the Eagle Lake turnover plots is considerably higher than that usually reported in the literature. Van Dyke (1972) counted 96 to 103 shoots/m² of *Carex* at Big Wall Lake. Bernard (1974) gives densities of 139 to 238 shoots/m² in Minnesota for *Carex rostrata*. At Eagle Lake shoot densities were as high as 596 shoots/m² in the spring (Fig. 3), but had dropped to 235 shoots/m² by September. The average maximum weight of a shoot was 6.21 g at Eagle Lake. It was only 4.3 g in Bernard's study of *C. rostrata* in Minnesota. It is the high shoot weights that

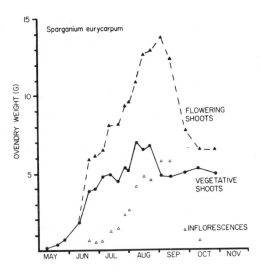

Fig. 6.   Average weights of flowering shoots, vegetative shoots, and inflorescences of *Sparganium eurycarpum* during the 1976 growing season at Eagle Lake.

are in part responsible for the high standing crop at the Eagle Lake turnover site. This is a very nutrient rich site receiving polluted water high in N and P. However, we believe the major reason that the plants grew more vigorously at the turnover sites was that the standing litter had been removed from these plots in October of the previous year. The removal of the litter stimulated the increased shoot density since nearby areas with litter only produced ≈ 136 shoots/m². Removal of litter also allowed soils to thaw earlier in the spring and increased the length of the growing season.

The average net primary productivity of the 5 emergents in the turnover study ranged from 4 $g \cdot m^{-2} \cdot day^{-1}$ for *Scirpus validus* to 23 $g \cdot m^{-2} \cdot day^{-1}$ for *Carex atherodes*. This was calculated by dividing their net primary production (Table 4) by the number of days required for their shoots to reach maximum weights (see Figs. 1 through 5). *Scirpus fluviatilis, Sparganium eurycarpum*, and *Typha glauca* had average net primary productivities of 6, 11, and 16 $g \cdot m^{-2} \cdot day^{-1}$, respectively. These values fall within the range reported for emergents in the literature (see Bernard, 1974).

Vegetative and Flowering Shoots Weights

The dates at which vegetative and flowering shoots of emergent species reach their average maximum weights are presented in Table 4 and, as an example, in Fig. 6 the average weight of flowering and vegetative shoots during the 1976 growing season is shown for *Sparganium eurycarpum*. Flowering shoots generally reach their maximum weight in July (*Scirpus fluviatilis, Carex rostrata*, and *Typha glauca*) and vegetative shoots in September (Table 4). With *Sparganium eurycarpum*, the flowering shoots reached their maximum on 2 September 1976 and vegetative shoots on 7 August 1976. This is because the seeds of *Sparganium* do not mature until late August, and they can account for > 40% of the weight of a flowering shoot (Table 4, Fig. 6).

The average maximum weights of flowering shoots are higher than those of vegetative shoots in *Sparganium eurycarpum* (98% higher) and *Typha glauca* (13% higher), but lower in *Carex rostrata* (35% lower) and *Scirpus fluviatilis* (24% lower). A change in the ratio of flowering to vegetative shoots could, then, have a significant impact on the standing crops of these emergents (see also Rieley and Jasnowski, 1972). For example, at Eagle Lake, in the turnover study (Table 4) only 13% of the shoots of *Sparganium eurycarpum* were flowering shoots. If, however, all of the shoots had been flowering shoots and all else had remained the same, the maximum standing crop would have been 1,638 g/m² instead of 1,054 g/m². The percentage of flowering stalks in a population of *Carex* is evidently a function of winter water levels (Bernard, 1975). If a change in environmental conditions can stimulate or inhibit flowering in other marsh species, which seems quite likely, future studies should pay more attention to the contribution of flowering and vegetative shoots to annual production. Of particular interest would be an examination of the impact of morphological differences between flowering and vegetation shoots on the leaf area of a stand and hence its photosynthesis.

ACKNOWLEDGMENTS

We thank the Iowa Conservation Commission for permission to work at Eagle Lake, and Larry Ewers for the use of his cabin. Paul J. Currier and Michael J. Krogmeier did the fieldwork on the general and selected surveys of Eagle Lake in 1975 and 1976. Dr. J. H. Zimmerman kindly identified the *Carex* spp. at Eagle Lake.

LITERATURE CITED

Auclair, A. N. D., Bouchard, A., and Pajaczkowski, J. (1976). Plant standing crop and productivity relations in a *Scirpus-Equisetum* wetland. *Ecology* 57, 941–952.

Bernard, J. M. (1974). Seasonal changes in standing crop and primary production in a sedge wetland and an adjacent dry old-field in central Minnesota. *Ecology* 55, 350–359.

Bernard, J. M. (1975). The life history of shoots of *Carex lacustris*. *Can. J. Bot.* 53, 256–260.

Bernard, J. M. and Bernard, F. A. (1973). Winter biomass in *Typha glauca* Godr. and *Sparganium eurycarpum* Engelm. *Bull. Torrey Bot. Club* 100, 125–127.

Bernard, J. M. and MacDonald, J. G., Jr. (1974). Primary production and life history of *Carex lacustris*. *Can. J. Bot.* 52, 117–123.

Clambey, G. K. (1975). A survey of wetland vegetation in north-central Iowa. Ph.D. Thesis, Iowa State University, Ames. 207 pp.

Currier, P. J., David, C. B., and van der Valk, A. G. (1977). Vegetative analysis of a wetland community at Eagle Lake marsh, Hancock County, Iowa. Proc. V Midwest Prairie Conference, Ames, Iowa. (*In press*).

Dykyjova, D. and Pribil, S. P. (1975). Energy content in the biomass of emergent macrophytes and their ecological efficiency. *Arch. Hydrobiol.* 75, 90–108.

Gleason, H. A. and Cronquist, A. (1963). Manual of the vascular plants of the Northeastern United States and adjacent Canada. Van Nostrand, New York.

Gorham, E. (1974). The relationship between standing crop in sedge meadows and summer temperatures. *J. Ecol.* **62**, 487–491.

Hayden, A. (1943). A botanical survey in the Iowa Lake Region of Clay and Palo Alto counties. *Iowa State College J. Sci.* **17**, 277–416.

Jervis, R. A. (1969). Primary production in the freshwater marsh ecosystem of Troy Meadows, New Jersey. *Bull. Torrey Bot. Club* **96**, 209–231.

Keefe, C. W. (1972). Marsh production: a summary of the literature. *Contrib. in Mar. Sci.* **16**, 163–181.

Květ, J. (1971). Growth analysis approach to the production ecology of reed-swamp plant communities. *Hydrobiologia* **12**, 15–40.

Mason, C. F. and Bryant, R. J. (1975). Production, nutrient content and decomposition of *Phragmites communis* Trin. and *Typha angustifolia* L. *J. Ecol.* **63**, 71–95.

Millar, J. B. (1973). Vegetation changes in shallow marsh wetlands under improving moisture regime. *Can. J. Bot.* **51**, 1443–1457.

Rieley, J. O. and Jasnowski, M. (1972). Productivity and nutrient turnover in mire ecosystems. I. Comparison of two methods of estimating the biomass and nutrient content of *Cladium mariscus* (L.) Pohl. *Oecol. Plant* **7**, 403–408.

Shaw, S. P. and Fredine, C. S. (1956). Wetlands of the United States. U.S.D.I. Fish and Wildlife Service Circular 39. 67 pp.

Spence, D. H. N., Campbell, R. M., and Chrystal, J. (1971). Productivity of submerged freshwater macrophytes. *Hydrobiologia* **12**, 169–176.

Stewart, R. E. and Kantrud, H. A. (1971). Classification of natural ponds and lakes in the glaciated prairie region. U.S.D.I. Fish and Wildlife Service, Resource Publication 92. 57 pp.

Stewart, R. E. and Kantrud, H. A. (1972). Vegetation of prairie potholes, North Dakota, in relation to quality of water and other environmental factors. U.S. Geol. Surv. Professional Paper 585-D. 36 pp.

Van Dyke, G. D. (1972). Aspects relating to emergent vegetation dynamics in a deep marsh, northcentral Iowa. Ph.D. Thesis, Iowa State University, Ames. 162 pp.

van der Valk, A. G. (1975). Floristic composition and structure of fen communities in northwest Iowa. *Proc. Iowa Acad. Sci.* **82**, 113–118.

van der Valk, A. G. (1976). Zonation, competitive displacement and standing crop of northwest Iowa fen communities. *Proc. Iowa Acad. Sci.* **83**, 50–53.

van der Valk, A. G. and Bliss, L. C. (1971). Hydrarch succession and net primary production of oxbow lakes in central Alberta. *Can. J. Bot.* **49**, 1177–1199.

van der Valk, A. G. and Davis, C. B. (1976a). Changes in composition, structure, and production of plant communities along a perturbed wetland coenocline. *Vegetatio* **32**, 87–96.

van der Valk, A. G. and Davis, C. B. (1976b). The seed banks of prairie glacial marshes. *Can. J. Bot.* **54**, 1832–1838.

van der Valk, A. G. and Davis, C. B. (1978). The role of seed banks in the vegetation dynamics of prairie glacial marshes. *Ecology, (In press).*

Walker, B. H. and Coupland, R. T. (1968). An analysis of vegetation–environment relationships in Saskatchewan sloughs. *Can. J. Bot.* **46**, 509–522.

Weller, M. W. (1975). Studies of cattail in relation to management for marsh wildlife. *Iowa State J. Research* **49**, 383–412.

Weller, M. W. and Fredrickson, L. H. (1974). Avian ecology of a managed glacial marsh. *Living Bird* **12**, 269–291.

Weller, M. W. and Spatcher, C. S. (1965). Role of habitat in the distribution and abundance of marsh birds. Iowa Agricultural and Home Economics Experiment Station Special Report No. 43. 31 pp. Ames, Iowa.

Westlake, D. (1963). Comparison of plant productivity. *Biol. Rev.* **38**, 385–425.

LIFE HISTORY ASPECTS OF PRIMARY PRODUCTION IN SEDGE WETLANDS

John M. Bernard

Department of Biology
Ithaca College
Ithaca, New York 14850

and

Eville Gorham

Department of Ecology and Behavioral Biology
University of Minnesota
Minneapolis, Minnesota 55455

Abstract   Sedge shoots differ in their life history; some such as *Carex lacustris* live for a maximum of 12–14 months whereas others such as *C. rostrata* live for $\approx$2 years. Mortality is very high and as many as 80–90% of shoots emerging do not complete the maximum possible life-span. Most sedge wetlands have maximum green standing crop values of < 1,000 g/m²; the highest reported value is 1,283 g/m². Such values indicate primary production in these wetlands to be <1,000 $g \cdot m^{-2} \cdot yr^{-1}$ when based on the harvest method. When life history of the shoots, especially mortality, is determined, productivity values are higher; *C. lacustris* production then approaches 1,600 $g \cdot m^{-2} \cdot yr^{-1}$.

Key words   Carex, *life history, primary production, population dynamics, sedge wetlands*.

INTRODUCTION

Many members of the genus *Carex* are large, semiaquatic macrophytes particularly common in temperate climates. They often form dense, almost monotypic stands along the borders of lakes, streams and bogs. Floating sedge mats frequently invade small ponds and may finally overspread them completely as "sedge meadows," some of which later evolve into sphagnum bogs. Even though they are common, widespread and locally important, they have not received as much study as reed swamp species such as *Typha* and *Phragmites*. This may be due in part to the difficult taxonomy and in part to the large numbers of *Carex* species in many local floras (Major, 1971). It may also be due to the fact that many sedges are not considered valuable as animal forage, though some *Carex* wetlands in Iceland have been managed for at least 1,000 years and *Carex lyngbyei*, *Carex nigra* and *Carex rostrata* are considered to be fine forage species for cattle (Ingvason, 1969). Hermann (1970) noted that many sedges were comparable to grasses in forage value in western United States rangelands.

Our interest has developed over the years since Pearsall and Gorham (1956) studied aboveground standing crop in single-species wetlands in England. That study indicated that sedge biomass was not as great as that in grass fens or reed swamps. Gorham and Somers (1973) continued work on sedge wetlands in Canada, studying aboveground primary production in particular, but noting the important role that species life history played in determining seasonal aspects of productivity. Since 1971 we have been jointly interested in the study of primary production in sedge wetlands with special reference to the role of species life history, population and nutrient dynamics, and the effects of environmental factors on ecosystem productivity.

Our purpose in this paper is to summarize some of the production data on *Carex* wetlands and where possible to relate these to aspects of species life history and environment.

METHODS

Our field methods have gradually changed from harvesting aboveground shoots and belowground organs in different quadrats throughout the year to the use of permanent reference quadrats. We consider the reference quadrat technique to be superior because it provides information on seasonal aspects of species life history such as emergence, mortality, flowering, etc., in addition to measurements of standing crop and productivity.

In this method, permanent quadrats of 50 × 50 cm are established and all living shoots within are measured for length and tagged with a number. Shoots outside the quadrat similar in length to shoots within are harvested, dried and weighed, and regressions of shoot weight *versus* shoot length are determined. The standing crop of tagged shoots within the quadrat is then determined on the basis of their length measurements. The same measurements and analyses are performed on each subsequent sampling date. When new shoots emerge within the permanent quadrat they also are tagged with a number and measured; those emerging during the same sampling period are treated as a cohort. Bernard and Solsky (1977) refined this basic method for their nutrient cycling study by successively tagging shoots outside the permanent quadrat so that those shoots harvested for nutrient analysis would be of exactly the same age as comparable shoots within the quadrat.

This method has been difficult to apply in belowground studies of standing crop and productivity and we have thus far continued to excavate quadrats. However, we do feel that the method may lend itself well to studies of root:shoot relationships and to determining some of the intrinsic causes of shoot mortality.

RESULTS AND DISCUSSION

Basic Life History

Shoots   Sedges may differ considerably in their life histories, as we have found in our work with *Carex lacustris* and *C. rostrata*. *Carex lacustris* shoots emerge in the summer beginning after 1 July and continuing until winter. Those shoots which exceed 50 cm in size die with the onset of winter; those <50 cm live through the winter frozen in ice and snow. Such overwintering shoots together with additional shoots that may emerge during winter and early spring constitute the population during the next summer. Flowering is normally uncommon, but may be extensive in certain years, probably because of flooding conditions the previous year (Bernard, 1975). All shoots that flower will die in August, whereas vegetative shoots have the capacity to survive until winter dieback, although not all do so.

Thus, *Carex lacustris* has 3 main categories of shoot life-span: (1) 4 to 5 months, emerging in July and August, growing >50 cm in height and dying in November; (2) 9–11 months, emerging from September to November, flowering in June and dying in August; and (3) 12–14 months, emerging from September to November, growing only as vegetative shoots and dying the following November. We consider any shoots emerging during winter or early spring to be a part of category (3), because we believe all shoots are formed during autumn, but some do not grow enough to emerge above the ground then, and those do not flower and die in August. Intermediate life-spans can, of course, result from early shoot mortality.

Carex rostrata, on the other hand, has a 2-year life-span. The shoots emerge in the summer beginning after 1 July as in *C. lacustris*, but all of them are capable of overwintering as green shoots and continuing growth the next year. Some shoots that do not appear aboveground before winter may emerge during the winter, or mostly in early spring. Both sets of shoots grow and mature during the summer, are joined by new shoots emerging after 1 July, again overwinter and emerge the next summer, then flower and die in August. *Carex rostrata*, therefore, has a maximum life-span which varies from about 1½ to 2 years, depending on whether new shoots emerge in late summer or early spring. As in the case of *C. lacustris*, we believe all shoots

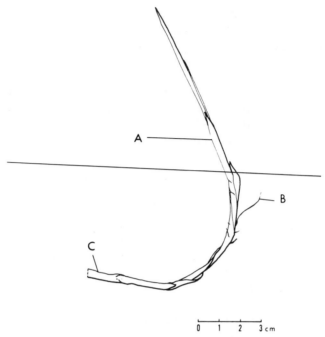

Fig. 1.   Drawing of a pressed specimen showing the early stage in development of shoots and roots of *Carex rostrata*. A represents a new shoot, B represents a new root, and C represents the horizontal rhizome from which the new shoot originates. The approximate location of the soil surface is indicated by the horizontal line.

emerging in winter or early spring to be members of the same cohort and, also similarly, mortality can shorten the life of many shoots well below the maximum life-span. Gorham and Somers (1973) noted that *C. aquatilis* has a life cycle similar to that of *C. rostrata*. Such a life strategy is apparently common in rhizomatous perennial species (Bernard, 1974, 1975; Westlake, 1971), and it is clearly more instructive to study such plants over their full life-span than on the more usual spring–autumn basis.

The life-spans of individual shoots are a part of the life of the entire "tiller clump" (Shaver and Billings, 1975) which may live considerably longer. For example, *C. aquatilis* shoots may not live >2 years, but the "tiller clump" of roots, rhizomes and stem bases where they emerge successively may live up to 4 years after aboveground growth ceases. Since individual rhizomatous tillers may live 4 to 7 years, there is a total life of "tiller clumps" of 5 to 8 (to 10) years according to Shaver and Billings (1975). Costello (1936) estimates that individual tussocks of *Carex stricta* persist for at least 50 years.

There is a striking difference in flowering strategy between *C. lacustris* and *C. rostrata*. As noted above, *C. lacustris* flowers are uncommon in most years, with almost all shoots remaining vegetative. An exception to this

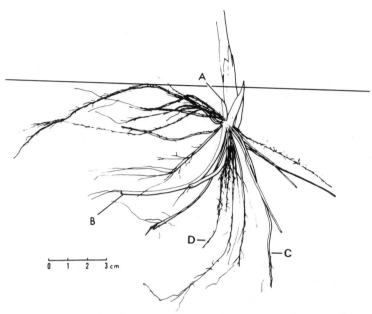

Fig. 2.   Drawing of a pressed specimen showing a later stage in development of shoots and roots in *Carex rostrata*. A represents a new axillary shoot, B represents the original horizontal rhizome, C represents a soil root, and D represents a fibrous root. The approximate location of the soil surface is indicated by the horizontal line.

occurred one summer following severe flooding, so that an environmental cause of flowering was suggested (Bernard, 1975). In *C. rostrata* all vegetative shoots which survive until they have matured will flower; the numbers of flowering shoots are, therefore, dependent on shoot mortality, which presumably depends in turn on environmental conditions.

Rhizomes and Roots   Early stages in the development of shoots and roots of *Carex rostrata* are shown in Fig. 1, which shows the beginning of a new rhizomatous shoot. The horizontal rhizome (C) begins growing from the parent plant about 1 July or later. It continues horizontal growth for a distance probably regulated at least in part by the density of the stand. It then turns vertically and becomes a new shoot (A) as it emerges above ground. Roots (B) begin to develop just about the time the new shoot emerges above ground. Figure 2 shows a later stage of development after a more extensive root system has developed. A new axillary shoot (A) has already begun to grow, possibly as early as late August, from the immature shoot.

There are 2 major types of roots formed by *C. rostrata*. One type is an unbranched root that grows almost straight down into the soil. Costello (1936) noted these same roots in *C. stricta* and called them "soil roots." Such soil roots may finally begin branching at their tip (C). The second type

Table 1.   Weights of Representative Young Shoots and Roots of *Carex rostrata*, 1–6 October 1971 (weights in grams)

Parameter	Rhizomatous shoots (n = 22)		Axillary shoots (n = 22)	
	$\bar{x}$	Range	$\bar{x}$	Range
Green shoot weight	0.74	0.2–1.6	0.63[a]	0.1–1.9
Weight of attached roots	0.26	0.0–0.7	0.08[b]	0.0–0.3
Total weight of shoots and roots	1.00	0.2–2.3	0.71	0.1–2.0
Weight ratio (root:shoot)	0.35	...	0.13	...

[a] Eight axillary shoots without roots averaged 0.25 g, whereas 14 axillary shoots with roots averaged 0.84 g.

[b] This figure includes shoots without roots, for the 14 axillary shoots with attached roots; these averaged 0.12 g.

is a nearly horizontal, fibrous root. Conway (1936) found both types of roots in the sedge *Cladium mariscus* and she noted that the soil roots seemed to be more common in waterlogged, anaerobic sites while the fibrous roots were more plentiful in aerated sites. Their location in the *C. rostrata* soils substantiates such a conclusion. Both root types can be seen in Fig. 2.

Preliminary examination in October 1971 of weights of representative, young, individual shoots and roots of *Carex rostrata* (Table 1) indicate that rhizomatous and axillary shoots are rather similar in size, 0.74 as compared to 0.63 g. However, 8 of 22 axillary shoots lacked roots and averaged 0.25 g, whereas the 14 axillary shoots with roots averaged 0.84 g. Only 1 of 22 rhizomatous shoots lacked roots, and it was the smallest shoot. Presumably the shoots without roots have emerged more recently than those with roots. The similarity in size of rhizomatous and axillary shoots is interesting, in view of the marked disparity in amounts of attached roots. The root:shoot ratio for rhizomatous shoots averages 0.35, whereas the ratio for axillary shoots averages 0.13. Including weight of shoot plus root, the rhizomatous total averages 1.00 g as compared to 0.71 g for the axillary total. If only the 14 axillary shoots with roots are considered, their total becomes 0.96 g, but the root:shoot ratio remains low at 0.14. Root:shoot ratio shows little relationship to shoot size in each type of shoot. However, the fact that axillary shoots with roots are considerably larger than those without suggests that each shoot depends largely, but not wholly, upon its own root system for essential nutrients.

If the differences noted above for rhizomatous and axillary shoots are common, then differences in the ratio of above- to belowground production in different stands could be partly owing to difference in their proportions of rhizomatous and axillary shoots. We intend to investigate this possibility.

Carex lacustris has a different root system in which individual roots are

Table 2. Age Structure of 180 *Carex rostrata* Shoots making up the Population in June 1974 in a Wetland in Central New York

Approximate date of emergence	No. shoots/m²	Percent of present population	Percent mortality from original
1972–1973			
Autumn–Spring	28[a]	16	83.0
1973			
26 July	16	9	33.3
31 August	60	33	none
9 October	48	27	none
7 December	4	2	none
1974			
Winter and early spring	24	13	none

[a] Refers to shoots in flower.

produced along the length of the rhizome rather than all being present at the base of the shoot.

Population Dynamics of Shoots

The life histories given earlier for *Carex lacustris* and *C. rostrata* are for maximum possible life-spans, but, as noted, few shoots survive for the entire period.

For example, a given year's population of *C. rostrata* includes all shoots present in the previous autumn and surviving the winter, plus new shoots emerging in winter or early spring. The age structure of a shoot population from New York in June 1974 is given in Table 2. Of the 180 shoots/m² present on that date, 28 (16%) were mature, flowering shoots with a life-span of ≈2 years (late summer or autumn of 1972 to July–August of 1974).

The percentage of mature shoots in the population is determined by the mortality which has occurred in the cohort year by year. Bernard (1974) noted that flowering shoots comprised ≈18% of a June population of *Carex rostrata* in Minnesota, while Gorham and Somers (1973) reported a figure of ≈40% in their study in Canada. The reasons for these mortality differences are unknown.

One hundred twenty-eight shoots, or 71% of the population present in June, had emerged the previous late summer or autumn, most of them in August and September. All of these shoots survived the winter except for some of the shoots emerging prior to 26 July, one third of which died. The fact that all shoots emerging after 26 July survived indicates that the winter season may not be important in the mortality of the younger shoots of this species.

Table 3. Total Number of Shoots of *Carex rostrata* Per Square Metre in a New York Wetland and the Cumulative Gains and Losses of Shoots throughout a Year, June 1973 to June 1974

Date	Emergence of shoots (cumulative gains)	Death of shoots (cumulative losses)	Total shoots/m²	Net change (%)	Net change (%/week)
		1973			
4 June	...	...	212	0	0
27 June	...	16	196	−7.5	−2.3
26 July	24	28	208	+6.1	+1.5
31 August	84	96	200	−3.8	−0.75
9 October	132	108	236	+18.0	+3.2
28 October	132	116	228	−3.4	−1.2
7 December	136	120	228	0	0
		1974			
18 June	160	192	180	−21.0	−0.76

Finally, 24 shoots (13%) of the 180/m² present in June 1974 emerged during the previous winter or in early spring of 1974. By 18 June all such shoots were mature and, on average, slightly shorter than those shoots which had emerged the previous summer and autumn.

The static view of age structure given in Table 2 is complemented by the more dynamic view of population gains and losses in Table 3, which also represents the stand of *Carex rostrata* in New York. The table shows the total number of shoots per square metre present on 4 June 1973 and the cumulative gains and losses over the following year. The number of shoots was relatively constant (180–236/m²); the largest differences between sampling dates during the growing season being the 18% increase between 31 August and 9 October after shoot emergence was well established, and the 7.5% decline between 4 June and 27 June, when some shoots were dying but none were emerging. Losses over winter amounted to 21%, and presumably included chiefly older shoots (cf. Table 2). When changes in shoot density are calculated on a weekly basis, the relative constancy of shoot numbers is even more apparent (Table 3). The largest weekly differences are only 3.2% increase and 0.76% decline during the summer and winter periods, respectively. Of special note is the fact that the death of all flowering shoots in the July–August period makes little difference in total shoot numbers since the number of flowering shoots is not large and they are being replaced then by new shoots just emerging. Also of note is the very high shoot turnover, 160 new shoots emerging/m² and 192 dying/m² during the 12-month period of study.

A complete analysis of shoot dynamics of *Carex lacustris* will not be presented, but large mortality rates during the growing season are also characteristic of this species. For example, in 1 permanent plot 32 shoots were tagged in early May. By early June, 12 were in flower and 4 had already died. Thus, a minimum of 50% of the shoots were not going to survive the entire growing season.

Significant annual shoot turnovers have been reported recently in a number of different species. Among them are *Glyceria maxima* (Mathews and Westlake, 1969; Westlake, 1971), 3 *Ranunculus* species (Sarukhán and Harper, 1973), *Danthonia caespitosa* and *Chloris acicularis* (Williams, 1970), and *Anthoxanthum odoratum* (Antonovics, 1972). The situation in regard to sexual reproduction in all these species, and in the *Carex* species thus far studied, was summed up by White and Harper (1970) who stated, "The commonest fate of plants under natural conditions is to die before reaching reproductive maturity." The causes of such mortality are not understood, but could include such factors as frost, insect damage, or internal competition of the kind noted by Haslam (1970) for *Phragmites*.

Standing Crop

Aboveground We have summarized aboveground standing crop values from different wetlands in a series of papers. The first (Bernard, 1973) noted that aboveground standing crops of sedges were lower in general than those of other wetland species such as *Typha* (Pearsall and Gorham, 1956; Penfound, 1956; Boyd, 1970; Boyd and Hess, 1970) *Phragmites* (Pearsall and Gorham, 1956; Björk, 1967), and *Glyceria* (Westlake, 1971). These species commonly have aboveground standing crops $>1,000$ g/m^2, whereas most *Carex* wetlands do not reach that figure. The only exceptions of which we are aware are *Carex lanuginosa* at 1,283 g/m^2 (J. M. Bernard and G. Howick, *personal observation*), *C. lacustris* at 1,037 g/m^2 (Bernard and MacDonald, 1974) and 1,145 g/m^2 (Bernard and Solsky, 1977) in New York, and *C. atherodes*, which exceeded 1,000 g/m^2 at 3 sites in northwestern Minnesota (Gorham and Bernard, 1975).

Bernard (1973) noted a decrease in standing crop with increasing latitude, the highest values from the more southerly sites averaging $\approx 1,000$ g/m^2, the lowest value from Swedish Lapland was 283 g/m^2 (Pearsall and Newbould, 1957). E. Gorham and M. G. Somers (*personal observation*) noted a similar decrease with increasing elevation. Gorham (1974) continued analyzing such relationships and correlated standing crop with summer temperature, particularly the highest monthly mean temperature.

Because of the nature of sedge life history, winter values of green standing crop are important to understand the productivity of these wetlands. The few data available indicate that green *Carex lacustris* shoots in winter weigh ≈ 180 g/m^2 (Bernard and MacDonald, 1974), *C. rostrata* ranges from 114 g/m^2 to 235 g/m^2 (Bernard, 1974; Bernard and Bernard, 1977; Gorham and Somers, 1973), *C. aquatilis* 40–50 g/m^2 (Gorham and Somers, 1973) and *C. lanuginosa* ≈ 25 g/m^2 (Bernard and Bernard, 1977).

Table 4.   Aboveground, Belowground, and Total Productivity in Sedge Wetlands[a]

Species	Author(s)	Location	Aboveground Yearly rate	Aboveground Daily rate (max)	Below-ground Yearly rate	Total Yearly rate
C. stricta	Jervis (1969)	New Jersey	1,492[b]	23	...	...
C. lacustris	Bernard and Solsky (1977)	New York	965	20.9	208	1,173
	Bernard and MacDonald (1974)	New York	857 (1,580)[c]	15.0 (20.3)[c]	161 161	1,018 (1,741)[c]
	Klopatek (1975)	Wisconsin	1,186	8.1	134	1,320
C. rostrata	Bernard (1974)	Minnesota	738	10.9	180	918
	Gorham and Somers (1973)	Alberta, Can.	515	6.0		
	J. M. Bernard and M. M. Hankinson (*personal observations*)	New York	540 (823)[c]			
C. aquatilis	Gorham and Somers (1973)	Alberta, Can.	340	4.0		

[a] All measurements are given in grams per square metre.
[b] Codominant, accounts for 47% of total aboveground standing crop shown in table.
[c] Corrected for severe shoot mortality throughout growing season (see discussion in text).

Belowground   The very scarce belowground standing crop values indicate winter maxima of $\approx$330 g/m² for *C. rostrata* in Minnesota (Bernard, 1974) while *C. rostrata* averaged 410 g/m² and *C. lacustris* 430 g/m² in New York (Bernard and Bernard, 1977). The midsummer minimum standing crops averaged about 140–200 g/m² less than the maximum winter values.

Net Primary Productivity

Table 4 summarizes productivity estimates from sedge meadows where careful seasonal measurements have been made. Net primary production aboveground declines with increasing latitude, the most southerly site in New Jersey yielding almost 1,500 $g \cdot m^{-2} \cdot yr^{-1}$. It should be noted, however, that the New Jersey site was a mixed stand of herbaceous marsh species and *C. stricta* contributed only 703 $g \cdot m^{-2} \cdot yr^{-1}$ with a maximum rate of 12 $g \cdot m^{-2} \cdot day^{-1}$.

Of special note are the 2 sets of figures for *C. lacustris* (Bernard and MacDonald, 1974) and *C. rostrata* (J. M. Bernard and G. Hankinson, *personal observation*). In each case, the first set of their data was obtained by subtracting the green standing crop in winter from the maximum green standing crop of summer, and did not take into account the dynamics of the shoot population throughout the growing season. The same is true of all sets of data presented by other authors. The second sets of data by Bernard and coauthors (Table 4) reflect shoot dynamics, especially shoot mortality during the growing season. It is apparent from these different estimates of annual harvest, 1,580 as compared to 857 g/m² for *C. lacustris*, and 823 as compared to 540 g/m² for *C. rostrata*, that productivity studies in sedge meadows require considerable attention to the seasonal dynamics of shoot populations.

The belowground production estimates in Table 4 range from 134–208 $g \cdot m^{-2} \cdot yr^{-1}$. These estimates coupled with the aboveground values give total production estimates ranging from approximately 900–1,300 $g \cdot m^{-2} \cdot yr^{-1}$ based on differences between seasonal maximum and minimum standing crop. The highest productivity shown, almost 1,750 $g \cdot m^{-2} \cdot yr^{-1}$ for *C. lacustris*, is based on the dynamics of shoot populations.

CONCLUSIONS

The life histories of species have a profound influence on their production processes, and we believe that year-round studies with due attention to life histories are essential to the accurate calculation of primary productivity in ecosystems. We see "natural history," therefore, as an integral part of quantitative ecosystem ecology. Two particular aspects of sedge life history stand out as critical. First, the very considerable green standing crop in winter must be reckoned with in many species. These species may have a competitive advantage, because green winter shoots can begin growth as soon as conditions become favorable, and they constitute the main aboveground organs that grow up to the time of maximum standing crop. Ignorance of a large winter standing crop would cause one seriously to overestimate net primary production. This winter standing crop also has high concentrations of certain nutrients (Bernard and Solsky, 1977; Klopatek, 1975), so nutrient studies will also benefit from a knowledge of winter condition.

A second extremely important aspect of sedge wetlands is the very high mortality rate, with as many as 80–90% of shoots dying before attaining their normal life-span. Striking underestimates of net production are possible unless such high levels of mortality are recognized. This mortality will also exert considerable influence upon nutrient cycling.

Significant areas of study remain. The lack of information on belowground organs is obvious and must be remedied before seasonal relationships between aboveground and belowground productivity can be understood. Relationships among shoots, including studies of the production and periodicity of axillary *versus* rhizomatous shoots, are essential not only to understand the development and life of "tiller clumps," but also to shed light on some of the intrinsic causes of mortality in these species. Extrinsic agents of mortality also deserve considerable attention. Although these are important areas of study in all ecosystems, we believe that sedges have many advantages for their investigation.

LITERATURE CITED

Antonovics, J. (1972). Population dynamics of the grass *Anthoxanthum odoratum* on a zinc mine. *J. Ecol.* **60**, 351–365.

Bernard, J. M. (1973). Production ecology of wetland sedges: The genus *Carex*. *Pol. Arch. Hydrobiol.* **20**, 207–214.

Bernard, J. M. (1974). Seasonal changes in standing crop and primary production in a sedge wetland and an adjacent dry old-field in central Minnesota. *Ecology* **55**, 350–359.

Bernard, J. M. (1975). The life history of shoots of *Carex lacustris*. *Can. J. Bot.* **53**, 256–260.

Bernard, J. M. and Bernard, F. A. (1977). Winter standing crop and nutrient contents in five central New York wetlands. *Bull. Torrey Bot. Club* **104**, 57–59.

Bernard, J. M. and MacDonald, J. G. Jr. (1974). Primary production and life history of *Carex lacustris*. *Can. J. Bot.* **52**, 117–123.

Bernard, J. M. and Solsky, B. A. (1977). Nutrient cycling in a *Carex lacustris* wetland. *Can. J. Bot.* **55**, 630–638.

Boyd, C. E. (1970). Production, mineral accumulation and pigment concentration in *Typha latifolia* and *Scirpus americanus*. *Ecology* **51**, 285–290.

Boyd, C. E. and Hess, L. W. (1970). Factors influencing shoot production and mineral nutrient levels in *Typha latifolia*. *Ecology* **51**, 296–300.

Björk, S. (1967). Ecologic investigations of *Phragmites communis*. *Folia Limnologica Scandinavia* **14**.

Conway, V. M. (1936). Studies in the autecology of *Cladium mariscus* R. BR. I. Structure and development. *New Phyt.* **35**, 177–205.

Costello, D. F. (1936). Tussock meadows in southeastern Wisconsin. *Bot. Gaz.* **97**, 610–647.

Gorham, E. (1974). The relationship between standing crop in sedge meadows and summer temperature. *J. Ecol.* **62**, 487–491.

Gorham, E. and Bernard, J. M. (1975). Midsummer standing crop of wetland sedge meadows along a transect from forest to prairie. *J. Minnesota Acad. Sci.* **41**, 15–17.

Gorham, E. and Somers, M. G. (1973). Seasonal changes in the standing crop of two montane sedges. *Can. J. Bot.* **51**, 1097–1108.

Haslam, S. M. (1970). The development of the annual population of *Phragmites communis* Trin. *Ann. Bot.* **34**, 571–591.

Hermann, F. J. (1970). Manual of the Carices of the Rocky Mountains and Colorado Basin. *USDA Ag. Handbook* **374**.

Ingvason, P. A. (1969). The golden sedges of Iceland. *World Crops* **21**, 218–220.

Jervis, R. A. (1969). Primary production in the freshwater marsh ecosystem of Troy Meadows, New Jersey. *Bull. Torrey Bot. Club* **96**, 209–231.

Klopatek, J. M. (1975). The role of emergent macrophytes in mineral cycling in a freshwater marsh. *In* "Mineral cycling in southeastern ecosystems" (F. G. Howell, J. B. Gentry and M. H. Smith, eds.), pp. 367–393. ERDA Symposium Series (CONF-740513).

Major, J. (1971). *Carex* for the ecologist. *Ecology* **52**, 539.

Mathews, C. P. and Westlake, D. F. (1969). Estimation of production by populations of higher plants subject to high mortality. *Oikos* **20**, 156–160.

Pearsall, W. H. and Gorham, E. (1956). Production ecology. I. Standing crops of natural vegetation. *Oikos* **7**, 193–201.

Pearsall, W. H. and Newbould, P. J. (1957). Production ecology. IV. Standing crops of natural vegetation in the sub-arctic. *J. Ecol.* **45**, 593–599.

Penfound, W. T. (1956). Primary production of vascular aquatic plants. *Limnol. Oceanogr.* **1**, 92–101.

Sarukhán, J. and Harper, J. L. (1973). Studies on plant demography: *Ranunculus repens* L., *R. bulbosus* L., and *R. acris* L. I. Population flux and survivorship. *J. Ecol.* **61**, 675–616.

Shaver, G. R. and Billings, W. D. (1975). Root production and root turnover in a wet tundra ecosystem, Barrow, Alaska. *Ecology* **56**, 401–409.

Westlake, D. F. (1971). Population dynamics of *Glyceria maxima*. *Hydrobiologia* **12**, 133–134.

White, J. and Harper, J. L. (1970). Correlated changes in plant size and number in plant populations. *J. Ecol.* **58**, 467–485.

Williams, O. B. (1970). Population dynamics of two perennial grasses in Australian semi-arid grassland. *J. Ecol.* **58**, 869–875.

PRIMARY PRODUCTION IN NORTHERN BOG MARSHES

R. J. Reader

Department of Botany and Genetics
University of Guelph
Guelph, Ontario, Canada N1G 2W1

Abstract Harvest-method estimates of primary productivity in 9 northern bog marshes (i.e., fens, minerotrophic peatlands, tundra, sedge meadows, etc.) ranged from 101 $g \cdot m^{-2} \cdot yr^{-1}$ to 1,026 $g \cdot m^{-2} \cdot yr^{-1}$ for aboveground plant parts and 141 to 513 $g \cdot m^{-2} \cdot yr^{-1}$ for belowground portions. No single taxon consistently made the greatest contribution to the production total. Environmental factors limiting the productivity of northern bog marshes include the length of the growing season, maximum summer air temperature, nutrient availability (especially N, P, K) and waterlogging. The relative importance of these factors apparently depends on the species composition and the location of the bog marsh.

Key words *Bog, enrichment, fertilization, nutrient, peat, primary production, wetland.*

INTRODUCTION

The aim of this paper is to summarize information currently available on the subject of primary production in northern bog marshes. The review is restricted to a consideration of primary production rates and environmental factors limiting primary productivity in northern bog marshes. Physiological aspects of the production process are not considered here.

To qualify as a northern bog marsh an ecosystem must possess the following 5 attributes. It must have (1) peat as the rooting medium for vegetation; (2) a water table above, or close to the peat's upper surface for at least a portion of the year; (3) an influx of mineral nutrients derived partially from precipitation and dust and partially from runoff from adjacent ecosystems; (4) vegetation consisting of nonvascular plants (especially bryophytes) and vascular plants (primarily herbaceous rather than woody) without a closed canopy of trees or shrubs; (5) a location $\geq 44°N$. Based on these criteria, the category of northern bog marsh would include marginal fens or laggs, minerotrophic peatlands or mires, sedge wetlands and tundra meadows at latitudes $\geq 44°N$ but would not include ombrotrophic peat bogs, bog forests, carrs, swamps, or heaths regardless of their locations.

Studies reporting primary productivity values for northern bog marshes had to meet 2 requirements to be included in this review. First, a study had to determine the productivity of all plant populations present in the bog marsh. This restriction was imposed since the objective of this paper was to evaluate the production potential of bog marsh plants as a trophic level, rather than to assess the productivity of individual bog marsh plant populations. Second, the methods used to derive production estimates had to be fully described. This restriction enabled an evaluation to be made of the importance of methodological differences between studies as a possible cause for variation evident in their production values.

Very few of the studies measuring primary production rates of bog marshes, and meeting the 2 requirements listed above, also included experimental or analytical procedures to quantify the relationship between total ecosystem primary productivity and environmental resource availability. However, some studies have monitored productivity changes by a limited number of bog marsh plant populations in response to artificially imposed changes in atmospheric or substrate conditions. The results of these experimental studies will be reviewed to point out some of the responses of bog marsh plants, in terms of primary production, to changes in environmental conditions. Bog marshes also experience naturally occurring changes in environmental conditions, e.g., seasonality, succession. However, it is not possible, in this review, to derive any generalizations about the effects of these natural environmental changes on the productivity of bog marsh plants because previous investigators have not measured or reported pertinent environmental data along with the primary production values.

This review considers first the methods used to measure bog marsh primary production, then compares northern bog marsh production rates, and finally considers the environmental factors which probably limit primary production in northern bog marshes.

METHODS

Both "gas-exchange" (Woodwell and Botkin, 1970) and "harvest" (Milner and Elfyn-Hughes, 1968) methods have been utilized to measure bog marsh primary productivity. Gas exchange techniques have been employed only for the estimation of primary productivity by single populations, e.g., Johansson and Linder (1975); Miller and Stoner (1976), rather than total ecosystems. For this reason, the results of the gas exchange studies are not discussed in this review. The harvest method offers several advantages over gas exchange for the estimation of total ecosystem primary productivity. A pair of clippers is the only piece of equipment needed to apply the harvest technique, while several pieces of relatively expensive equipment (e.g., infrared gas analyzer), are required to monitor photosynthetic rates using gas exchange techniques. It is also very difficult to extrapolate accurately from short-term (minute or hour), small scale (individual leaf or plant) gas exchange measurements to longer term (day or year) or larger scale (square metre or hectare) production values. This extrapolation problem can be minimized easily in the harvest method by simply adjusting the size of the area sampled and the sampling periodicity to achieve the desired level of resolution. However, even estimates of total ecosystem primary productivity derived from the application of the harvest method are likely to contain errors. Undetected losses of new growth resulting from incomplete extraction of belowground plant parts, natural senescence, or consumer activities will all cause primary production to be underestimated. On the other hand, primary production will be overestimated if translocation of photosynthate from old to new growth is not accounted for, or if new growth cannot be accurately distinguished and separated from old growth (e.g., annual radial increment in woody stems or root stocks). Every production study conducted to date using the harvest method includes at least 1 of the abovementioned errors, and most studies contain all of these errors to some degree. It is important to realize that the primary production values discussed in the following sections are not free from errors and, therefore, may indicate only the order of magnitude of bog marsh primary productivity with any accuracy. It should also be noted that the production potential of a bog marsh is likely a dynamic quantity and that the production values compared in this paper represent estimates of that production potential for single, usually unrelated points in a successional time frame.

All of the aboveground productivity values (grams of dry matter per square metre per year) presented in this paper were derived by simply subtracting the minimum plant weight recorded in periodic harvests of the standing crop from the maximum plant weight recorded. In most instances the productivity values were determined for each of the plant species or types (e.g., vascular *versus* nonvascular) present in the ecosystem. Total belowground productivity (grams per square metre per year) was estimated with the following equation:

$$\text{Total belowground productivity} = \left(\frac{\text{belowground standing crop}}{\text{aboveground standing crop}}\right) \times \left(\text{total aboveground productivity}\right).$$

This method of calculating belowground productivity may give fair results when values obtained at the end of the growing season are utilized, but the strong asymmetry of aboveground and belowground development would produce a serious error in the estimation of belowground production early in the growing season. Difficulties associated with the extraction of belowground plant parts and the identification of new growth have made it impossible to obtain more realistic, direct estimates of belowground productivity. The ecosystem's total primary production was derived by summing the values for the individual components. I made no attempt to standardize the values presented in the original studies, even though the frequency and intensity of sampling varied considerably among the studies, as did the analytical methods, e.g., drying temperature, correction factors for death losses, etc. The number of assumptions involved in any standardization procedures would have outnumbered assumptions already included in the sampling and analytical techniques.

RESULTS

The best estimates of primary productivity currently available for northern bog marsh ecosystems are summarized in Table 1. In most studies only aboveground productivity was reported, presumably because of the difficulties involved in the extraction of roots or the identification of new root growth. The 5 values given for belowground production are really indicative only of the magnitude of root production to be expected in bog marshes, i.e., 141–513 $g \cdot m^{-2} \cdot yr^{-1}$. Productivity values for aboveground plant components ranged from 101 to 1,026 $g \cdot m^{-2} \cdot yr^{-1}$, with 7 of the 9 values falling somewhere between 101 and 530 $g \cdot m^{-2} \cdot yr^{-1}$.

No single species or genus consistently made the greatest contribution to total primary production at the 9 sites. Not only does this fact make it difficult to compare productivity totals for the different bog marshes, but it also means that it will not be possible to use any single species or genus as an "indicator" of total bog marsh production. The only similarity between the 9 marshes was that vascular plants made a greater contribution to total aboveground production than did nonvascular plants (the Dempster study area of Wein and Bliss [1974] was an exception).

DISCUSSION OF RESULTS AND CONCLUSIONS

The data presented in Table 1 indicate that bog marsh productivity values are relatively low compared to the values reported for the other types of marshes discussed in this volume and that productivity can vary considerably between sites. Low productivity is most obviously a reflection of the northern locations of these marshes. The length of the growing season (average daily air temperature >5°C) ranged from 55 days at Barrow, Alaska (71°N) to 210 days at Houghton Lake, Michigan (44°N). Gorham (1974) found a positive correlation between peak plant standing crop and maximum monthly summer air temperature for a latitudinal series of sedge meadows. Presumably a similar relationship exists for other wide-ranging genera of bog marsh plants. The diverse composition of the bog marshes included in Table 1 prevents any meaningful evaluation from being made of the relationship between air temperature (or site latitude) and the primary productivity of selected bog marsh genera or the entire flora.

Gorham's analysis also suggested that mineral nutrient availability was an important factor influencing sedge standing crop. Results of experimental

Table 1. Summary of Annual Production Estimates for Northern Bog Marsh Vegetation

Author, location, and site	Species	Primary productivity ($g \cdot m^{-2} \cdot yr^{-1}$) Above-ground	Below-ground
Reader and Stewart (1972)	*Chamaedaphne calyculata*	227	
Elma, Manitoba, Canada	*Carex rostrata*	116	
(49°53'N, 95°54'W)	*Calamagrostis canadensis*	48	
lagg	*Salix bebbiana*	219	513
	Salix serissima	11	
	Aulacomnium palustre	36	
	Hypnum pratense	31	
	Other species	338	
	Total	1,026	513
Bernard (1974)	*Carex rostrata*	738	180
Princeton, Minnesota, USA	*Equisetum fluviatile*		
(45°25'N, 93°40'W)	*Sagittaria latifolia*	42	17
sedge wetland			
	Total	780	197
Richardson et al. (1976)	*Chamaedaphne calyculata*	187	
Houghton Lake, Michigan, USA	*Betula pumila*	151	
(44°20'N, 84°50'W)	Herbs	3	
leatherleaf-bog birch			
	Total	341	
		D	EC
Wein and Bliss (1974)	*Eriophorum vaginatum*	27	28
Yukon territory and Alaska	*Ledum palustre*	4	9
Dempster (D)	*Vaccinium vitis-idaea*	2	12
(64°45'N, 138°21'W)	*Carex* spp.	1	10
and Eagle Creek (EC)	*Vaccinium uliginosum*	0	9
(65°26'N, 145°30'W)	*Empetrum nigrum*	0	1
	Betula nana	4	2
	Rubus chamaemorus	3	3
	Andromeda polifolia	.2	1
	lichens	1	3
	mosses	127	32
	Total	169	110
Tieszen (1972)	*Carex aquatilis*	25	
Barrow, Alaska, USA	*Dupontia fischeri*	21	
(71°20'N, 156°39'W)	*Eriophorum angustifolium*	29	
wet meadow	*Calamagrostis holmii*	3	
	Eriophorum scheuchzeri	10	
	Poa arctica	5	
	Other species	8	
	Total	101	

Table 1.   Continued

Author, location, and site	Species	Primary productivity ($g \cdot m^{-2} \cdot yr^{-1}$)	
		Above-ground	Below-ground
Forrest and Smith (1975)	*Eriophorum angustifolium*	96	24
Moor House, Westmoreland, UK	*Trichophorum caespitosum*	394	187
(54°30′N, 3°W)	*Sphagnum papillosum*	35	...
Cottage Hill A	Other species	5	?
	Total	530	>211
Wielgolaski (1975)	*Salix lapponum*		
Stigstuv, Norway	*Salix herbacea*		
(60°18′N, 7°41′E)	*Carex nigra*	252	410
wet meadow	Other vascular	173	...
	Nonvascular		
	Total	425	410
Doyle (1973)	*Schoenus nigricans*	29	66
Glenamoy, Ireland	*Molinia caerulea*	38	75
(54°15′N, 9°45′W)	*Calluna vulgaris*		
blanket bog	*Erica tetralix*	47	?
	Myrica gale		
	Eriophorum angustifolium		
	Narthecium ossifragum	11	?
	Scirpus sp.		
	Sphagnum spp.	50	...
	Total	175	>141

field studies conducted with several different bog marsh plants (Table 2) substantiate the idea that low nutrient availability can limit their productivity. Phosphorus, K, and N were the 3 elements most often found to be limiting bog marsh plant growth, although the relative importance of these elements as limiting factors changed from site to site. For example, the productivity of cotton grass (*Eriophorum vaginatum*) was most limited by K availability at bog marshes in Breconshire and Cardiganshire, UK (Goodman and Perkins, 1968a, 1968b) and by P at Smaland, Sweden (Tamm, 1954), while neither K nor P was found to be limiting at Moor House, UK (Gore, 1961a, 1961b). Gore (1963) found that N was a more important factor limiting purple moor grass (*Molinia caerulea*) productivity at Moor House, than was either P or K. The outcome of nutrient enrichment experiments conducted in the field is often influenced by other site factors, such as substrate pH and water level. The solubility, and hence availability, of mineral nutrients is pH dependent (Truog, 1946). Consequently, the results of a nutrient enrichment experiment could be very different when conducted in a series of bog marshes differing in the pH of their substrate. Waterlogging and the accompanying effects of poor aeration influence nutrient availability both directly, by impairing the ability of roots to absorb nutrients, and indirectly by re-

Table 2. Summary of the Treatment and Findings of Nutrient Enrichment Experiments Conducted in Northern Bog Marshes

Author	Species involved	Location	Nutrient added	Effect on primary production
Tamm, 1954	*Eriophorum vaginatum*	Smaland, Sweden	50 kg/ha P	9-fold increase in aboveground dry weight 16 months after treatment applied
Gore, 1961*a*, *b*	Transplanted *Eriophorum vaginatum* and *Molinia caerulea*	Moor House, UK	2,500 or 5,000 kg/ha Ca, 24.4–244 kg/ha P, or 600 kg/ha N plus P	No significant increase in the dry weight of aboveground shoots after 1 year
Gore, 1963	Transplanted *Molinia caerulea*	Moor House, UK	2,500 or 5,000 kg/ha Ca, 24.4–244 kg/ha P, or 600 kg/ha N plus P	Significant increase in the shoot and root dry weight of plants in response to N after 2nd year, but no effect of P or P plus N
Goodman and Perkins, 1968*a*, *b*	*Eriophorum vaginatum*	Breconshire and Cardiganshire, UK	0–25X where X is 12.6 kg/ha K or 9.88 kg/ha P or 14.4 kg/ha Mg or 40 kg/ha Ca or 20 kg/ha N	Tiller production rate was increased by nutrients in the following order: $K \geqslant P >$ mixture $\geqslant N > Ca >$ **Mg**
Goodman, 1963, 1968	*Eriophorum angustifolium*	Aberystwyth, Wales, UK	0–50, 213 kg/ha Ca	Tiller production increased (up to 3×) at intermediate concentrations during 5-year period following application
Sheikh, 1969	Transplanted *Molinia caerulea*	Hampshire, UK	0–54 kg/ha N and K and P	Shoot and root dry weight doubled after 2 growing seasons

Table 2. Continued

Author	Species involved	Location	Nutrient added	Effect on primary production
Haag, 1974	*Eriophorum* spp.., *Carex* spp.	Northwest Territories, Canada	0–200 kg/ha N or P	Nitrogen almost doubled production measured 3 months after treatment, but P did not increase production. Nitrogen plus P had a greater effect than N alone.
Richardson et al., 1976	*Chamaedaphne calyculata*, *Betula pumila*	Houghton Lake, Michigan, USA	0–1,108 kg ha^{-1} yr^{-1} solution containing P, carbonate, N, Na, Fe	No effect on aboveground dry weight up to 18 months following initial application
	Carex spp.	Houghton Lake, Michigan, USA	4,432–11,080 kg ha^{-1} yr^{-1} solution containing P, carbonate, N, Na, Fe	Up to 25% greater aboveground standing crop 4 months following initial application

stricting root growth to upper portions of the peat substrate. Laboratory experiments conducted by Loach (1968) and Gore and Urquhart (1966) have demonstrated the detrimental effects of waterlogging on nutrient uptake and primary productivity of *Eriophorum vaginatum* and *Molinia caerulea*. The apparently conflicting results of field experiments conducted with *Eriophorum vaginatum* (Table 2) may be due to the effects of pH and/or water level differences at the field sites.

Only 2 field studies involving nutrient enrichment have been conducted in North American bog marshes. Haag (1974) found that the addition of N or N plus P caused an immediate (i.e., after 3 months) increase in the primary productivity of a bog marsh dominated by species of *Eriophorum* and *Carex* in the Northwest Territories. Richardson et al. (1976) also recorded an immediate increase in the aboveground standing crop of a sedge wetland in Michigan in response to the addition of simulated sewage effluent. An earlier experiment, on the other hand, produced no significant change in the aboveground weight of 2 woody bog marsh plants (i.e., *Chamaedaphne calyculata* and *Betula pumila*) found in the same wetland. The long-term effects of nutrient enrichment on the productivity of these woody plants has yet to be evaluated.

It should be obvious from the brevity of this review that further experimentation, both in the field and in the laboratory, will be necessary to draw firm conclusions concerning the effects of substrate conditions and the climatic regime on primary production in northern bog marshes. The only facts apparent at the present time are that individual bog marshes vary considerably both in their productivity and in their response to nutrient enrichment and that the northern climate restricts the production potential of bog marshes.

LITERATURE CITED

Bernard, J. M. (1974). Seasonal changes in standing crop and primary production in a sedge wetland and an adjacent old-field in central Minnesota. *Ecology* 55, 350–359.

Doyle, G. J. (1973). Primary production estimates of native blanket bog and meadow vegetation growing on reclaimed peat at Glenamoy, Ireland. *In* "Primary production and production processes, tundra biome" (L. C. Bliss and F. E. Wielgolaski, eds.), pp. 141–151. Tundra Biome Steering Committee, Edmonton and Stockholm.

Forrest, G. I., and Smith, R. A. H. (1975). The productivity of a range of blanket-bog types in the Northern Pennines. *J. Ecol.* 63, 173–202.

Goodman, G. T. (1963). The role of mineral nutrients in *Eriophorum* communities. I. The effect of added ground limestone upon growth in an *E. angustifolium* community. *J. Ecol.* 51, 205–221.

Goodman, G. T. (1968). II. The effects of added ground limestone upon the availability and uptake of inorganic elements in an *E. angustifolium* community. *J. Ecol.* 56, 545–564.

Goodman, G. T., and Perkins, D. F. (1968*a*). III. Growth response to added inorganic elements in two *E. vaginatum* communities. *J. Ecol.* 56, 667–683.

Goodman, G. T., and Perkins, D. F. (1968*b*). IV. Potassium supply as a limiting factor in an *E. vaginatum* community. *J. Ecol.* 56, 683–696.

Gore, A. J. P. (1961a). Factors limiting plant growth on high-level blanket peat. I. Calcium and phosphate. *J. Ecol.* **49**, 399–402.

Gore, A. J. P. (1961b). II. Nitrogen and phosphate in the first year of growth. *J. Ecol.* **49**, 605–616.

Gore, A. J. P. (1963). III. An analysis of growth of *Molinia caerulea* (L.) Moench. in the second year. *J. Ecol.* **51**, 481–491.

Gore, A. J. P., and Urquhart, C. (1966). The effects of waterlogging on the growth of *Molinia caerulea* and *Eriophorum vaginatum*. *J. Ecol.* **54**, 617–633.

Gorham, E. (1974). The relationship between standing crop in sedge meadows and summer temperature. *J. Ecol.* **62**, 487–491.

Haag, R. W. (1974). Nutrient limitation to plant production in two tundra communities. *Can. J. Bot.* **52**, 103–116.

Johansson, L-G., and Linder, S. (1975). The seasonal pattern of photosynthesis of some vascular plants on a subarctic mire. *In* "Fennoscandian Tundra Ecosystems". (F. E. Wielgolaski, ed.), pp. 194–200. Springer-Verlag, New York.

Loach, K. (1968). Relations between soil nutrients and vegetation in wet-heaths. II. Nutrient uptake by the major species in the field and in controlled conditions. *J. Ecol.* **56**, 117–127.

Miller, P. C., and Stoner, W. A. (1976). A model of stand photosynthesis for the wet meadow tundra at Barrow, Alaska. *Ecology* **57**, 411–430.

Milner, C., and Elfyn-Hughes, R. (1968). "Methods for the measurement of the primary production of grasslands." I. B. P. Handbook No. 6. Blackwell, Oxford.

Reader, R. J., and Stewart, J. M. (1972). The relationship between net primary production and accumulation for a peatland in southeastern Manitoba. *Ecology* **53**, 1024–1037.

Richardson, C. J., Wentz, W. A., Chamie, J. P. M., Kadlec, J. A., and D. L. Tilton. (1976). Plant growth, nutrient accumulation and decomposition in a central Michigan peatland used for effluent treatment. *In* "Freshwater wetlands and sewage effluent disposal" (D. L. Tilton, R. H. Kadlec, and C. J. Richardson, eds.), pp. 77–117. University of Michigan, Ann Arbor, Michigan.

Sheikh, K. H. (1969). The effects of competition and nutrition on the interactions of some wet-heath plants. *J. Ecol.* **57**, 87–99.

Tamm, C. O. (1954). Some observations on the nutrient turnover in a bog community dominated by *Eriophorum vaginatum* L. *Oikos* **5**, 189–194.

Tieszen, L. L. (1972). The seasonal course of aboveground production and chlorophyll distribution in a wet arctic tundra at Barrow, Alaska. *Arctic and Alpine Res.* **4**, 307–324.

Truog, E. (1946). Soil reaction influence on availability of plant nutrients. *Soil Sci. Soc. Amer. Proc.* **11**, 305–308.

Wein, R. W., and Bliss, L. C. (1974). Primary production in arctic cotton grass tussock tundra communities. *Arctic and Alpine Res.* **6**, 261–274.

Wielgolaski, F. E. (1975). Primary productivity of alpine meadow communities. *In* "Fennoscandian Tundra Ecosystems" (F. E. Wielgolaski, ed.), pp. 121–128. Springer-Verlag, New York.

Woodwell, G. M., and Botkin, D. B. (1970). Metabolism of terrestrial ecosystems by gas exchange techniques: the Brookhaven approach. *In* "Analyses of Temperate Forest Ecosystems" (D. E. Reichle, ed.), pp. 73–85. Springer-Verlag, New York.

THE ROLE OF HYDROLOGY IN FRESHWATER WETLAND ECOSYSTEMS

J. G. Gosselink and R. E. Turner

Center for Wetland Resources
Louisiana State University
Baton Rouge, Louisiana 70803

Abstract   This communication discusses the key role of the hydrologic regime as a controller of wetland ecosystems. The source, velocity, renewal rate, and timing of the water in a wetland ecosystem directly controls the spatial heterogeneity of wetlands and the nutrient, O_2, and toxin load of the sediments. These secondary factors in turn control or modify such ecosystem characteristics as species composition and richness, primary productivity, organic deposition and flux, and nutrient cycles. Wetlands are classified according to their hydrodynamic regime, and trends in ecosystem response along a hydrodynamic gradient are discussed briefly.

Key Words   *Diversity, hydrology, nutrient cycles, nutrients, organic flux, O_2, primary production, toxins, water renewal, water velocity, wetlands.*

INTRODUCTION

Freshwater wetlands are found in many kinds of physical regimes, and include such diverse ecosystems as the extensive raised and blanket bogs of the northern Canadian shield, sedge meadows in Minnesota, freshwater tidal marshes on the Atlantic and Gulf coasts, and marshes of high-energy rivers such as the Mississippi and the Atchafalaya. Much of the Everglades is a marshy sea and the shores of many lakes are laced with pocket marshes and fringe marshes. All these wetlands have at least one thing in common: they are flooded frequently enough so that the roots of the emergent vegetation exist in an anaerobic environment. (The term "emergent vegetation" is used to include mosses and other nonvascular plants such as sphagnum, vascular grasses and herbs, and woody trees such as black spruce, cypress, and tupelo.)

On the other hand, different wetlands may vary widely in other ecosystem attributes, including characteristic biota, species richness, productivity, the amount of organic matter accumulating in the sediments, and in the degree of coupling with adjacent ecosystems.

One can consider any specific wetland type to occupy a certain niche. This niche is one instance of the environmental mosaic that results from the interaction of climatic, geologic, hydrologic, and biologic processes. It supports certain species at a particular level of productivity. To the extent that the hydrologic regime distinguishes emergent wetlands from aquatic and terrestrial systems, it is the primary determinant of all wetland systems. Yet solid quantitative information about the hydrodynamic characteristics of different wetlands is surprisingly difficult to find. In this paper we discuss, in a conceptual way, the relationship between hydrodynamics and wetland ecosystem characteristics. Specifically we identify several attributes of the hydrologic regime that seem to be important variables, we show that these attributes directly influence or modify a range of other secondary abiotic parameters (pH, nutrient flux, dissolved O_2, etc.), and that these secondary parameters in turn determine the biotic response. Finally, we complete the loop by relating changes in the hydrologic regime to environmental changes resulting from the biota.

The insights and examples used draw heavily from European and Canadian literature about inland bogs, mires, and fens. These are, in general, fairly low-energy examples (in the sense of water velocity, flooding frequency) of the broad range of wetlands that exist, and we attempt to extend and amplify concepts from these systems to wetlands in general. (Fig. 1).

CONCEPTUAL MODEL

Figure 1 shows a simple model that allows us to set bounds to the discussion and to define components. Climatic influences such as temperature and radiation may have an overriding influence on the type of species that develop in a wetland, but because the hydrologic interactions with wetland ecosystems are our primary interest, we limit the discussion to local effects.

Hydrologic Parameters

The following attributes of the hydrologic regime (**1**, in Fig. 1) seem to us of most importance to the biota:

The SOURCE determines ionic composition, oxygen saturation, and toxin load. The VELOCITY affects turbulence and the ability of the water to carry suspended particulate matter. The RENEWAL RATE describes the frequency of replacement of the water. It depends on water depth (volume),

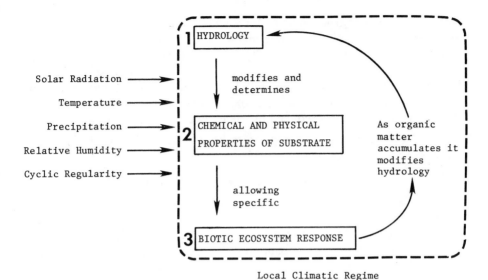

Fig. 1. General conceptual model of the role of hydrology in wetland ecosystems.

frequency of inundation, and velocity, and it is one of the most difficult hydrodynamic parameters to measure. The TIMING, that is, the frequency of inundation (daily, seasonal) and its regularity or predictability, influences the potential for system succession and maturation (Margalef, 1968).

Chemical and Physical Properties

Four chemical and physical properties of the substrate (2, in Fig. 1) that are strongly influenced by the hydrologic regime, and which appear to limit ecosystem development in flooded systems are:

Water Under most conditions of plant growth water is a limiting commodity. However, freshwater emergent plants must often cope with long periods of flooding. For this reason, the major effect of water on wetland plants is secondary; that is, it does not itself directly limit plant growth. Rather its influence can be traced through secondary responses related to O_2 availability in the root zone.

Nutrients The necessity for, role of, and limitations placed on plant growth by inorganic nutrients are too well documented to require amplification.

Toxins By definition, toxins have a controlling role in ecosystems. They can be natural toxins, as salt or H_2S, or human artifacts such as pesticides.

Oxygen Availability The flooded condition of wetland soils results in an anoxic environment, and this in turn leads to a whole set of chemical differences compared to oxidized soils (Patrick and Mikkelsen, 1971). For example, there is a tendency for micronutrients to be more readily available; toxic materials such as H_2S accumulate in the substrate; O_2 is limiting to the roots and plants must develop alternate methods of respiring or of acquiring O_2. Surprisingly, there has been relatively little experimental study of the effects of anoxic soil conditions on marsh plant growth, but the analyses of flooded rice and marsh soils by Patrick and Mikkelsen (1971) and by Armstrong (1975), for example, show that the anoxic environment is a stress factor and that growth in general is reduced by highly anaerobic conditions.

Spatial Heterogeneity   This appears to be a major determinant of species richness in ecosystems (Jacobs, 1975).

ECOSYSTEM RESPONSE TO HYDROLOGY

In the discussion that follows we address selected ecosystem characteristics (3, in Fig. 1), as they are controlled or modified within wetlands by the hydrologic regime. These characteristics are: species composition and richness, primary productivity, organic deposition and flux, and nutrient cycling.

Species Composition and Richness

A major factor influencing species richness is spatial heterogeneity, since the greater the spatial heterogeneity, the greater the number of niches, and the more opportunity for a successful invasion by a species (Jacobs, 1975). The hydrologic regime is not the only factor affecting spatial diversity, but it is a major one. First, flooding waters provide a vehicle for the movement of materials, either dissolved or suspended, which is absent in terrestrial environments. This may have the effect of minimizing the spatial diversity because of uniform mixing. Therefore, wetlands subject to sheet flow by flooding waters tend to be quite uniform and to have large areas of monospecific stands. Examples are the large reed (*Phragmites communis*) marshes at the mouth of the Mississippi River, the sawgrass wetlands of the Florida Everglades, and blanket bogs of Minnesota. On the other hand, the hydrologic regime can contribute to elevational and substrate differences, which are a chief source of species diversity in wetlands (Hinde, 1954). As rising waters crest over stream banks, current velocity is reduced, resulting in a gradient of elevation and sediment grain size. The secondary effect is typical plant zonation, with different species occurring at different elevations. Diversity generally seems to increase with elevation (Fig. 2) and therefore is a function of flooding duration and depth. Data from Heinselman (1970) indicate that plant species richness increases with increasing water velocity (and probably renewal rate) in northern Minnesota peatlands (Table 1). Thus, the hydrologic regime can lead to either uniformity or to diversity, depending on the regime of a specific wetland landscape. Whether diversity increases through time is often determined by whether the developing biota have a reciprocal effect on the hydrologic regime (Gorham, 1957). These successful effects will be discussed later.

Table 1. The Relationship between the Hydrologic Regime and Species Richness in the Northern Minnesota Peatlands[a]

	Species present						Comments
	Tree	Shrub	Field herbs	Grasses and ferns	Ground layer	Total	
1. Rich swamp forest	6	16	28	11	10	71	Good surface flow: minerotrophic
2. Poor swamp forest	3	14	17	12	5	51	Downstream from 1; not adapted to strong water flow
3. Cedar string bog and fen	3	10	10	12	4	39	Better drainage than 2
4. Larch string bog and fen	3	9	9	12	4	37	Similar to 3; sheet flow
5. Black spruce feather moss forest	2	9	2	2	10	25	Gentle water flow on semiconvex template
6. Sphagnum bog	2	8	2	1	7	20	Isolated, little standing water
7. Sphagnum heath	2	6	2	2	5	17	Wet, soggy, and on convex template

[a] The data are from Heinselman, 1970. Note the increasing number of vegetation types as the velocity, frequency, and duration of the hydrologic flow per square metre increases.

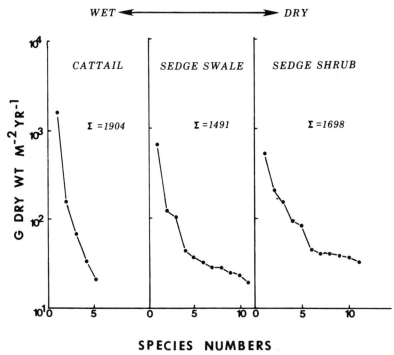

SPECIES NUMBERS

Fig. 2. The relationship between the individual species and their importance in the community ranked in terms of their annual production. Note that the differences in the production of the dominant and of the entire community and species richness going from wet to drier environments (after Jervis, 1969).

Primary Productivity

Primary productivity may be limited at times by any of the critical substrate parameters listed above. Water stress is not usually a factor (but see Pigott, 1969, and Dunn et al., 1976), but the lateral and vertical spread of ombrotrophic bogs seems to be controlled by the availability of water (Gorham, 1957). The timing or seasonality of the rain input may also be critical (Kulczynski, 1949).

The availability of dissolved nutrients for plant growth is a function of concentration (i.e., source) of the nutrients and of renewal. Gorham (1957), for instance, states that the total flux of nutrients into a wetland is more important than the instantaneous concentration, and Newbould (1960) has shown that the nutrients available in the water in the vicinity of the plant roots at any one time are < the amount taken up by the plant during the growing season. In the smooth cordgrass (*Spartina alterniflora*) marshes of coastal Massachusetts, N is limiting in inland marshes but not in streamside marshes, presumably because the frequent flushing of the latter results in a larger total N supply (Valiela et al., 1975).

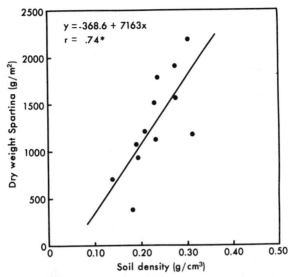

Fig. 3. The relationship of sediment density to smooth cordgrass (*Spartina alterniflora*) biomass in a salt marsh (from Delaune et al., 1975): asterisk (*) indicates that regression is significant at the 95% probability level.

Dissolved nutrient availability is also influenced indirectly by the degree of anoxia of the substrate, since many nutrients, especially micronutrients and PO_4, become soluble under anaerobic conditions (Patrick and Mikkelsen, 1971).

The availability of nutrients bound to particulate materials is also strongly influenced by hydrodynamic considerations, particularly the velocity of flooding waters. Perhaps the best example of this is the recent report by Delaune et al. (1976) that the end-of-season biomass of smooth cordgrass is directly related to the density of the substrate (Fig. 3). Substrate density is a reflection of the amount of inorganic sediment input compared to organic content. The higher the flooding velocity, the greater the sediment input, and the more vigorous the plant growth. If one assumes that streamside locations sustain higher velocities than inland ones, then the same relationship between velocity and productivity is shown also in the well-known "edge effect," the stimulation of production along stream banks (Kirby and Gosselink, 1976, and Smalley, 1959, for salt marshes; Buttery et al., 1965, for freshwater marshes).

In addition to nutrients, the source of water is also a source of toxins. Probably the most ubiquitous of these are salts, usually associated with sea water. There is an extensive literature to show that salt inhibits growth of all but a few true halophytes (Phleger, 1971; Seneca, 1972). The importance of the renewal rate is illustrated clearly with this chemical. In many coastal areas subject to regular flooding, saline marshes support a vigorous flora. However, at sites subject to infrequent tidal flooding (for instance, at the spring tide elevation) salt is concentrated as a result of evapotranspiration, and all vegetation is killed (Gosselink et al., 1971). The regular renewal of the water at lower elevations prevents salt accumulation.

Other toxins, such as herbicides and pesticides, are also carried into wetlands by flooding waters. In these cases also the frequency and velocity of flooding are important in determining the total amount available to wetland flora and fauna.

Under saturated conditions, with low renewal rates, the depletion of O_2 in the substrate leads to a number of chemical changes, which together have a significant effect on plant productivity. In the first place, apparently only those plants flourish that have evolved specialized tissues to allow diffusion of O_2 from the leaves through the stem into the roots (Armstrong, 1975). Even so, the anaerobic environment of the roots seems a significant stress. The effect may be indirect, as through accumulation of toxic sulfides (Joshi et al., 1975), or solubilization and absorption of high concentrations of micronutrients (Patrick and Mikkelsen, 1971). The results of highly anaerobic conditions have been described without the exact mechanism being known. Harms (1973), for example, found that tupelo seedlings grew best under aerobic conditions, moderately well when continuously flooded with moving water, and poorest with stagnant flooding. This effect is related to depth and duration of flooding (renewal rate) and also to water velocity directly. In one of the few careful quantitative studies of hydrodynamics of wetlands, Sparling (1966) showed that the O_2 content of bog waters was directly related to its velocity. At velocities >1 cm/s, the waters remained agitated and saturated with O_2. Below 0.4 cm/s, O_2 replenishment was equal to that of diffusion into stationary water; because of sediment and root respiration, O_2 then tends to be depleted.

Organic Deposition and Flux

Wetland systems are generally net producers; that is, primary production exceeds consumption. The fate of this excess production is strongly influenced by the hydrologic regime. At one extreme are depression bogs that accumulate most of their primary production as peat (Bellamy and Reilly, 1966). At the other extreme, Teal (1962) estimated an organic export from highly flushed salt marshes of ≈45% of net primary production. There is little quantitative information about the role of hydrology in this process. However, Odum and de la Cruz (1967) did measure organic export from a Georgia tidal salt marsh. The rate of total particulate and total organic export was directly proportional to volumetric flow rates. Gosselink et al. (1977) showed that as the frequency of inundation of coastal marshes decreases, the inorganic concentration and particle size of the marsh sediments decreases, and organic concentration increases; that is, flooding frequency is directly related to inorganic silt input and to organic export (Fig. 4). Whigham and Simpson (1975) also found in freshwater marshes in New

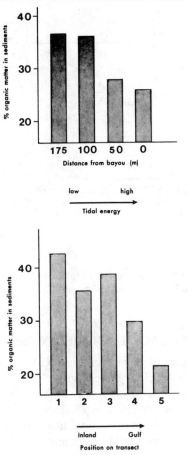

Fig. 4. The relationship of hydrologic energy to organic concentration in salt marsh sediments. Tidal energy decreases from the edge of the bayou inland and from the Gulf inland.

Jersey that soil organic concentration increased on a gradient from actively flooded stream banks to less actively flooded inland high marshes.

Nutrient Cycling

The nutrient load in flooding waters depends on the flux volume of water and its concentration (source). Thus, ombrotrophic bogs are nutrient-poor because the source is rainwater, whereas the water supply to minerotrophic bogs is often rich in minerals, depending on the substrate (Bay, 1967). This influences not only the loading of nutrients to the wetland but often the acidity. Waters with low dissolved solids tend to become acid in contact with peats, whereas mineral-rich waters are much less acid, especially if they percolate over a calcium carbonate rock (Gorham, 1957). Saline waters are highly buffered and even in reducing sediment seldom are much below pH 7.

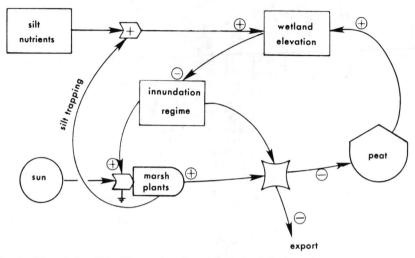

Fig. 5. The relationship of interactions between the hydrologic regime, marsh plants, and sedimentation characteristics in a hypothetical marsh.

FEEDBACK LOOPS (SUCCESSION)

The classical examples of bog succession (Gorham, 1957), emphasize the feedback loop through which organic accumulation changes the hydrologic regime by raising the wetland elevation or blocking and diverting earlier waterflows. Figure 5 illustrates the reciprocal effects between wetland ecosystems and hydrology. Plant production in the marsh yields organic matter that can become deposited in sediment as peat or exported. Whether it is exported or deposited depends to some extent on the inundation regime. That is, a high-energy hydrologic regime will carry much of the production out of the marsh (e.g., Mason and Bryant, 1975), but there are at least 2 feedback loops that modify this response. In the first place, the vegetation of the marsh acts as a silt trap, particularly at the edge (e.g., Buttery et al., 1965); this tends to increase the sedimentation rate and raise the elevation of the marsh. Second, as the marsh elevation increases, the frequency and depth of flooding decrease so that less organic production is exported and is instead deposited as peat. This deposition results in further increases in marsh elevation. These 2 factors tend to cause a wetland to continue to grow upward until it reaches the upper level of the flooding water; or in the case of ombrotrophic bogs, the elevation above which it does not stay saturated. The buildup of biotic material increases peat production at the expense of inorganic inputs. Heinselman (1970) has discussed the situation in the northern Minnesota peatlands. In this case, the surface vegetation was shown to delineate the subsurface hydrology. Peat accumulation was found to alter the hydrology, since it acted as a barrier to percolation of surface waters.

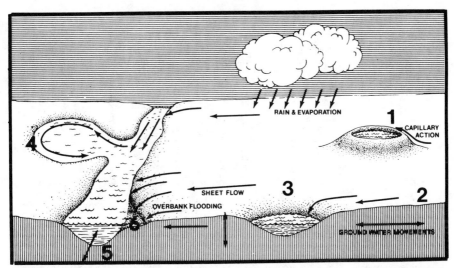

Fig. 6. A schematic representation of 6 types of freshwater marsh environments, and their hydrologic regime. The numbers correspond to the description of each in Table 2.

1. Raised-convex (ombrogenous bogs, Gorham, 1957)
2. Meadow (blanket bogs, DuRietz, 1949; lacustrine bogs)
3. Sunken minerotrophic (fen, DuRietz, 1949; carr, Gorham, 1957)
4. Lotic (fen, carr; reed swamp, Gorham, 1957)
5. Tidal
6. Lentic (riverine)

Another feedback loop through the biota occurs with nutrients. As peat deposition increases or silt trapping increases, the marsh builds upward. The increase in elevation, and in total standing biomass aboveground and peat underground, tend to close the nutrient cycle and less flux occurs across the marsh boundaries.

Although the modification of the hydrologic regime and simultaneous biotic changes may produce dramatic instances of vertical succession (ombrotrophic bogs from minerotrophic wetlands), or horizontal lakeshore succession from open water through grassed and wooded wetlands to raised bogs (Gorham, 1957), many more energetic wetland systems seem to be arrested in immature stages by the pulsed inundation regime. Odum (1971) terms this "pulse stability" of fluctuating water-level ecosystems and cites coastal tidal marshes and the Florida Everglades as examples. In Margalef's (1968) terms, these systems cannot mature because the flooding waters continually dilute them, preventing the accumulation of biological information.

CLASSIFICATION OF WETLAND SYSTEMS ON A HYDRODYNAMIC ENERGY GRADIENT

These considerations of the importance of hydrology in wetlands suggest that the classification of wetlands along a hydrodynamic energy gradient

Table 2. The Major Hydrodynamic Characteristics of Different Marsh Types Shown in Fig. 6

Marsh type	Water inputs to marsh				Type of water flow				Outputs from marsh			Hydro–pulse
	Capillarity	Precipitation	Upstream	Downstream	Capillary	Subsurface	Surface sheet flow	Overbank flooding	Percolation	Evapotranspiration	Downstream runoff	
1. Raised–convex	+	+			+	+			+	+		Seasonal
2. Meadow	+	+	v.little (+)				+			+	v. little	Seasonal
3. Sunken–convex	+	+				+	v. slow			+		Seasonal
4. Lotic	+	+				+	+	+		+	+	Seasonal
5. Tidal	+	+		+		+	+	+		+	+	Tidal
6. Lentic	+	++				+	+	+		+	+	Variable/seasonal

could be a useful approach to understanding functional relationships. Figure 6 catalogs freshwater wetlands into 6 classes based on source and velocity of water flow. In general flow rate, or other indications of hydrologic energy such as renewal time or frequency of flooding, increases from raised convex wetlands to lentic and tidal wetlands. The major sources of water and types of flow are summarized in Table 2. If a classification of wetlands on a hydrodynamic energy gradient is of functional value, it should correspond with gradients in ecosystem attributes. We therefore consider briefly the 4 system attributes discussed above to see if any consistent trends appear when different wetland classes are compared.

During classical successional processes, hydrologic flows are diverted and reduced by the developing biota. According to ecosystem theory (e.g., Odum, 1971; Margalef, 1968), diversity should increase with maturity. This would mean that diversity should be highest in perched bogs and lowest in high-energy wetlands. The examples of Heinselman (1970) portrayed in Table 2, and Bay (1967), indicate the reverse. Bay compared perched bogs (raised-convex wetland), to one in contact with the groundwater supply (sunken wetland) and reported much higher vegetation diversity on the latter site, which he related to a richer nutrient supply. These examples suggest that the opportunities for niche differentiation increase with the hydrologic energy of the wetland. Whether this relationship holds true for extremely high-energy freshwater tidal and lentic wetlands is not clear. For example,

Table 3. Comparison of Cypress Swamp Productivity as Related to Water Flows[a]

Stagnant (cypress domes)	NPP (g)
Stagnant (cypress domes, Florida)	192
Cypress dome and riverine combined (Withlacoochee State Forest, Florida)	600
Very slowly flowing water (Okefenokee Swamp, Georgia)	692
Riverine-undrained-edge strand (Big Cypress Swamp, Florida)	1170
Semiriverine, seasonal flooding (des Allemands Swamp, Louisiana)	1140

[a] From Conner and Day, 1976.

freshwater marshes at the mouth of the Mississippi River are often monotypic (data from Chabreck, 1972) compared to the diverse freshwater marshes found in less energetic situations elsewhere along the Louisiana coast.

Productivity is so strongly influenced by many factors that generalizations may be impossible to make. Low-energy wetlands tend to be concentrated in northern areas, whereas southern wetlands are usually lentic, lotic, or tidal. Thus, comparisons of different wetland types are confounded by different temperatures and growing seasons. In addition, productivity measurements have not been standardized, and large fluctuations in estimates result from use of different techniques. Comparisons of single species growing in different hydrologic regimes within a limited geographical area may provide some insight into the role of hydrology in production. Whigham and Simpson (1977) compared production of *Zizania aquatica* from tidal and nontidal sites. Considering only those in New Jersey, tidal sites were more productive than nontidal ones. This correlation between the energy of the flooding regime and productivity is reinforced by a comparison of productivity of cypress swamps in the southeastern United States (Conner and Day, 1976). Productivity appears to be positively correlated with the water-flow regime (Table 3).

Accumulation of organic matter and peat development are characteristic of low-energy wetlands. Raised sphagnum bogs and lotic freshwater marshes are alike in this respect. The rate of peat accumulation depends on the production rates as well as on the duration and depth of surface flooding, because oxidation of the substrate occurs rapidly upon exposure. The proportion of primary production that is deposited as peat in different wetlands is difficult to determine. It is expected that as flooding energy increases, the proportion of production exported also increases. Thus deposition and export are inversely related to one another and to flooding energy.

On a continuum from low-energy, ombrotrophic bogs to lentic and tidal wetlands internal cycling of nutrients should decrease and dependence on external sources increase. Nutrient budgets have not been calculated for many marsh systems. Nevertheless, much information can be pieced together. Perched bogs are limited to precipitation for nutrient input, so are usually very nutrient-poor. Most nutrients in these wetlands are bound in organic form. Bay (1967) reports 2 to 4 times higher concentrations of nutrients in interstitial waters of forested bogs in contact with the regional water table (sunken wetland?) than in a perched bog. At the other extreme in tidal salt marshes, Hopkinson and Day (1977) estimate that nitrogen fluxes across salt marsh boundaries in coastal Louisiana are ≈75% of that recycled within the marsh. This is a very open system.

CONCLUSIONS

If this analysis has any validity, it suggests that in order to build a complete functional picture, we should pay much more attention to the hydrologic regime. For instance, continuous recordings of water level require little time for installation and maintenance, and yield frequency, duration, regularity, and depth of inundation—important parameters that can also be used as indexes of velocity. Experiments should be designed with the hydrodynamic regime in mind so that appropriate sites are chosen to show a range of hydrodynamic inputs and an effort is made to separate different hydrodynamic parameters.

ACKNOWLEDGMENTS

This work is a result of research sponsored by the Louisiana Sea Grant Program, a part of the National Sea Grant Program maintained by the National Oceanic and Atmospheric Administration of the U.S. Department of Commerce (Grant 2-35231), and by the United States Geological Survey (Grant 14-08-0001-G-234).

LITERATURE CITED

Armstrong, W. (1975). Waterlogged soils. In "Environment and Plant Ecology" (J. R. Etherington, ed.), pp. 181–218. J. Wiley and Sons, New York.
Bay, R. R. (1967). Ground water and vegetation in two peat bogs in northern Minnesota. Ecology 48, 308–310.
Bellamy, D. J., and Reilly, J. (1966). Some ecological statistics of a "miniature bog." Oikos 18, 33–40.
Buttery, B. R., Williams, W. T., and Lambert, J. M. (1965). Competition between Glyceria maxima and Phragmites communis in the region of Surlurgham Broad. J. Ecol. 53, 183–195.
Chabreck, R. H. (1972). Vegetation, water, and soil characteristics of the Louisiana coastal region. Louisiana State University Agricultural Experiment Station Bulletin No. 664. 72 pp.
Conner, W. H., and Day, J. W., Jr. (1976). Productivity and composition of a baldcypress-water tupelo site and a bottomland hardwood site in a Louisiana swamp. Amer. J. Bot. 63, 1354–1364.
Delaune, R. D., Patrick, W. H. Jr., and Brannon, J. M. (1976). "Nutrient transformations in

Louisiana salt marsh soils." Louisiana State University Center for Wetland Resources, Baton Rouge. Sea Grant Publ. No. LSU-T-76-009.

Dunn, E. L., Giurgevich, J. R., and Antlfinger, A. E. (1976). Intrinsic limitations to photosynthesis in Southeastern salt marsh species. *Bull. Ecol. Soc. Amer.* **57**, 40 (abstract).

DuRietz, G. E. (1949). Huvudenheter och huvudgranser i svensk myrvegetation. *Svensk bot. Tidskr.* **43**, 274–309 (English summary).

Gorham, E. (1957). The development of peat lands. *Quart. Rev. Biol.* **32**, 145–166.

Gosselink, J. G., Reimold, R. J., Gallagher, J. L., Windom, H. L., and Odum, E. P. (1971). "Spoil Disposal Problems for Highway Construction through Marshes." Institute of Ecology, University of Georgia, Athens.

Gosselink, J. G., Hopkinson, C. S. Jr., and Parrondo, R. T. (1977). Marsh plant species, Gulf Coast area. Volume I. "Production marsh vegetation." U.S. Army Corps of Engineers Tech. Rept. D-77. Louisiana State University, Center for Wetland Resources, Baton Rouge, La. (*In press*).

Harms, W. R. (1973). Some effects of soil type and water regime on growth of tupelo seedlings. *Ecology* **54**, 188–193.

Heinselman, M. L. (1970). Landscape evolution, peatland types, and the environment in the Lake Agassiz Peatlands Natural Area, Minnesota. *Ecol. Monogr.* **40**, 235–261.

Hinde, H. P. (1954). Vertical distribution of salt marsh phanerogams in relation to tide levels. *Ecol. Monogr.* **24**, 209–225.

Hopkinson, C. S., and Day, J. W. Jr. (1977). A model of the Barataria Bay salt marsh ecosystem. *In* "Ecosystems Modeling in Theory and Practice: An Introduction with Case Histories" (C. A. S. Hall and J. W. Day Jr., eds.). J. Wiley and Sons, New York.

Jacobs, J. (1975). Diversity, stability, and maturity in ecosystems influenced by human activities. *In* "Unifying Concepts in Ecology" (W. H. van Dobben and R. H. Lowe-McConnell, eds.), pp. 187–207. Dr. W. J. B. V. Publishers, The Hague.

Jervis, R. A. (1969). Primary production in the freshwater marsh ecosystem of Troy Meadows, N. J. *Bull. Torrey Bot. Club* **96**, 209–231.

Joshi, M. M., Ibrahim, I. K., and Hollis, J. P. (1975). Hydrogen sulfide effects on the physiology of rice plants and relation to straighthead disease. *Phytopathology* **65**, 1165–1170.

Kirby, C. J., and Gosselink, J. G. (1976). Primary production in a Louisiana Gulf coast salt *Spartina alterniflora* marsh. *Ecology* **57**, 1052–1059.

Kulczynski, S. (1949). "Peat bogs of Polesie." Mem. Acad. Sci. Cracovie, Classe Sci. Math et nat, B. No. 15.

Margalef, R. (1968). "Perspectives in Ecological Theory." University of Chicago Press, Chicago, Illinois.

Mason, C. F., and Bryant, R. J. (1975). Production, nutrient content, and decomposition of *Phragmites communis* Trin and *Typha angustifolia* L. *J. Ecol.* **63**, 71–96.

Newbould, P. J. (1960). Report on the ecology of Hartland bog and fen, Dorset, given to the British Ecological Society. *J. Ecol.* **48**, 762.

Odum, E. P. (1971). "Fundamentals of Ecology." 3rd ed. W. B. Saunders Co., Philadelphia, Pa.

Odum, E. P., and de la Cruz, A. A. (1967). Particulate organic detritus in a Georgia salt marsh–estuarine ecosystem. *In* "Estuaries" (G. H. Lauff, ed.), pp. 383–388. Publ. 83, AAAS, Washington, D.C.

Patrick, W. H. Jr., and Mikkelsen, D. S. (1971). Plant nutrient behavior in flooded soils. "Fert. Technol. and Use." 2nd ed. Pp. 187–215. Soil Sci. Soc. of Am., Madison, Wisconsin.

Phleger, C. E. (1971). Effect of salinity on growth of a salt marsh grass. *Ecology* **52**, 908–911.

Pigott, C. D. (1969). Influence of mineral nutrition on the zonation of flowering plants in coastal salt marshes. *In* "Ecological Aspects of Mineral Nutrition in Plants" (I. H. Rorison, ed.), pp. 25–35. 9th Symp. British Ecol. Soc.

Seneca, E. D. (1972). Seedling responses to salinity in four dune grasses from the outer banks of North Carolina. *Ecology* **53**, 465–471.

Smalley, A. E. (1959). The role of two invertebrate populations, *Littorina irrorata* and *Orchelimum fidicinium*, in the energy flow of a salt marsh ecosystem. Ph.D. Dissertation, University of Georgia, Athens, 126 pp.

Sparling, J. H. (1966). Studies on the relationship between water movement and water chemistry in mires. *Can. J. Bot.* **44**, 747–758.

Teal, J. M. (1962). Energy flow in the salt marsh ecosystem of Georgia. *Ecology* **43**, 614–624.

Valiela, I., Teal, J. M., and Sass, W. (1975). Production and dynamics of salt marsh vegetation and the effect of experimental treatment with sewage sludge. I. Biomass, production, and species composition. *J. Applied Ecol.* **12**, 973–981.

Whigham, D. F., and Simpson, R. L. (1975). Ecological studies of the Hamilton marshes. Progress report for the period June 1974 to January 1975. In-house publication, Biol. Dept. Rider College, Lawrenceville, New Jersey (mimeo). 185 pp.

Whigham, D. F., and Simpson, R. L. (1977). Growth, mortality, and biomass partitioning in freshwater tidal wetland populations of wild rice (*Zizania aquatica* var. *aquatica*). *Bull. Torrey Bot. Club* **104**, 347–351.

PRIMARY PRODUCTION PROCESSES: SUMMARY AND RECOMMENDATIONS

Armando A. de la Cruz

P. O. Drawer Z
Mississippi State University
Mississippi State, Mississippi 38762

Wetlands represent a transition zone between terrestrial and aquatic systems with their photosynthetic products and by-products serving as a major basis for interaction among these ecosystems. Although still limited, information on primary production of freshwater wetlands has steadily increased during the last several years. Studies reported in this section focus on production in total communities (Reader, 1978; van der Valk and Davis, 1978; Whigham et al. 1978) and the production ecology of individual wetland species (Bernard and Gorham, 1978).

Aboveground primary production data have been obtained by several methods: (1) measuring peak biomass, (2) maximum minus minimum biomass, (3) disappearance methods, (4) multiple harvests, and (5) gas exchange. Aboveground net annual productivity calculated from maximum minus minimum biomass ranged from 102 to 530 g/m² for northern bogs (Reader, 1978) to values of 731 to 2,852 g/m² for prairie glacial marshes (van der Valk and Davis, 1978). Bernard and Gorham (1978) and Whigham et al. (1978) have shown, however, that the range of values cited above are underestimates of actual aboveground production and it appears that production estimates may increase as much as twofold when adjusted for mortality during the growing season.

In contrast to aboveground production, little data currently exist on belowground productivity in freshwater wetlands. Available estimates of belowground production for freshwater wetlands (Table 1) are comparable to minimum values for *Spartina alterniflora* wetlands in North Carolina

Table 1. Annual Belowground Net Primary Productivity of Wetland Communities Summarized by Investigators from Various Published Sources

Community	Net belowground production ($g \cdot m^{-2} \cdot yr^{-1}$)	Reference
Freshwater		
Northern bog wetlands	211–513	Reader (1978)
Prairie glacial marshes	253–640	van der Valk and Davis (1978)
Sedge meadows	134–208	Bernard and Gorham (1978)
Freshwater tidal marshes	160–223	Whigham et al. (1978)
Freshwater riverine marshes	296–710	Klopatek and Stearns (1977)
Typha spp. marshes	1,300	Keefe (1972)
	371–954	Keefe (1972)
Brackish water		
Juncus roemerianus marsh	1,360	de la Cruz and Hackney (1977)
Saline water		
Spartina alterniflora marshes		
Short form	460	Stroud (1976)
	2,500	Valiela et al. (1976)
Tall form	503	Stroud (1976)
	3,500	Valiela et al. (1976)

(Stroud, 1976) but much lower than belowground estimated production reported by Valiela et al. (1976) for the same species and for *Juncus*-dominated wetlands studied by de la Cruz and Hackney (1977).

The papers by Bernard and Gorham (1978) and Gosselink and Turner (1978) suggest other important factors that should be considered during productivity studies. Bernard and Gorham (1978) provide data that clearly show species' life histories should be studied because they have a profound influence on production processes and will considerably affect estimates of primary production. The presence of overwintering green shoots, for example, may contribute to maximum standing biomass and cause an overestimation of net production. Conversely, high mortality rates during the growing season will result in underestimates of net production. This situation was documented by Bernard and Gorham (1978) who observed that 84% of the *Carex rostrata* shoots died before completing a normal life-span, and by Whigham et al. (1978) who observed a high leaf mortality rate of 19% in freshwater tidal wetlands. This rate is comparable to the 10–25% rate which I observed for *Juncus roemerianus* in Mississippi wetlands.

Gosselink and Turner (1978) considered how physical parameters, particularly those related to hydrology of wetlands, affect wetland production. Their paper conveys an urgent message to at least include observations on water depth and frequency, duration and regularity of inundation. Rate of water flow may be important in the supply of nutrients (Newbould and

Gorham, 1956; Newbould, 1960; Ingram, 1967), in aeration of peats (Ingram, 1967) and in determining vegetation composition (Newbould, 1960). Reader (1978) discusses other environmental factors that affect primary production including geographic latitude, temperature, nutrient availability and water-logging.

Clearly, to understand fully primary production processes, studies must consider both the hydromechanics and nutrient chemistry of the system. Additionally, for meaningful comparison of production data among wetland types, there is a need to consider a common set of physical, chemical and biological parameters. One can, therefore, envision future productivity studies that routinely encompass the following parameters:

1. *Geographical Location and Hydrologic Regime* Gosselink and Turner (1978) consider the importance of various physical factors that affect wetland productivity. Measurement of air and soil temperatures is a simple matter and should be done routinely. Water level gauges will provide data on 4 parameters of the hydrologic regime (depth and frequency, duration and regularity of inundation) and should be regularly employed in wetland studies.

2. *Soil Regime* Reader (1978) showed that the availability of mineral nutrients in soil is an important factor influencing biomass production. The investigation of nutrient chemistry of wetland substrates is essential to explain, in part, the production variability observed among different stands. Field experiments of nutrient modification will provide clues not only to which elements limit plant productivity but also to the concentrations and combination of nutrients that will maximize a plant's potential to convert radiant energy into biomass. At a minimum, primary production studies should include data on nutrient composition of wetland substrates.

3. *Community Type and Stand History* Detailed vegetational descriptions of wetland communities may seem unimportant and indeed have often been neglected, particularly in almost monotypic wetlands. If community composition is not detailed, there can be erroneous comparisons of production data. Eville Gorham (*personal observation*) has pointed out that it is uncertain whether wetlands of diverse species composition are more or less productive than essentially monotypic stands. This seems to be a question of considerable theoretical interest and may have some practical significance in terms of secondary production. Van der Valk and Davis (1978) have clearly demonstrated that multiyear studies are necessary to estimate accurately production in areas where there are cyclic changes in vegetation brought about by changing environmental conditions in the wetland. Accurate characterization of primary production in freshwater wetlands demands consideration of both community composition and stand history.

4. *Life History* As discussed by Bernard and Gorham (1978) life his-

Table 2. Examples of Suggested Primary Productivity Procedure for Sampling Aboveground Materials in Freshwater Wetlands

Growth forms	Harvest procedure	Plot size	Plot shape	Sampling interval	Replicates (N)	Method reference
Tall grasses (e.g., *Phragmites*, *Spartina cynosuroides*)	Counted and measured	0.25–1.0 m²	Square	Monthly to twice per month	20	Mason and Bryant (1972)
Short grasses (e.g., *Distichlis*, *Panicum* and *Spartina patens*)	Clip	0.1–0.25 m²	Circular	Monthly to twice per month	10	Wiegert and Evans (1964)
Sedge—rush forms (*Carex, Scirpus, Juncus*)	Counted and measured	0.25 m²	Square	Twice per month	10	Bernard and Gorham (1978)
Broad-leaved emergents (*Typha, Nuphar, Peltandra*, and other annuals and perennials)	Clip	0.25 m²	Square	Twice monthly to a month	6	Whigham et al. (1978)

tory studies are imperative to obtain estimates of primary production that are as close as possible to true net production. Information on mortality, natality and phenology of plants or plant parts should be obtained from permanent sampling plots and incorporated in the productivity value estimated from harvest techniques.

5. *Extrinisic Factors* The primary productivity of a wetland is also influenced by a number of biological factors such as grazers, pathogenic organisms, parasites and toxins (i.e., exocrines) secreted by certain biotic components of the system. Some of these biological regulators of plant productivity may be among the most interesting bases of interactions of the wetlands. Unfortunately, they are the most difficult parameters to study and offer a true challenge to future investigations of primary productivity.

Current investigators recognize that variability in sampling procedure and the differences in the method and formula employed in calculating production values are primary reasons why comparisons of productivity data are unreliable. The desire for standardized procedures has been expressed by some workers and shunned by others, but a certain amount of uniformity in methods should be tolerated. The deciding factor in sampling logistics rests on the vegetative morphology, density and growth pattern of the wetland community to be sampled. Consciousness of the possible damage field experiments may do to the wetland has led some researchers (e.g., van der Valk and Davis, 1978; de la Cruz and Hackney, 1977) to use stratified random sampling from more or less restricted zones on preselected grids or from alternate points along a transect line through a uniform vegetation stand. Milner and Hughes (1968) recommend sampling at least 2 m away from plots harvested during the previous collection.

Recommended sampling procedures for aboveground production are summarized in Table 2. Because turnover rate is an important factor, use of disappearance techniques (Wiegert and Evans, 1964), life history methods (Bernard and Gorham, 1978) or single plant procedures (Mason and Bryant, 1975; Williams and Murdock, 1972) is recommended if they can be employed with statistical confidence. The simple maximum biomass or biomass increment methods are also useful but they will yield underestimates of actual net production.

The need for standardized techniques to estimate belowground productivity is acute. The limited estimates of belowground biomass have generally been determined from periodic increases in standing crop or maximum minus minimum biomass methods. Fitted periodic regression curves have also been used to calculate root productivity from monthly biomass change (de la Cruz and Hackney, 1977). The major problem with these procedures is the complete separation of living from dead roots as Stroud (1976) has pointed out. In sampling belowground biomass, root morphology, rooting depth, wetland type and patterns of vegetation must be considered. These

Table 3. Examples of Suggested Primary Productivity for Sampling Belowground Materials in Freshwater Wetlands

Growth forms	Harvest procedure	Core size	Core shape	Sampling interval	Replicates (N)	Method reference
Sedge—rush marsh (Carex, Scirpus, Juncus)	Corer	10 cm diam, 30 cm deep	Cylinder	Monthly	19	de la Cruz and Hackney (1977)
Tall grass (Phragmites, Spartina)	Digging	0.5 m², 20 cm deep	Cube	Monthly	10	Davis and van der Valk (1978)
Annuals in freshwater tidal wetlands	Corer 10 cm deep	5 cm diam	Cylinder	Monthly	10	Whigham et al. (1978)
Nonclonal perennials in freshwater tidal wetlands (Pontederia, Peltandra, Sagittaria)	Digging	Complete extraction		Monthly	10	Whigham et al. (1978)
Clonal perennials in freshwater tidal wetland, (Typha, Acorus, Scirpus, Sparganium)	Digging	0.25 m²	Square	Monthly	10	Whigham et al. (1978)

factors will determine the size and shape of the sampling device, number of samples, location of samples with respect to major plant clumps and the time interval of sampling. Suggested procedures for estimating belowground production in herbaceous dominated wetlands are given in Table 3. Additional information on sampling belowground biomass is provided by Gallagher (1974), Stroud (1976), Valiela et al. (1976), and de la Cruz and Hackney (1977).

The few studies on energy and nutrient contents of plant organs (e.g., Boyd, 1969a, 1969b, 1970; de la Cruz and Poe, 1975; Mason and Bryant, 1975) should be supplemented in current and future investigations so that productivity values can be conveniently expressed in these terms. Because standardization of techniques is desirable, drying and ashing procedures must be considered. Drying time and temperature (usually ranging from 80° to 110°C) to achieve constant weight should be determined for each plant species. If plants are to be analyzed for nutrients, drying at 50°C for 120 hours is recommended to minimize volatilization. Other analytical procedures including standard methods provided by the instrument manual and those found in standard methods publications (e.g., AOAC, APHA, IBP Handbooks) could probably benefit from comparison and intercalibration.

In conclusion, the study of primary production processes in herbaceous dominated freshwater wetlands has intensified in the past several years. A substantial body of data has accumulated on aboveground biomass in these wetlands suggesting that some of them may be among the most productive naturally occurring ecosystems. However, few accurate estimates of net annual productivity, either above- or belowground, exist for freshwater wetlands. Likewise, it is becoming clear that hydrology and soil chemistry are extremely important, but little adequate information is available on exactly how these factors influence the production process. To understand more fully the primary production process in freshwater wetlands and to permit comparison of wetland types, future studies should employ standardized techniques and must consider hydrology, soil regimes, community type (and stand history), species life history and several extrinisic biological factors. As production processes are better elucidated, we will improve our ability to assess natural and artificial perturbations of freshwater wetlands.

ACKNOWLEDGMENTS

I thank Eville Gorham for his valuable suggestions and critical review of this manuscript.

LITERATURE CITED

Bernard, J. M., and Gorham, E. (1978). Life history aspects of primary production in sedge wetlands. In "Freshwater Wetlands: Ecological Processes and Management Potential" (R. E. Good, D. F. Whigham and R. L. Simpson, eds.), pp. 39–51. Academic Press, New York.

Boyd, C. E. (1969a). The nutrient value of three species of water weeds. *Econ. Bot.* **23**, 123–127.

Boyd, C. E. (1969b). Production, mineral nutrient absorption and biochemical assimilation by *Justicia americana* and *Alternanthera philoxeroides*. *Arch. Hydrobiol.* **66**, 139–160.

Boyd, C. E. (1970). Amino acid, protein, and caloric content of vascular aquatic macrophytes. *Ecology* **51**, 902–906.

de la Cruz, A. A., and Hackney, C. T. (1977). Energy value, elemental composition and productivity of belowground biomass of a *Juncus* tidal marsh. *Ecology* **58**, *(In press)*.

de la Cruz, A. A., and Poe, W. E. (1975). Amino acid contents of marsh plants. *Estuar. Coast. Mar. Sci.* **3**, 243–246.

Gallagher, J. L. (1974). Sampling macro-organic matter profiles in salt marsh plant root zones. *Soil Sci. Soc. Am. Proc.*, **38**, 154–155.

Gosselink, J. G., and Turner, R. E. (1978). The role of hydrology in freshwater wetland ecosystems. *In* "Freshwater Wetlands: Ecological Processes and Management Potential" (R. E. Good, D. F. Whigham and R. L. Simpson, eds.), pp. 63–78. Academic Press, New York.

Ingram, H. A. P. (1967). Problems of hydrology and plant distribution in mires. *J. Ecol.* **55**, 711–724.

Keefe, C. W. (1972). Marsh production: A review of the literature. *Contrib. Mar. Sci.* **16**, 163–181.

Klopatek, J. M., and Stearns, F. W. (1977). Primary productivity of emergent macrophytes in a Wisconsin marsh ecosystem. *Am. Midl. Nat.* *(In press)*.

Mason, C. F., and Bryant, R. J. (1975). Production, nutrient content and decomposition of *Phragmites communis* Trin. and *Typha angustifolia* L. *J. Ecol.* **63**, 71–95.

Milner, C., and Hughes, R. E. (1968). Methods for the measurement of primary production in grassland. IBP Handbook No. 6, Blackwell Sci. Publ., Oxford and Edinburg.

Newbould, P. J., and Gorham, E. (1956). Acidity and specific conductivity measurements in some plant communities of the New Forest valley bogs. *J. Ecol.* **44**, 118–128.

Newbould, P. J. (1960). The ecology of Cranesmoor, a New Forest valley bog I. The present vegetation. *J. Ecol.* **48**, 361–383.

Reader, R. J. (1978). Primary production in northern bog marshes. *In* "Freshwater Wetlands: Ecological Processes and Management Potential" (R. E. Good, D. F. Whigham and R. L. Simpson, eds.), pp. 53–62. Academic Press, New York.

Stroud, L. M. (1976). Net primary production of belowground material and carbohydrate patterns of two height forms of *Spartina alterniflora* in two North Carolina Marshes, Ph.D. Dissertation, North Carolina State Univ., Raleigh, N.C.

Valiela, I., Teal, J. M. and Persson, N. Y. (1976). Production and dynamics of experimentally enriched salt marsh vegetation: belowground biomass. *Limnol. Oceanogr.* **21**, 247–254.

van der Valk, A. G., and Davis, C. B. (1978). Primary production of prairie glacial marshes. *In* "Freshwater Wetlands: Ecological Processes and Management Potential" (R. E. Good, D. F. Whigham and R. L. Simpson, eds.), pp. 21–37. Academic Press, New York.

Whigham, D. F., McCormick, J., Good, R. E., and Simpson, R. L. (1978). Biomass and primary production in freshwater tidal wetlands of the Middle Atlantic Coast. *In* "Freshwater Wetlands: Ecological Processes and Management Potential" (R. E. Good, D. F. Whigham and R. L. Simpson, eds.), pp. 3–20. Academic Press, New York.

Wiegert, R. G., and Evans, F. C. (1964). Primary production and disappearance of dead vegetation on an old field in southeastern Michigan. *Ecology* **45**, 49–63.

Williams, R. B., and Murdock, M. B. (1972). Compartmental analysis of the production of *Juncus roemerianus* in a North Carolina salt marsh. *Chesapeake Sci.* **13**, 69–79.

PART II

Decomposition Processes

DECOMPOSITION OF INTERTIDAL FRESHWATER MARSH PLANTS

William E. Odum and Mary A. Heywood

Department of Environmental Sciences
University of Virginia
Charlottesville, Virginia 22903

Abstract This paper summarizes the existing information concerning decomposition of marsh plants which grow in the regularly flooded portions of freshwater tidal marshes. The 6 species investigated *(Peltandra virginica, Nuphar luteum, Pontederia cordata, Bidens laevis, Zizania aquatica, Sagittaria latifolia)* all exhibit rapid rates of decomposition. Typically, 40–50% of the original ash-free dry weight is lost from litterbags within 10 days and 70–80% within 60 days. These marsh plants appear to decay more rapidly and produce larger quantities of dissolved organic matter than the salt-tolerant plants from higher salinity, coastal marshes. There are indications that they contain more N and that the pattern of N change during decomposition differs from that of the higher salinity plants. The leaves of *Peltandra virginica* contain initially between 2–3% N on an ash-free dry weight basis. During decomposition this may increase to 4–5.5% N during the first 20 days and then decline to 3–3.8% after 50 days. The C:N ratio follows a similar, but inverse pattern.

Key words Bidens laevis, C:N ratio, decomposition, freshwater tidal marsh, marsh plants, N change, Nuphar advena, Nuphar luteum, Peltandra virginica, Pontederia cordata, Sagittaria latifolia, Zizania aquatica.

INTRODUCTION

Although tidal freshwater marshes cover large tracts of the coastal zone of eastern North America, particularly between southern Virginia and northern New Jersey, only a modest research effort has been directed to these environments. This is in marked contrast to the massive literature which exists for saline tidal marshes. As a result, such basic ecosystem processes as primary and secondary production, decomposition, nutrient cycling, and succession are poorly understood for the tidal freshwater marsh ecosystem.

The purpose of this paper is to examine the preliminary data which exist for one of these processes, decomposition of tidal freshwater plants. Our objectives are: (1) to summarize existing research results, (2) attempt to synthesize general principles, (3) to compare these preliminary results and principles with the extensive literature on saline tidal marshes, and (4) suggest directions for future research.

We have chosen to restrict our discussion to those marsh plants which occur primarily in the regularly flooded portion of the marsh. These include arrow arum *(Peltandra virginica)*, spatterdock or yellow pondlily *(Nuphar luteum)*,[1] wild rice *(Zizania aquatica)*, pickerel weed *(Pontederia cordata)*, arrowhead *(Sagittaria latifolia)*, and bur marigold *(Bidens laevis)*. Certainly, these are not all of the species which occur in this zone, but they are apparently the only ones for which published decomposition results exist. Plants which are found in the irregularly flooded sections of the tidal freshwater marsh, such as the common reed *(Phragmites communis)* and the cattail *(Typha latifolia)* have been omitted from this discussion. For further information on these thoroughly studied species, see Mason and Bryant (1975).

METHODS

Original data presented in this paper (Heywood, 1977; W. E. Odum *(personal observations)* are based on the following methodology. Decomposition rates were estimated using nylon bags with 2-mm mesh. Forty grams of air-dried, green plant leaves were placed in each bag. Four replicates were removed at intervals, the plant material washed thoroughly, ovendried at 105°C for 48 hours, weighed, and ground in a Wiley mill to 20-mesh size. Replicate 1-g samples were ashed in a muffle furnace at 550°C for 12 hours. Litterbags were placed in the freshwater headwaters of Ware Creek, a tributary of the York River, Virginia. For more detailed description of methods, see Heywood (1977).

Total N was determined by a modified Kjeldahl technique (Jackson, 1958) and by the indol phenol blue method for ammonia using a Technicon Autoanalyzer® I (Anonymous, 1971). Organic carbon was assumed to be equal to 45% of the ash-free dry weight. This was confirmed by analyses of 12 samples on a Perkin-Elmer CHN™ analyzer which gave values ranging from 43–48% and mean of 45%. Analysis of variance was used to test the data for significance.

Methodological Difficulties Decomposition studies of tidal freshwater marsh plants are hampered by several factors which may not be as bothersome in other environments. Foremost among these is the problem of high sedimentation rates. Typically, these marshes are characterized by a sub-

---

[1] *Nuphar luteum* equivalent to *N. advena*, Beal (1977); eds. note.

strate composed of fine, unconsolidated sediments. These sediments shift about constantly in response to wind, tide, and precipitation-generated perturbations. As a result, litterbags quickly become infiltrated and covered with fine, sticky sediments. In addition to covering the litterbags and rapidly creating anaerobic conditions, these fine sediments adhere to every surface and fissure in the litter and are almost impossible to remove completely.

To circumvent this problem we have found it necessary: (1) to deploy litterbags whenever possible at sites where they are well flushed by tidal action, (2) to scrub and sonicate litter samples thoroughly, and (3) report decomposition rates on an ash-free basis. Even with these precautions a large error factor may be introduced into the results. This is particularly likely to happen when litterbags are placed on the surface of the high marsh in poorly flushed marsh ponds. The most consistent results are obtained from samples suspended in the water column of well-flushed tidal creeks.

A second problem concerns the growth state at which marsh plant tissues should be harvested for litterbag studies. In higher salinity marshes it is usually possible to harvest standing "dead" material which has not been appreciably decomposed. In the tidal freshwater marsh this is virtually impossible because the initial stages of decomposition proceed so rapidly. For a species such as *Peltandra virginica* there is no way to harvest undecomposed dead leaves. The only apparent solution is to harvest living green leaves which appear to be approaching death. This is, of course, artificial and possibly misleading because these leaves contain compounds which the plant would probably translocate back into the root system. For comparative purposes, therefore, it is important to harvest plant tissues from different species at approximately the same growth stage and as near death as possible as evidenced by a color change from green to yellow green.

A similar problem common to other environments concerns the season when litterbag material is harvested. Most marsh species vary in nutrient content seasonally. Leaves of *Peltandra virginica*, for example, may contain as much as 3% N (ash-free dry weight) early in the summer but only 2% or less in the late summer. Obviously, litterbag material for an experiment should be gathered at one location at one time; each batch of material which is gathered at different times or from different locations should be analyzed separately for elemental composition.

RESULTS AND DISCUSSION

Decomposition Rates

From the standpoint of decomposition, the best studied tidal freshwater marsh plant is arrow arum, *Peltandra virginica*. A typical set of decomposition curves for this species is shown in Fig. 1. There are 2 factors that should be emphasized. First, the initial rate of decomposition of this species is extremely high; this indicates a considerable loss of soluble compounds and a high rate of dissolved organic matter (DOM) production. After the initial 30

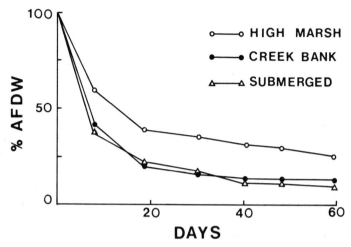

Fig. 1. The percentage of the original ash-free dry weight (AFDW) of leaves of *Peltandra virginica* remaining in litterbags placed on the high marsh (irregularly flooded), creek bank (flooded twice daily) and permanently submerged in the water column. Each data point represents 4 replicates.

to 40 days, the decomposition rate slows dramatically so that it requires from 150–400 days for all of the *Peltandra virginica* tissue to disappear from the litterbags. Second, there are significant differences in decomposition rates depending upon the sites where the litterbags are placed. For example, submerged leaves decompose more rapidly ($P \leq .01$) than those placed upon the marsh surface. There are several factors which may contribute to the higher rates of decomposition of submerged leaves including better access for detritivores, more constant physical conditions for decomposer bacteria and fungi, a greater availability of dissolved nutrients, and a more suitable environment for rapid leaching.

Other investigators have found a similar pattern of rapid decomposition for *Peltandra virginica*. Walker and Good (1976) found 20% remaining after 40 days on a regularly flooded marsh surface. Whigham and Simpson (1976) report high rates of decomposition with <40% of the original material remaining after 90 days.

Because these field results indicated an initial rapid loss of soluble compounds from *Peltandra virginica*, we designed a laboratory experiment to examine soluble losses over very short time intervals. Weighed pieces of air-dried, yellow leaves were placed in flasks of freshwater situated on a shaker table, individual flasks were removed at 1-day intervals, and the detritus filtered out of the water, dried and weighed. The results (Fig. 2) detail the extremely rapid loss of weight during the first 9 days of immersion culminating in a 50% loss of dry weight. For comparison, pieces of green cordgrass, *Spartina alterniflora*, were treated in a similar manner and

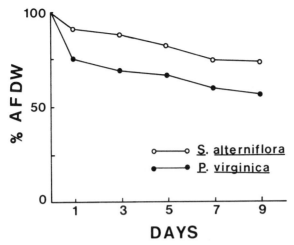

Fig. 2. The percentage of the original ash-free dry weight (AFDW) of green leaves of *Peltandra virginica* and *Spartina alterniflora* retained on filter paper after immersion in freshwater in 250-ml flasks at 20°C. Each data point represents 4 replicates.

showed a significantly ($P \leq .01$) slower loss of weight. This suggests that much of the primary production of tidal freshwater marshes is released rapidly, probably as dissolved organic matter, to surrounding aquatic ecosystems.

Comparison with Saline Species To confirm our initial impression of a more rapid rate of decomposition for tidal freshwater marsh plants when compared to more saline species, we deployed litterbags of *Peltandra virginica* and *Spartina alterniflora* at the same sites in both freshwater and estuarine marshes. The results from the freshwater site (Fig. 3) show a significantly greater decay rate ($P \leq .01$) for *Peltandra virginica*. Almost identical results were obtained from an estuarine marsh (salinity = 25‰).

Other Freshwater Marsh Species Our preliminary data on 4 additional species of tidal freshwater marsh plants (Fig. 4) indicate relatively rapid rates of decomposition. All 4 species exhibit an initial rapid loss of weight (40% or more in the 1st month) followed by a more gradual decline until all material disappears from the litterbags after 150–350 days. In all cases these rates of decay are significantly higher ($P \leq .05$) than for the higher salinity species *Spartina alterniflora* and *Juncus roemerianus* deployed at the same sites.

D. F. Whigham and R. L. Simpson *(personal communication)* have data for *Zizania aquatica* which show a 60% loss in weight after 90 days for litter on the high marsh, but only 35% loss for inundated litter during the same time period. This lack of agreement with our data may be due to either differences in physical and chemical conditions at the sites or may be be-

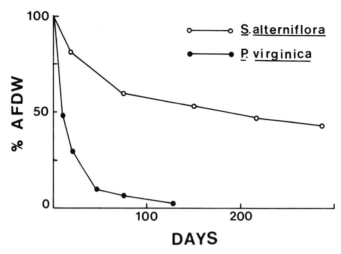

Fig. 3. The percentage of the original ash-free dry weight (AFDW) of green leaves of *Peltandra virginica* and *Spartina alterniflora* remaining in submerged litterbags. Each data point represents 4 replicates.

cause their data are expressed in terms of dry weight while ours are expressed as ash-free dry weight thus excluding adsorbed inorganic sediment particles. D. F. Whigham and R. L. Simpson *(personal communication)* also investigated *Bidens laevis* and found a rapid rate of decomposition similar to that shown in their data for *Peltandra virginica* (75% loss of weight after 90 days at a regularly flooded site).

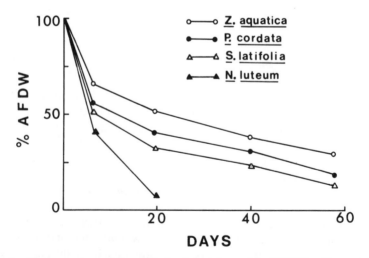

Fig. 4. The percentage of the original ash-free dry weight (AFDW) of leaves of *Zizania aquatica*, *Pontederia cordata*, *Sagittaria latifolia* and *Nuphar luteum* remaining in submerged litterbags. Each data point represents 4 replicates.

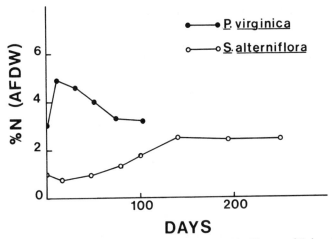

Fig. 5. The relative N content (ash-free dry weight, AFDW) of leaves of *Peltandra virginica* and *Spartina alterniflora* from submerged and intertidal litterbags. Each data point represents 6 replicates.

Nitrogen Changes During Decomposition

The only tidal freshwater marsh species for which we have data concerning N changes during decomposition is *Peltandra virginica* (Fig. 5). Submerged and creek bank litter follows a trend of initial relative N increases followed by subsequent declines. Plant material which was originally 2.9% N (ash-free dry weight) increased to 4.0–5.5% N after 10–20 days and then declined to 3.0–3.8% N after 50 days. The C:N ratio declined initially from 15.5 in fresh *Peltandra virginica* leaves to 7.6–10.6 after 29 days and rose to 12.3–15.2 after 50 days.

In contrast, litter placed on the irregularly flooded high marsh rose steadily in N content from 2.9% to 6.1% after 20 and 6.5% after 50 days. The C:N ratio dropped from 15.5 to 9.2 after 10 days and reached 7.0 after 50 days. We believe that this contrasting pattern can be explained in part by the rapid colonization of the high marsh litter by autotrophic bacteria and nitrogen-fixing blue-green algae, both of which were observed qualitatively on the detrital material.

R. Walker and R. E. Good (*personal communication*) followed N changes in decaying *Peltandra virginica* for 1 month at 3 sites on the Mullica River, New Jersey. Their results showed an increase from initial values of 1.76–2.05% N per gram of ash-free dry weight to values of 3.10–3.80% after 24 days. Two of the sites demonstrated subsequent declines while the litter at the third site rose slightly in N content. At 1 site, the N content dropped from 3.24% to 3.15% between days 24 and 45 while at the other site it dropped from 3.80% to 3.43% between days 24 and 31. Both changes were not statistically significant ($P > .01$).

To summarize, *Peltandra virginica* appears to exhibit an initial increase in relative N content with a drop in the C:N ratio followed by a decline in N and rise in the C:N ratio between days 20 and 40. This pattern does not apply to the high marsh where there is a continual, gradual increase in N. As Heywood (1977) has pointed out, this type of N loss is typical of plants which have a high initial nitrogen content in contrast to species such as *Spartina alterniflora* or *Spartina cynosuroides* which typically contain less than 1.5% N as live or standing dead material, but increase steadily during decomposition and ultimately reach values of 2–3% N after several months of decay.

CONCLUSIONS

Although there are few research data on decomposition of intertidal freshwater marsh plants, certain preliminary conclusions can be drawn. First, these plants appear to decompose more rapidly than species from estuarine and higher salinity marshes. There are also indications (M. Dunn and B. Nash, *personal communication*) that the tissues of freshwater marsh plants are much more readily attacked by detritivores (such as amphipods and isopods) and more easily macerated than tissues from *Spartina* species growing at higher salinities. Both observations can be explained, at least in part, by the chemical composition of the freshwater marsh plants. In addition to their relatively high N content, preliminary analyses suggest that they contain low amounts of lignin and cellulose, a combination which evidently leads to a rapid rate of decomposition. Although this rate may be influenced by physical variables such as salinity and temperature, in all cases it should be more rapid than for the higher salinity marsh plant species with their more refractive tissues.

Second, the production of DOM from tidal freshwater marshes may be higher per unit of primary production than from saline and estuarine marshes. This suggests that the DOM pathway of energy flow (dead plant → DOM → bacteria → consumer) may be important in the tidal river ecosystems where these marshes occur.

Finally the pattern of N change during decomposition of tidal freshwater marsh plants appears to differ from that of saline marsh plants. In either case the C:N ratio of aged detritus is similar (between 12 and 18).

Obviously, there are numerous and varied possibilities for future research on decomposition in tidal freshwater marshes. In addition to more detailed decomposition studies of the 6 species mentioned in this paper, there are 20 to 30 other species which occur in significant quantities to merit some study. Research to date has been limited to stems and leaves. Walker and Good (1976) found that the underground standing crop of a species such as *P. virginica* may be at least 10× higher than that present aboveground. What happens to this material when it dies? Much of the subsequent decomposition must occur under anaerobic conditions. Studies of root decomposition appear to be essential.

Finally, critical questions center on the problem of detritus utilization and secondary production. How important is the DOM pathway? Is freshwater plant material more easily consumed and assimilated by detritivores than material from more saline marshes? Which marshes are dependent upon, or at least utilize, detritus from these marshes? In short, how important are tidal freshwater marshes as the base of a detritus food web?

LITERATURE CITED

Anonymous. (1971). Methods for chemical analysis of water and wastes. U.S. Environmental Protection Agency, U.S. Government Printing Office, Washington, D.C.

Beal, E. O. (1977). A manual of marsh and aquatic vascular plants of North Carolina. Technical Bulletin No. 247. North Carolina Agricultural Exp. Station. Raleigh, North Calolina.

Heywood, M. A. (1977). The effects of nutrient enrichment on the decomposition of *Spartina cynosuroides* and *Peltandra virginica*. Masters Thesis. University of Virginia.

Jackson, M. L. (1958). Soil chemical analysis, Prentice Hall, Inc., Englewood Cliffs, New Jersey.

Mason, C. F. and Bryant, R. J. (1975). Production, nutrient content and decomposition of *Phragmites communis* Trin. and *Typha angustifolia* L. *J. Ecol.* **63,** 71–95.

Walker, R. and Good, R. E. (1976). Vegetation and production for some Mullica River-Great Bay tidal marshes. *Bull. N.J. Acad. Sci.* **21,** 20 (abstract).

Whigham, D. F. and Simpson, R. L. (1976). Sewage spray irrigation in a Delaware River freshwater tidal marsh. *In* "Freshwater Wetland and Sewage Effluent Disposal" (D. L. Tilton, R. H. Kadlec and C. J. Richardson, eds.), pp. 119–144. The University of Michigan, Ann Arbor, Michigan.

LITTER DECOMPOSITION IN PRAIRIE GLACIAL MARSHES[1]

C. B. Davis and A. G. van der Valk

Department of Botany and Plant Pathology
Iowa State University
Ames, Iowa 50011

Abstract   Data from permanent quadrats show that shoot death begins in early June for *Carex atherodes*, in late July for *Scirpus validus* and *Scirpus fluviatilis*, and in September for *Sparganium eurycarpum* and *Typha glauca*. Studies of standing and fallen litter were conducted on these 5 emergent species. Losses from the standing litter compartment are primarily the result of fragmentation that occurs during the first winter and spring. Leaching from the standing litter may also remove nutrients in the first few weeks after shoot death. In the fallen litter, K, Na, and Mg were released very rapidly, while N, P, and Ca were released more slowly. In contrast, Al and Fe were accumulated by fallen litter in all species. Release and uptake of nutrients by fallen litter varied from species to species, from site to site for the same species, and seasonally. *Scirpus validus* fallen litter lost the most weight followed, in turn, by *Sp. eurycarpum*, *T. glauca*, and *C. atherodes* litter. Fallen litter was submersed only in the spring and early summer. The amount of time fallen litter was submersed varied from species to species in the marsh. Flow models for N, P, K, and Ca summarize the available information on the seasonal flux of these nutrients in the standing and fallen litter compartment of prairie glacial marshes.

Key words   *Biomass*, Carex, *decomposition, Iowa, litter, marsh, nutrient release*, Scirpus, Sparganium.

INTRODUCTION

One of the major wetland complexes in North America occupies an area extending from central Iowa, through western Minnesota and the Dakotas,

[1] Journal Paper No. J-8792 of the Iowa Agriculture and Home Economics Experiment Station, Ames, Iowa. Project 2071.

and well into the Canadian Provinces of Alberta, Manitoba, and Saskatche-wan. This region is underlain by deep till deposits left behind by retreating glaciers. The till surface is pockmarked with millions of poorly drained de-pressions of various sizes. As a result, this area is referred to as the "prairie pothole region," and the wetland communities that develop in these potholes are called prairie glacial marshes. These marshes range in size from <0.1 ha to well over 1,000 ha.

In 1850, Iowa alone had nearly 300,000 ha of wetlands (Shaw and Fredine, 1956), most of which occurred in the glaciated north-central and northwest-ern parts of the state. With the encouragement of both federal and state governments, drainage of these pothole wetlands began in earnest before 1900. By 1906, ≈65,000 ha had been drained, and by 1922, an additional 140,000 ha of wetland had been "reclaimed" for agriculture. Today, <30,000 ha of wetland remain in Iowa (Shaw and Fredine, 1956). Most of the remain-ing wetlands are state-owned prairie glacial marshes set aside for public hunting.

Interest in protecting these remaining marshes has resulted mainly from their value as breeding and feeding sites for migratory waterfowl. Recently, however, interest in the potential value of these marshes as nutrient sinks for nonpoint sources of agriculture runoff has arisen. This interest prompted the Iowa Agriculture and Home Economics Experiment Station to fund a long-term interdisciplinary study of nutrient cycling at 2 prairie glacial marshes in north-central Iowa, Goose Lake and Eagle Lake.

In this paper, we summarize the information obtained to date on litter production and decomposition in prairie glacial marshes in our continuing study. Specifically, we review the available information on the production of standing litter, the transfer of material from the standing to the fallen litter compartment, and the seasonal uptake and release of nutrients from standing and fallen litter. We then synthesize what is known about the flux of N, P, K, and Ca in the litter compartments of prairie marshes in a series of simple input-output models.

The studies were carried out at Goose Lake in Hamilton County and at Eagle Lake in Hancock County, Iowa. Both lakes lie within the boundaries of the Wisconsin glacial drift sheet and are typical prairie glacial marshes. They are classified as Type IV wetlands (Stewart and Kantrud, 1971).

A survey of the water chemistry of Iowa lakes situated within the Wiscon-sin glacial drift sheet (Bachmann, 1965) revealed that these lakes are typical hard water lakes of the bicarbonate type (Hutchinson, 1957). Total alkalinity averaged 160 ppm $CaCO_3$; total hardness averaged 199 ppm $CaCO_3$; Ca content averaged 35 ppm; Mg content averaged 27 ppm; and SO_4 content averaged 46 ppm. Our measurements at Goose Lake show it to have a total alkalinity of between 160 and 170 ppm; a total hardness of between 160 and 180 ppm; a Ca content of ≈130 ppm; a Mg content of between 30 and 50 ppm; and a SO_4 content of ≈12 ppm. No water chemistry data are yet available for Eagle Lake, but it also is a hard water lake. Both lakes receive sizable inputs from agricultural runoff.

Goose Lake is privately owned and managed by a local hunting club. It has a surface area of ≈65 ha and a maximum depth of 1.5 m. Water depth is controlled by a dam at the southern end of the marsh. Air temperatures range from near 37°C during the summer to −30°C in midwinter. Annual precipitation is ≈800 mm, approximately half of which falls during the spring. Dominant emergent macrophyte species are *Typha glauca* Godr., *Scirpus fluviatilis* (Torr.) Gray, *Scirpus validus* Vahl., and *Sagittaria latifolia* Willd. (see Weller and Spatcher, 1965). Nomenclature follows Gleason and Cronquist (1963). Eagle Lake is owned and managed for hunting by the Iowa Conservation Commission. It has a surface area of ≈360 ha and a maximum depth of 1 m. Water depth is controlled by a dam at the northern end of the lake. Air temperatures and precipitation patterns are similar to those at Goose Lake. Dominant emergent macrophyte species are *Typha glauca*, *Scirpus fluviatilis*, *Scirpus validus*, *Sagittaria latifolia*, *Carex atherodes* Spreng., and *Sparganium eurycarpum* Engelm.

METHODS

Changes in shoot density in *C. atherodes*, *Sc. validus*, *Sc. fluviatilis*, *Sp. eurycarpum*, and *T. glauca* during the 1976 growing season were determined from weekly counts made on three 1 × 1 m permanent quadrats located in homogeneous stands of each species.

Standing litter was sampled by harvesting ten 1 × 1 m quadrats of each species. Quadrats were located at 10–15 m intervals along transect lines located in each stand sampled. Culms were clipped at the substrate level except during the winter when they were clipped at ice level. Stumps of these culms were collected after the spring thaw. Litter was separated by age (1974 litter, 1975 litter) in the field and returned to the laboratory for analysis.

The rate of decomposition of fallen litter was measured by deploying nylon mesh bags containing leaf and culm material on the bottom of the marsh. Fresh standing litter was collected immediately after the first killing frosts in October of 1974 at Goose Lake and 1975 at Eagle Lake. The litter for each species was thoroughly mixed and ovendried to constant weight at 80°C. In the Goose Lake study, 50 g ovendry weight of fresh *Typha glauca* and 20 g ovendry weight of fresh *Scirpus fluviatilis* litter were placed in 3-mm nylon mesh bags measuring 20 × 40 cm. In the Eagle Lake study, 50 g were again used for *Typha glauca*, but 30 g were used for the other 3 species. Numbered plastic tags were inserted into the bags for identification. Thirty-six bags of each species were distributed randomly in 3 adjacent 2 × 2 m quadrats, 12 bags per quadrat. The bags were deployed on 3 October 1974 at Goose Lake and on 15 November 1975 at Eagle Lake. Bags were anchored to stakes by nylon cord. One bag from each quadrat was collected at each sampling time.

At each collection time, the samples were carefully washed to remove silt, clay, and small invertebrates, then were dried to constant weight at 80°C. This temperature was selected to minimize the volatilization of chemical constituents. After being weighed, the samples were mixed, subsampled, ground in a 40-mesh Wiley mill, and stored in glass jars at 5°C. Mineral analysis was carried out at the Soils Laboratory of the University of Wisconsin. Total N was determined by semimicrokjeldahl procedures, and P, K, Na, Ca, Mg, Al, and Fe were determined by multispectral analysis (Genson et al., 1976).

RESULTS AND DISCUSSION

The Standing Litter Compartment

Shoot Death and the Duration of the Standing Litter Stage   Material enters the standing litter stage with the onset of leaf and shoot death (**see** van der Valk and Davis, 1978, figs. 1–5). Shoot death in *Carex atherodes* begins in early May and is almost linear throughout the growing season. This pattern evidently is caused by competition among the densely packed *C. atherodes* shoots. May and June deaths result from self-shading caused by differential growth rates among spring shoots. July, August, and early September deaths were restricted predominantly to new, and therefore small, shoots that developed during mid and late summer. A few *C. atherodes* shoots were still alive in early November, having survived several severe frosts.

Shoot death in *Scirpus fluviatilis* and *S. validus* does not begin until late July and progresses rapidly throughout the rest of the growing season. The decrease in *S. fluviatilis* density resulted from the death of the fruiting shoots and the smaller vegetative shoots. Losses of *S. validus* shoots were caused primarily by a serious infestation of dipteran larvae. Seventy percent of the *S. validus* shoots and 77% of the *S. fluviatilis* shoots died before the first killing frost in late September.

The third pattern of shoot death is found in *Sparganium eurycarpum* and *Typha glauca* shoots. In each of these species, 80% of the shoots alive at peak density remained alive until killed by the late September and early October frosts. Almost all of the late summer deaths were fruiting shoots.

An additional factor contributing to the development of the standing litter compartment is the death of leaves or portions thereof. The older leaves of *Carex atherodes*, *Sparganium eurycarpum*, and *Scirpus fluviatilis* often die well before the rest of the shoot shows any signs of senescence. The tips of *Typha glauca* leaves also may die several weeks before the rest of the shoot dies.

Thus, the timing of initiation and the pattern of development of the standing litter compartment varies considerably from species to species. Our

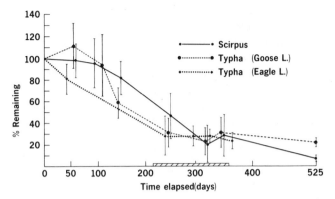

Fig. 1. Change in dry weight of standing litter of *Scirpus fluviatilis* and *Typha glauca* at Goose Lake and *T. glauca* at Eagle Lake. 100% is equal to standing live crop on 15 August. Shoot death occurred at day 35. Growing season is indicated by the crosshatched bar. Vertical bars at each sample point indicate (+) or (−) one standard deviation.

findings and those of others (Mason and Bryant, 1975; Bayly and O'Neill, 1972*a*, *b*) suggest that future studies of the standing litter stage of decomposition should begin during the growing season.

Timing and Patterns of Litter Fall   As with the timing and pattern of shoot death, the rates at which materials are transferred from the standing litter compartment to the fallen litter compartment or to the soil—water environment vary from species to species, time to time, and site to site. The major processes that function in the breakdown of the standing litter are leaching, fragmentation, and toppling (Davis and van der Valk, 1978).

At least some transfer of fresh litter from the standing to the fallen compartment occurred during the growing season in all 5 species studied. Because *Typha* shoots generally do not begin to die until late summer or early fall, such transfers were modest in *Typha* stands, occurring mainly near the end of the growing season when newly-killed leaf tips fragmented and fell into the water. In *Carex atherodes*, *Sparganium eurycarpum*, and *Scirpus fluviatilis* leaves died throughout the growing season and dead leaves often drooped sufficiently to bring them into contact with the water. For the most part, these transfers also seemed minor. Of the 5 species studied, only *Scirpus validus* had a significant amount of litter fall preceding the autumn freezes. Dead *S. validus* shoots, especially the thinner ones, are easily bent and broken by winds accompanying summer thunderstorms and the tops of these broken shoots may fall into the water. Only a small percentage of the 1976 peak standing crop, however, fell to the water before the killing frosts of late September.

The seasonal pattern of litter fall after the first killing frost is illustrated in Fig. 1 for *Scirpus fluviatilis* and *Typha glauca* at Goose Lake and for *T.*

glauca at Eagle Lake. Losses were caused almost entirely by fragmentation of the standing litter during winter and spring storms. *Typha glauca* and *Scirpus fluviatilis* litter confined in nylon mesh bags suspended 1 m above the water surface at Goose Lake during the 525-day period of this study lost dry weight at a much slower rate than did unconfined standing litter. Standing *S. fluviatilis* litter lost 79% of its peak dry weight after 525 days; suspended *S. fluviatilis* litter lost only 16%. Standing *Typha glauca* litter lost 93% of its peak dry weight while suspended *T. glauca* litter lost only 44% during this period (Davis and van der Valk, 1978). These differences in dry-matter loss reflect the importance of fragmentation in the breakdown of the standing litter. As a result, suspended litterbags cannot be used to estimate the rates at which litter and nutrients are transferred from the standing to the fallen compartment.

Because the sampling of *Scirpus fluviatilis* and *Typha glauca* stands at Goose Lake occurred during the same period and the stands sampled are adjacent to one another, comparison of the pattern and rates of litter fall in the 2 species is appropriate (see Fig. 1). The standing litter of both species remained relatively intact throughout the fall of 1974. During January and February 1975, the dry weight of standing *Scirpus fluviatilis* litter decreased sharply. These plants have long, thin leaves that become very brittle when dry and are easily broken by strong winter winds. Standing *S. fluviatilis* litter had a 50% decrease in dry weight by the first of April (175 days), at which time only a few leaves remained on the more resistant culms. *Typha glauca* leaves, however, are less susceptible to fragmentation and were lost more gradually. Standing litter losses in the *T. glauca* stand did not reach 50% until early June. After June 1975, the new 1975 shoots and the ensuing crop of new standing litter afforded the remaining culms of 1974 standing litter some protection from the winds. Stumps of these culms may remain standing for 2 or more years before microbial activity (which rots their bases) topples them. Microbial activity is undoubtedly also a factor in the breakdown of the aerial portions of standing litter, but such activity proceeds slowly. Stalks generally topple into the water before microbial decomposition can have much impact (Davis and van der Valk, 1978).

Although fragmentation due to wind, rain, sleet, and snow is the primary cause of litter fall in prairie glacial marshes, other factors can be equally effective in particular situations. During periods of high water, muskrat populations often grow explosively (Weller and Spatcher, 1965; Weller and Fredrickson, 1974). Their feeding and lodge-building activities can destroy large areas of emergent vegetation, bypassing the standing litter stage by introducing felled culms directly into the fallen litter compartment. In drier years, deer and domestic cattle are able to enter the marsh. The deer have only minor impact, but a herd of hungry cattle feeding on fresh shoots can knock down a large volume of standing litter. Fires also may destroy large areas of standing litter. Burned stumps and buried charcoal at Eagle Lake

Table 1. A Summary of Weight Loss and Changes in Nutrient Content (% Dry Weight) of the Fallen Litter from 5 Marsh Macrophyte Species at Goose and Eagle Lakes

Species	Dry wt loss (%) after lapse of ( ) days	Percent of dry wt expressed as: initial %/net gain or [net loss]								50% dry wt loss reached in ( ) days
		N	P	K	Na	Ca	Mg	Al	Fe	
S. fluviatilis[a]	39 (525)	.98/[.25]	.204/.059	.397/[.262]	.200/[.166]	.421/.648	.159/.035	.014/.236	.023/.188	525[c]
T. glauca[a]	53 (525)	1.26/[.81]	.242/[.111]	1.14/[.99]	1.61/[.85]	.346/[.246]	.796/[.756]	.020/.117	.025/.099	325
T. glauca[b]	51 (330)	.46/.04	.061/[.009]	.194/[.116]	1.01/[.97]	1.78/[.85]	.242/[.153]	.005/.004	.004/.015	325
S. eurycarpum[b]	67 (330)	.56/0	.008/.001	.891/[.802]	.093/[.076]	1.86/[.73]	.320/[.141]	.007/.031	.005/.051	250
C. atherodes[b]	20 (330)	.34/.09	.048/.068	.213/[.041]	.010/.032	.212/.187	.092/.015	.007/.136	.009/.129	330[c]
S. validus[b]	84 (330)	.36/[.02]	.055/[.015]	.401/[.361]	.062/[.050]	.360/.004	.168/[.103]	.003/.005	.004/.012	225

[a] Goose Lake sample for *Scirpus* and *Typha*, 15 August (Davis and van der Valk, 1978).
[b] Eagle Lake sample for *Typha*, *Sparganium*, *Carex* and *Scirpus*, 15 November (Davis and van der Valk, 1978).
[c] *Scirpus fluviatilis* and *Carex rostrata* litter had not lost 50% dry weight by the end of these studies.

attest to the past occurrence of fires. The effects of muskrats and fire on mineral cycling in prairie glacial marshes have never been studied.

Standing litter also was sampled at Eagle Lake beginning in the fall of 1975. Because of high water levels in the spring, we were able to obtain complete data only for *Scirpus validus* and *Typha glauca*. Standing litter of *S. validus* fell very rapidly in the fall and was gone by November 1975. The weight-loss curve for standing *T. glauca* litter at Eagle Lake is similar to the curve for *T. glauca* at Goose Lake (Fig. 1).

The Fallen Litter Compartment

Table 1 summarizes the changes in dry weight and nutrient content of the 5 emergent macrophyte species studied at Goose and Eagle lakes. The Goose Lake study began in October 1974 and was concluded after 525 days. The Eagle Lake study started in November 1975 and is still in progress. Eagle Lake data presented here are for the first 330 days of the study. Initial values for the Goose Lake study were for peak standing crop (15 August 1974), whereas the initial values for the Eagle Lake study are for the first litter sample in the fall (15 November 1975).

The content of Al and Fe increased in all species studied. These increases were especially noticeable in *Scirpus fluviatilis* at Goose Lake and in *Carex atherodes* at Eagle Lake. *Scirpus fluviatilis* also accumulated P, Ca, and Mg. *Carex atherodes* accumulated all 8 of the nutrients studied. *Scirpus fluviatius* and *Carex atherodes* are morphologically similar. The ability of *S. fluviatilis* and *C. atherodes* to accumulate nutrients more effectively than do the other species studied probably reflects structural differences between these 2 species and the other 3 for which data were available. Corrections for weight loss during decomposition (Davis and van der Valk, 1978) indicate that these accumulations represent actual uptake from the environment.

An analysis of the nutrient-release data in Table 1 reveals that there are basically 3 nutrient-release patterns for these species. The most rapidly lost nutrients were K, Na, and Mg. Those held most tenaciously were Al and Fe. Nitrogen, P, and C had intermediate release rates. No 2 species studied had identical patterns of nutrient release. Even the 2 *Typha glauca* stands differed considerably. These differences are not surprising because of the wide range of initial concentrations among the 7 species and the differences in the site conditions both within and between the marshes. For instance, increases of Mg and Na in *Scirpus fluviatilis* and *Carex atherodes* may have been primarily a function of low initial concentrations. Site characteristics such as duration of standing water also can affect the pattern of nutrient loss. The fallen litter of *Scirpus fluviatilis* and *Carex atherodes* was submerged only during the spring. Although this litter did remain moist during the summer, it was not subjected to continuous leaching as was the continuously submersed fallen litter of *Typha glauca* and *Scirpus validus*. *Sparganium eurycarpum* litter was submersed from November 1975 until July 1976, creating a third

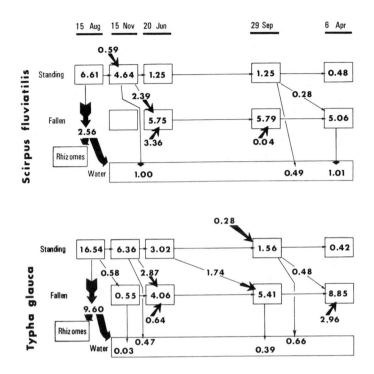

Fig. 2. Nitrogen content, flow, and release during decomposition of standing and fallen litter of *Scirpus fluviatilis* and *Typha galuca* at Goose Lake (g/m²); 1974–1976.

set of conditions. The lack of summer precipitation eliminated the potential for leaching in unsubmersed fallen litter at Eagle Lake in 1976. However, unsubmersed, fallen *Scirpus fluviatilis* litter at Goose Lake was subjected to leaching by rainstorms in June and October 1975.

Initial N content and the amount of dry weight lost by the end of the studies seem to be positively correlated (Table 1). Tissues rich in N are probably more suitable substrates for microbial populations responsible for litter decomposition than are N-poor tissues. At Goose Lake, *Typha glauca* lost more weight than did *Scirpus fluviatilis*. At Eagle Lake, *Sparganium eurycarpum* lost more weight than did *Typha glauca* and *T. glauca* lost more weight than did *Carex atherodes*. The lone exception to this pattern is *Scirpus validus* litter at Eagle Lake. The initial N content of *S. validus* litter was only slightly higher than that of *C. atherodes*. Yet *S. validus* had the greatest weight loss while *C. atherodes* had the least. This unexpectedly high weight loss for *S. validus* was undoubtedly a function of the small amount of structural tissue in *S. validus* shoots.

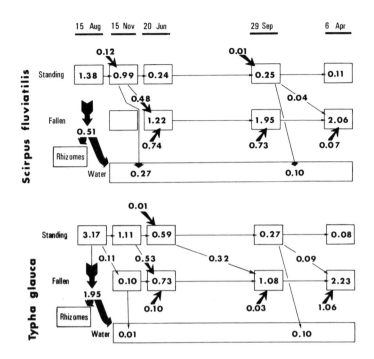

Fig. 3. Phosphorus content, flow, and release during decomposition of standing and fallen litter of *Scirpus fluviatilis* and *Typha glauca* at Goose Lake (g/m²); 1974–1976.

Nutrient Fluxes in Standing and Fallen Litter

Estimates of the contribution made by decomposing macrophyte litter to the nutrient cycles at Goose Lake were obtained by converting data on changes in the dry weight and nutrient content of decomposing *Scirpus fluviatilis* and *Typha glauca* litter to absolute quantities (grams per square metre) (see **Appendix** for formulas used). Initial quantities (15 August) are for living plants at peak standing crop. All other quantities are for litter. These values are illustrated for N, P, K, and Ca in Figs. 2 through 5.

Nitrogen   Figure 2 reveals that losses of N between 15 August and 15 November 1975 were considerable in both *Scirpus* (39%) and *Typha* (58%). These losses were caused by a combination of translocation to the rhizomes preceding shoot death and leaching from the living and dead shoots. The relative importance of these 2 processes is unknown. Further refinement of our nutrient-release estimates must await further study of these early transfers.

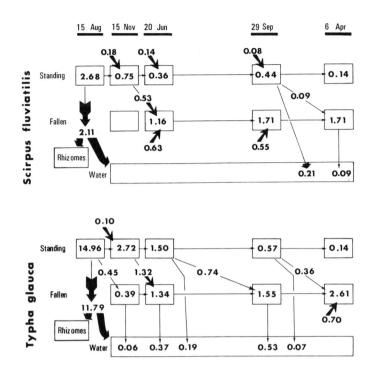

Fig. 4. Potassium content, flow, and release during decomposition of standing and fallen litter of *Scirpus fluviatilis* and *Typha glauca* at Goose Lake (g/m²); 1974–1976.

The transfer of N and other nutrients from the standing to the fallen litter compartment was caused by fragmentation and occurred mainly during the first winter and spring. Leaching of N from the standing litter also occurred mostly in the winter–spring periods. During the period between 15 November and late March, temperatures in northern Iowa are below freezing most of the time, and precipitation is mostly in the form of snow. Under these conditions, nutrient release from standing and fallen macrophyte litter was probably negligible. Therefore, releases of N (and the other nutrients) that are indicated in Fig. 2 as having occurred between 15 November and 20 June undoubtedly took place mainly during May and June.

During the summer, *Typha* standing litter had net gains in N content. This familiar phenomenon undoubtedly is the result of microbial activity. The apparent accumulation of all 4 nutrients by *Scirpus* standing litter during the early fall of 1974 may indicate that the standing live crop increased between 15 August 1974 and the time of shoot death, that microbial uptake occurred in the standing litter, or that a combination of the 2 occurred. The concentra-

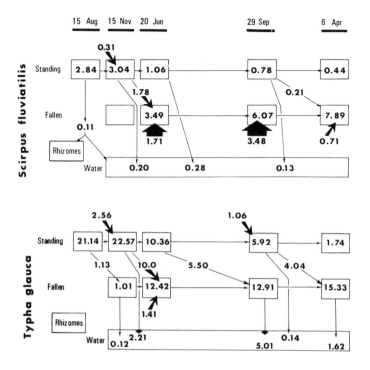

Fig. 5. Calcium content, flow, and release during decomposition of standing and fallen litter of *Scirpus fluviatilis* and *Typha glauca* at Goose Lake (g/m²); 1974–1976.

tions of N in the fallen litter were quite variable, but in all instances, there was a net gain of N in this compartment. After 1 year, *Typha* had released or translocated 97.5 kg N/ha, whereas *Scirpus* had gained 4.3 kg N/ha.

Phosphorus   Figure 3 shows that early leaching and translocation removed 37% of the initial P content from *Scirpus* shoots before 15 November 1974. *Typha* lost 65% of its P during the same period. As with N, translocation and leaching were the major causes of these early losses. Phosphorus losses from the standing litter of both species between 15 November and 20 June were caused mainly by fragmentation. Smaller losses followed the second winter season. Transfer of P from the standing to the fallen *Typha* litter continued during the summer, but standing *Scirpus* litter actually accumulated a small amount of P during this period.

Fallen *Scirpus* litter had a large increase in P during the summer of 1975. Fallen *Typha* litter also had a slight increase in P content during this period. These increases probably resulted from microbial uptake, silt contamination, or a combination of both processes (Davis and van der Valk, 1978). After 1 year, *Typha* had accumulated 18.2 kg P/ha and *Scirpus* had accumulated 8.2 kg P/ha.

Potassium Figure 4 illustrates the great mobility of K. Initial translocation and leaching removed 79% of the K from standing *Scirpus* litter and 79% of the K from standing *Typha* litter by November 1974. Subsequent releases of K from the standing litter were small.

A different pattern emerged for the fallen litter. Fallen *Scirpus* litter had a large net increase in K during the 1st year, but had a slight loss during the second winter and spring. Fallen *Typha* litter, by contrast, lost K during the 1st year, but accumulated it during the second winter and spring. After 1 year, *Scirpus* had released or translocated 5.3 kg K/ha, and *Typha* had released or translocated 128.4 kg K/ha.

Calcium Figure 5 illustrates the relative immobility of calcium. Initial losses occurred only in *Scirpus*, and these were minor (4%). The only appreciable releases from the standing litter occurred in *Typha*, 10% between 15 November 1974 and 20 June 1975. Fallen *Scirpus* litter increased considerably in Ca content; fallen *Typha* litter lost Ca. The increases observed in *Scirpus* litter probably resulted from precipitation of $CaCO_3$ from the alkaline waters of Goose Lake. The loss of Ca from *Typha* litter under the same conditions cannot yet be explained.

After 1 year, *Scirpus* had accumulated 49.1 kg Ca/ha, whereas *Typha* had released or translocated 23.1 kg Ca/ha. These losses accounted for only 11% of the Ca initially available in the newly formed *Typha* litter.

CONCLUSIONS

The preliminary models in Figs. 2 through 5 provide one with a first approximation of the patterns of nutrient flow through the litter compartments of these *Typha glauca* and *Scirpus fluviatilis* stands. One can see that initial losses due to leaching and translocation can be great, as in K, P, and N, and that later losses from the standing litter result mainly from the bulk transfer of nutrients incorporated in falling litter. Certain nutrients (N, Ca) tend to accumulate in the fallen litter, and other nutrients tend to be readily leached from fallen litter. These models also reveal the seasonal patterns in the uptake and release of nutrients and the tremendous variability in the patterns of litter decomposition between species.

The study of the ecological dynamics of prairie glacial marshes is in its infancy. Further study is needed on several problems including determining the patterns and rates of shoot death during the growing season. Very few data exist on the increase in new standing litter resulting from death of individual leaves and small shoots. Almost nothing is known about death rates and decomposition rates of rhizomes and roots. What roles are played by herbivores, parasites, and pathogens in the formation and breakdown of litter and the release of nutrients? What impact does the accumulation of standing and fallen litter have on the productivity of a site? These questions require answers.

The contribution made by litter decomposition to the marsh nutrient cycles is only vaguely understood. Very little is known of the fate of nutrients released during litter decomposition. What are the turnover times for these nutrients? To what extent are nutrients immobilized by being translocated from the shoots to the rhizomes before the shoots die?

Finally, there is a need to examine the impact that the prevailing continental climate has on patterns and rates of litter decomposition in prairie glacial marshes. What effect does the extreme cold of the prairie winters have on the breakdown of litter and the release of nutrients? How do nutrient cycling patterns change during the 5- to 20-year drawdown cycles characteristic of prairie glacial marshes described by Weller and Spatcher (1965)? Cycling patterns will change as the species composition and vegetation density of marshes change during these cycles. Long-term studies are needed to monitor and intepret these changes.

ACKNOWLEDGMENTS

We thank the Iowa Conservation Commission for permission to work at Eagle Lake, and the owners of Goose Lake for permission to work there. We are also indebted to Paul J. Currier and Michael J. Krogmeier who did most of the sampling of standing litter and to Mr. Lawrence Ewers for allowing us to use his cabin at Eagle Lake.

LITERATURE CITED

Bayly, I. L., and O'Neill, T. A. (1972a). Seasonal ionic fluctuations in a *Phragmites communis* community. *Can. J. Bot.* **50**, 2103–2109.
Bayly, I. L., and O'Neill, T. A. (1972b). Seasonal ionic fluctuations in a *Typha glauca* community. *Ecology* **53**, 714–719.
Bachmann, R. W. (1965). Some chemical characteristics of Iowa lakes and reservoirs. *Proc. Iowa Acad. Sci.* **72**, 238–243.
Davis, C. B., and van der Valk, A. G. (1978). The decomposition of standing and fallen litter of *Typha glauca* and *Scirpus fluviatilis*. *Can. J. Bot.* (*In press*).
Genson, J. J., Liegel, E., and Schulte, E. E. (1976). Wisconsin soil testing and plant analysis procedures. No. 6 Soil Fertility Series. Department of Soils Science. University of Wisconsin, Madison.
Gleason, H. A., and Cronquist, A. (1963). "Manual of Vascular Plants of Northeastern United States and Adjacent Canada." D. Van Nostrand Company, Princeton, New Jersey.
Hutchinson, G. E. (1957). "A Treatise on Limnology." Vol. I. John Wiley and Sons, Inc. New York.
Mason, C. F., and Bryant, R. J. (1975). Production, nutrient content, and decomposition of *Phragmites communis* Trin. and *Typha angustifolia* L. *J. Ecol.* **63(1)**, 71–95.
Shaw, S. P., and Fredine, C. G. (1956). Wetlands of the United States. U.S. Fish Wildl. Serv. Circ. 39.
Stewart, R. E., and Kantrud, H. A. (1971). Classification of natural ponds and lakes in the glaciated prairie region. U.S. Bur. Sport Fish. Wildl. Res. Publ. 92.

van der Valk, A. G., and Davis, C. B. (1978). Primary production of prairie glacial marshes. *In* "Freshwater Wetlands: Ecological Processes and Management Potential" (R. E. Good, D. F. Whigham and R. L. Simpson, eds.), pp. 21–37. Academic Press, New York.

Weller, M. W., and Fredrickson, L. H. (1974). Avian ecology of a managed glacial marsh. *Living Bird* 12, 269–291.

Weller, M. W., and Spatcher, C. S. (1965). Role of habitat in the distribution and abundance of marsh birds. Iowa Agric. Home Econ. Exp. Sta. Spec. Rep. 43. Ames, Iowa.

APPENDIX

Conversion of biomass and nutrient content to absolute quantities (from Davis and van der Valk, 1978)

1. that rates of litter decomposition, in both the standing and fallen litter compartments, are constant during the time intervals between sampling dates, and

2. that litter falling into the water at any time instantaneously acquires the same nutrient content and decomposition rate as that of litter present in the fallen litter compartment since the beginning of the study.

$$E_s = (S)(C_s) \tag{1}$$

where:

E_s = amount of any nutrient in the standing compartment (living or dead) on any date (g/m²)

S = the biomass (living or dead) in the standing compartment on that date (g/m²)

C_s = concentration of any nutrient in the biomass (living or dead) on that date (%)

$$E_t = (\Delta S)\,(\bar{C}_s) \tag{2}$$

where:

E_t = amount of any nutrient transferred from the standing litter compartment to the fallen compartment during any time interval (g/m²)

ΔS = change in standing biomass during that time interval (g/m²)

$\bar{C}_s$ = mean concentration of any nutrient in the standing litter compartment during that time interval (%)

$$T_t = (\Delta S)(1 - R_f/2) \tag{3}$$

where:

T_t = amount of standing litter that falls into the fallen litter compartment during a given time interval and remains in the fallen litter compartment at the end of that time interval (g/m²)

R_f = fallen litter decomposition rate during that time interval (%)

$$F_t = F_{t-1}(1 - R_f) \tag{4}$$

where:

F_t = amount of litter in the fallen litter compartment, at the beginning of a time interval, that remains in the fallen litter compartment at the end of the time interval (g/m²)

F_{t-1} = amount of litter in the fallen litter compartment (g/m²) at the beginning of a time interval

$$F = T_t + F_t \tag{5}$$

where:

F = total amount of fallen litter present at the end of a time interval (g/m²)

$$E_f = F(C_f) \tag{6}$$

where:

E_f = amount of any element in the fallen litter compartment at the end of a time interval (g/m²)

C_f = concentration of any element in the fallen litter at the end of any time interval (%)

DECOMPOSITION IN NORTHERN WETLANDS

Jim P. M. Chamie

Botany-Biology Department
South Dakota State University
Brookings, South Dakota 57007

and

Curtis J. Richardson[1]

School of Natural Resources
The University of Michigan
Ann Arbor, Michigan 48109

Abstract   Northern wetland decomposition processes were reviewed and litterbag data obtained from 4 commonly occuring plant groups in a central Michigan peatland from 1973 to 1975. Air-dried leaves, small stems and large stems of *Salix* spp., *Carex* spp., *Betula pumila* and *Chamaedaphne calyculata* were confined in nylon mesh bags (1-mm mesh) for periods up to 590 days. The effects of aerobic and anaerobic conditions on decomposition were determined via dry weight losses, rates of decay and nutrient concentration changes for N, P, K, Ca and Mg. Percentages of dry weight remaining after nearly 2 years were always >58% of the initial weight for any plant group. The 99% turnover time for most of the vegetation components in the wetland exceeded 10 years. An estimate of organic peat turnover time at the ecosystem level was 5,000 to 10,000 years. Anaerobic decomposition occurred more slowly 15 cm below the peat—water interface than at the interface. For most species and their parts, nutrient concentration losses occurred in P, K and Mg while increases occurred in N and Ca. Elemental mobility followed the order: K > Mg > P > N > Ca.

Key words   *Decomposition, marshes, Michigan, nutrient losses, peatlands, plant breakdown, wetlands.*

[1] Present address: School of Forestry and Environmental Studies, Duke University, Durham, North Carolina 27706.

INTRODUCTION

This paper mainly presents decomposition data obtained from a field study conducted on a wetland[2] in central Michigan from April 1973 to June 1975. Objectives of the study were to determine the effects of aerobic and anaerobic decomposition upon: dry weight losses, rates of decay, and nutrient concentration changes for 4 commonly occurring plant groups. Decomposition rate parameters and turnover times were also estimated.

Background and Rates of Decomposition Decomposition, with accompanying weight losses, results from a combination of respiration, leaching, comminution loss, plus physical removal. However, addition of material may occur along with growth of microorganisms into, through and out of decaying vegetation samples (Heal and French, 1974). The rate of decomposition of organic matter determines the availability of nutrients necessary for growth of microorganisms. This relationship is much more valid during the later stages of weight loss, especially in wetlands, where leaching may account for 50–75% of the weight loss during the 1st year (Romarov, 1968). The later stages of decomposition rely almost completely upon the activity of microorganisms (Remezov, 1961).

Nykvist (1959) showed that leaching occurred almost immediately under both aerobic and anaerobic conditions with most of the water-soluble organic substances being released within 6 to 12 months. He also showed that a positive correlation existed between decomposition of litter and amount of water-soluble organic substances present. Sorenson's (1974) studies have also shown that decay rates in the laboratory were faster than those of native soil organic matter and that the rate of soil organic decomposition increased 12–30% following repeated drying-rewetting treatment.

Jenny et al. (1949) formulated nonconstant integrative parameters, known as k and k', to denote the decomposition rate of vegetation under steady state conditions. Simply stated, k' equals the percentage loss of the original weight over a specified period of time with a single year usually used. The term k equals the $-\ln(1 - k')$ or the negative natural logarithm of the percentage weight remaining after a designated period of time. Both parameters, k and k', measure the effectiveness of the decomposer organism, plus leaching and physical loss due to fragmentation.

The values for k range from 0.0001 to 4.0 for most ecosystems (Olson, 1963). The lower values correspond to wetlands, while the higher values correspond to tropical forests. In many decomposition studies, the assumption of a relatively constant rate loss has been made with data fitted to a negative exponential curve. Decomposition rates of wetlands fit or follow this negative exponential decay model when weight losses are measured over long periods of time generally exceeding 1 year (Chamie, 1976).

---

[2] The term "wetland" is used generically to include marshes, bogs, fens, peatlands, and potholes whenever general statements are applicable.

Role of Microorganisms   Generally the total number of microorganisms is higher in organic soils, especially peat, than in nonorganic soils due to the higher proportion of available carbon from organic matter (Dawson, 1956). The high concentration of C in organic soils and litter provides the substrate necessary for rapid decomposition when N and P are freely available. These higher microorganisms' numbers tend to create higher levels of microbial activity than is generally estimated (Küster, 1968).

Even more difficult to interpret than physical and chemical processes are the specific roles that microorganisms play in the decomposition of different stages of organic matter. In wetlands measurements of seasonal and vertical variability of microorganisms are further compounded by aerobic—anaerobic metabolic differences (Latter and Cragg, 1967). As a result, the measurement of total microbial activity is now being favored. This activity may include many of the following reactions: oxidation-reduction, volatilization, solubilization, precipitation, and the production of organic and insoluble chelates (Miller, 1973).

Aerobic and Anaerobic Conditions   Waterlogged soils, especially organic soils, develop a very thin oxidized layer a few millimetres thick in the surface and may, if good drainage exists, even have some local oxidized areas in the remaining reduced strata (Aomine, 1962). In general, however, the surface layer undergoes aerobic respiration, while the remaining areas undergo anaerobic respiration and fermentation. Anaerobic decomposition is much slower and is done by obligate anaerobic and some facultative anaerobic bacteria.

Anaerobic bacteria function most efficiently at near neutral pH and low redox potential values < -200 mV adjusted to a pH of 7 and given as Eh_7 (Aomine, 1962). Generally a soil zone with an Eh_7 exceeding $+200$ mV is considered aerobic and will support nitrification, but a zone < -200 mV is considered anaerobic and will support denitrification (Tusneem and Patrick, 1972). It has been shown that aerobic activity is very limited 3 cm below the waterlogged peat surface (Dickinson, 1974). The presence of such gases as CH_4, H_2 and H_2S indicates anaerobic conditions (Lähde, 1969). Other substances can also accumulate under anaerobic conditions such as short-chain organic acids, alcohols, amines and mercaptans.

Periodic or even constant flooding of a soil's surface, characteristic of wetlands, leads to an overall decrease in the activity of soil fauna (Tusneem and Patrick, 1972). The surface layer of peat has been shown to lack many common forest soil faunal elements and unable to support them if they are introduced (Dickinson, 1974). In peat, for example, nematodes and earthworms are usually absent due to the unavailability of O_2 (Kozlovskaya, 1963). In addition most fungi are obligate aerobes, and as most waterlogged wetland ecosystems are anaerobic, fungi numbers are usually very low and greatly outnumbered by bacteria and actinomycetes (Brock, 1966).

Review of Decomposition Methods   Various techniques have been used to study plant decomposition in northern wetlands. Field studies on the decomposition of litter or selected vegetation placed in or on the litter layer have been performed using methods which directly or indirectly measure the microbial activity in soils. These techniques include: (1) enzyme assay methods which measure biological activities as opposed to numbers and biomass of microorganisms (Macfadyen, 1970; Skujins, 1973); (2) thermal measurements detected by galvanometers, thermistors, etc. (Mina, 1962); (3) ATP-assay for total numbers and biomass of microorganisms (Ausmus and Edwards, 1972; Parkinson et al., 1971); (4) gas exchange or respiration primarily as O_2 consumption or CO_2 evolution (Chapman, 1971; Sorokin and Kadota, 1972; Ivarson, 1977); and (5) rate of substrate disappearance from mesh bags and other materials (Boyd, 1970; Mason and Bryant, 1975; Rosswall, 1973; van Cleve, 1971). The first 3 techniques although better measurements, have not proven suitable on waterlogged wetland soils. Respiration measurements have a great disadvantage for waterlogged organic soils because they usually measure only aerobic processes and exclude anaerobic processes which are very important in many wetland systems.

The litterbag technique using nylon or fiberglass mesh bags containing known amounts of selected vegetation has been successfully used in forests (Shanks and Olson, 1961), tundra (Heal and French, 1974), grasslands (Clark, 1970) and wetlands (Boyd, 1970, Reader and Stewart, 1972). Mesh bags permit organic matter quickly to become part of the natural litter layers and give reasonable estimates of decay rates if the mesh openings are large enough to permit macrofaunal entry but small enough to prevent large losses of vegetation from bags.

METHODS

Houghton Lake Study Area   The wetland studied consists of 716 ha located 2.3 km southwest of Houghton Lake, Michigan (44°20′N, 84°50′W) at an elevation of 346 m. Because the organic soils (histosols) in the area are categorized as Rifle peat and Houghton muck with underlying Newton loamy sand, the study site shall be referred to as a peatland.

The peatland study site supports 2 distinct vegetation types. One called the sedge-willow (S-W) community includes predominantly sedges (*Carex* spp.) and willows (*Salix* spp.). The second is the leatherleaf—bog birch (L-B) community consisting of mostly *Chamaedaphne calyculata* (L.) Moench and *Betula pumila* L., respectively. The L-B community, which also contains small areas of sedge–willow, comprises 19% of the total peatland. Standing water is present during most of the year, but the peatland is often dry during late summer and early autumn. The L-B cover type is usually drier than the S-W cover type. For a further site description, its productivity, peat–water and nutrient characteristics, see Chamie (1976), Richardson et al. (1976), Richardson et al. (1978) and Wentz (1976).

Experimental Design   Large representative areas were chosen in each of the 2 communities. Sixteen 6 × 6 m plots were systematically placed approximately 6–10 m from one another in each cover type area. Plots were marked using metal conduit and heavy twine to prevent disturbance. All plots were numbered then divided into 20 subplots (1 × 1 m) which were systematically numbered and randomly selected for analyses. All experiments utilized full factorial, completely crossed designs.

Decomposition of Vegetation   The following 10 categories of plant parts were used in a litterbag study: sedge leaves (mostly *Carex lasiocarpa* Ehrh., *Carex aquatilis* Wahl, *Carex comosa* Boott, *Carex oligosperma* Michaus and *Carex rostrata* Stokes); leaves, small stems ≤2.0 mm in diameter and large stems ≥12.0 mm of willows (mostly *Salix pellita* Anders, *Salix discolor* Muhl. and *Salix pedicellaris* Pursh.); leaves, small stems ≤2.0 mm and large stems ≥3.0 mm in diameter of leather-leaf; and leaves, small stems ≤2.0 mm and large stems ≥12.0 mm in diameter of bog birch.

All samples were first dried 2–3 weeks to constant air-dry weights, then ovendried at 85°C to a constant weight (1–2 days longer). Oven-drying revealed that moisture content of air-dried vegetation ranged from 10–15%. A regression equation relating air-dry to ovendry weight was calculated from representative samples from each category: $Y = -1.06 + 1.22(X)$, where X = ovendried weight and Y = air-dry weight, $r^2 = 0.9956$, and $s_{y \cdot x} = 0.6236$. This equation was found applicable to all peatland plants in this study.

Approximately 1,300 bags (20 × 20 cm in size) were made out of nylon cloth with a 1-mm mesh. This size opening permitted maximum movement of all but the largest decomposer organisms and minimized possible vegetation loss from the bags due to fragmentation. The bags contained 15–40 g of vegetation air-dried to a constant weight with leaves usually comprising <20 g because of their volume. All bags were identified with a plastic label, sewn closed, and weighed prior to placement in the field.

On 30 September 1973 all bags were placed in nonadjacent subplots within each of the large plots. Half the bags were placed on the surface of the peat soil, while the other half were placed 15 cm below the peat's surface or the soil–water interface. It was hypothesized that aboveground placed bags would undergo aerobic decomposition, while those belowground would primarily undergo anaerobic decomposition. All belowground material was placed first by making an opening in the peat with a sharp spade then individually placing each bag as horizontal as possible and at least 20 cm from similar litter bag material placed aboveground. All paired bags were marked by wire flags to facilitate retrieval at a later date.

Sixteen bags from each group were randomly chosen for collection at 215, 265, 320 and 370 days of exposure in the field, then 8 bags were collected from each group at 590 days of exposure.

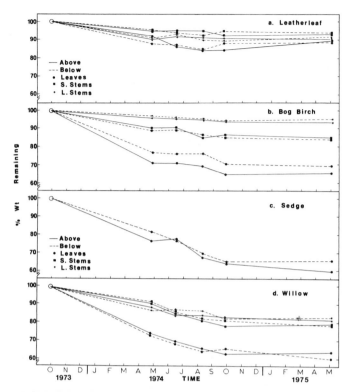

Fig. 1. Percentages of original weight remaining in leaves, small and large stems of (a) leatherleaf, (b) bog birch, (c) sedge, and (d) willow placed in aboveground and belowground positions (September 1973). All points represent a mean of 8 samples, except May 1975 which had only 4 samples. Standard errors about the means never varied more than 13% (leaves of all species), 15% (small stems), and 20% (large stems).

Laboratory Analysis   All collected material was dried at 85°C to a constant weight, then weighed to the nearest 0.01 g. Randomly selected portions of each bag's contents were ground using a Wiley Mill to pass through a no. 20 mesh screen. All samples were stored in 60 cm³ plastic vials until nutrient analysis could be performed. Total nitrogen was determined using the semimicro Kjeldahl procedure (Black, 1965). Total phosphorus analysis was done utilizing spectrophotometry following wet acid digestion of the ground samples, which consisted of a perchloric and nitric acid system (Behan and Kinraide, 1973). Sodium, Mg, K and Ca were determined by atomic absorption spectrophotometry following standard atomic absorption techniques (Perkin-Elmer, 1973).

Statistical Analysis   All two-way analyses of variance were made utilizing a factorial program, which can accommodate unequal observations per cell and multiple-way layouts. All one-way analyses of variance were performed using the MIDAS statistical package (Fox and Guire, 1974). Simple and complex contrasts were determined using a posteriori multiple comparison tests.

A critical level of $\alpha = .10$ was initially set for all analyses of variance. This level was chosen based on the inherent variability of the populations, laboratory procedures used in analyses, and levels previously used on other projects. Representative sets of results from each type of analysis were examined to determine if the assumptions of the models were being violated. In each case all assumptions were met.

RESULTS AND DISCUSSION

Weight Loss

Differences in weight loss were shown (Fig. 1) to be significant ($P < .10$) with respect to season and depth (aboveground and belowground) for some components. Significant differences in weight loss ($P < .10$) occurred for all plant parts over time except for leatherleaf stems. Significant differences ($P < .10$) by position were only noted in willow small stems and the leaves of leatherleaf and bog birch (Fig. 1).

Closer examination of all 10 plant groups indicated that a general encompassing statement concerning aboveground vs. belowground weight losses could not be made. This may be due to the decay rate differences between species and their plant parts, the accuracy of methods employed, and the dry summer conditions of 1973 which may have produced aerobic conditions in depths where belowground samples were placed. No significant differences were obtained for the interaction of any of the factors for any of the 10 groups of vegetation.

The percentage of dry weight remaining for any of the 10 groups was always >58% of their initial dry weights following 20 months field exposure (Fig. 1a–d). However, few changes occurred after 1 year for all groups. Leatherleaf leaves and stems had the smallest losses, 6–16%, during the entire study. Reader and Stewart (1972) found that leaves of ericaceous shrubs in Manitoba, Canada peatlands lost 7–38% of their original dry weight during the 1st year.

The weight losses for leaves of sedge, willow and bog birch ranged from 18–29%, 26–36% and 31–41% at 215, 370 and 590 days, respectively. Stems from all species lost weight more slowly than their corresponding leaves. Willow and bog birch stems never lost more than 21% and 16%, respectively. The high concentration of cellulose, hemicellulose and lignin in woody stems (Epstein, 1972) probably accounted for this much slower rate of decomposition.

The loss rates for the leaves of our non-ericaceous species were less than those reported at other sites. For example, Hodkinson (1975) found that after 1 year *Salix* spp. leaves confined in mesh bags placed in a beaver pond in Alberta, Canada lost ≈67% of their initial weight, while *Juncus tracyi*

Table 1. Comparison of Fractional Weight Loss [k'] and k [k = −ln(1 − k')] to 10 Groups of Vegetation Placed Aboveground after 215, 370 and 590 Days in the Houghton Lake, Michigan Wetland

| | Time and decomposition rates | | | | | |
| | 215 days | | 370 days | | 590 days | |
Species and part	k'	k	k'	k	k'	k
Sedge (leaves)	.230	.261	.361	.448	.417	.540
Willow (leaves)	.271	.316	.368	.459	.353	.435
Willow (small stems)	.107	.113	.219	.247	.205	.229
Willow (large stems)	.116	.123	.166	.182	.171	.188
Leatherleaf (leaves)	.089	.093	.154	.167	.109	.115
Leatherleaf (small stems)	.066	.068	.089	.093	.072	.075
Leatherleaf (large stems)	.101	.107	.088	.092	.116	.123
Bog birch (leaves)	.295	.350	.372	.465	.322	.389
Bog birch (small stems)	.103	.109	.136	.146	.133	.143
Bog birch (large stems)	.054	.056	.079	.082	.082	.086

leaves lost 39%. Latter and Cragg (1967) reported that the dry weight loss from leaves of *Juncus squarrosus* placed in nylon nets was about 20–25% after 1 year and 50% after 2.5 years. In England, Mason and Bryant (1975) found that *Typha angustifolia* and *Phragmites communis* confined in mesh bags and placed on the surface of mud lost 10–20% of their initial dry weight during the 1st month. Burkholder and Bornside (1957) reported that 50% of the dry weight of *Spartina* was lost in the first 6 months. Rosswall (1973) showed that winter leaching rates were a major part of the weight losses of many species in wet tundra areas.

Rates of Decomposition

The decomposition rate parameters, k and k', for our peatland given in Table 1 also show a higher weight loss in the first 6- to 12-month period with little, if any, change occurring thereafter to most groups. The weight loss for leatherleaf stems was negligible during the 590 day study period. Sedge showed the highest decay rate followed by leaves of bog birch and willow. Leatherleaf leaves were considerably more resistant to breakdown than other nonwoody material.

Regressions comparing values of k and k' against time of decomposition at 215, 370 and 590 days were calculated from transformed data. The highest correlations ($r^2 = 0.7311$) and best relationships followed a negative exponential curve. This was also found by Gosz et al., 1973; Hodkinson, 1975; and Olson, 1963. Because the negative exponential curve reaches an asymptotic level, this curve can be closely equated to biological half-life or turnover rates. A parameter calculated from this exponential relationship gives

Table 2. Decomposition Rate Parameters Including Turnover Times for Leaves, Small Stems and Large Stems of Sedge, Willow, Bog Birch and Leatherleaf in the Houghton Lake, Michigan Wetland

Part and species	Stem diam (mm)	k-value (370 days)	Turnover times (years)[a]		
			Half-life	95%-life	99%-life
Leaves					
Sedge	...	.448	1.55	6.70	11.17
Willow	...	.459	1.51	6.54	10.89
Bog birch	...	.465	1.49	6.45	10.75
Leatherleaf	...	.167	4.15	17.96	29.94
Small stems					
Willow	≤2.0	.247	2.81	12.15	20.24
Bog birch	≤2.0	.146	4.75	20.55	34.25
Leatherleaf	≤2.0	.093	7.45	32.26	53.76
Large stems					
Willow	≥12.0	.182	3.81	16.48	27.47
Bog birch	≥12.0	.082	8.45	36.59	60.98
Leatherleaf	≥3.0	.092	7.53	32.61	54.35

[a] Half-life = 0.693/k, 95%-life = 3/k, 99%-life = 5/k (Olson, 1963).

the time required for 50% of the original weight of organic matter to disappear. Therefore, solving: $0.50 = e^{-kt}$, $t = 0.693/k$. Similarly, the time required for 99% to disappear would be: $t = 5/k$ from solving $0.99 = e^{-kt}$ (Olson, 1963).

The k value at 1 year was used to extrapolate turnover times (Table 2). The 99% turnover time for most of the vegetation components in our peatland exceeded 10. Leatherleaf, the species most resistant to decay, requires in excess of 30 and 50 years for 99% decomposition in leaves and stems, respectively. Willow was the woody species least resistant to decay.

To estimate the long-term rates and the turnover time at the ecosystem level, a new k value was derived by dividing the annual litterfall by the total accumulated organic matter. The annual litterfall in the L-B community equals ≈ 140 g/m² (Chamie, 1976). The upper cubic metre of ovendried peat weighs between 135 to 255 kg (J. Chamie and W. A. Wentz, *personal communication*). If this value is assumed to be the total accumulated organic matter, k values of 0.0001 to 0.0005 are obtained. These k values are much smaller than those determined using short-term decomposition bag data studies and indicate composite vegetation turnover times (5/k) of 5,000 to 10,000 years for the ecosystem. Because our peatland has a large amount of accumulated organic matter, these k values may more closely reflect the true k rates for the entire system.

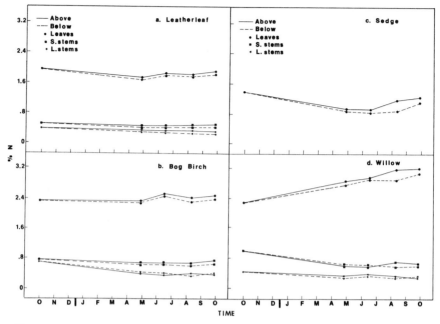

Fig. 2. Changes in N concentrations in leaves, small and large stems of (a) leatherleaf, (b) bog birch, (c) sedge, and (d) willow placed in aboveground and belowground positions (September 1973). Points equal the mean of 8 samples. Standard errors about the means never varied more than 3% (leaves of woody species), 7% (large stems and sedge leaves), and 8% (small stems).

Elemental Changes

Initial concentrations of elements in plant tissue vary from species to species. However, loss rates of a particular element from different types of vegetation generally are the same because of the way specific elements are incorporated in organic matter (Epstein, 1972). Loss of nutrients through leaching depends on the type of vegetation, its physical condition, the respiration rates, and the presence of aerobic and anaerobic conditions.

Boyd (1970) felt that leaching of nutrients should proceed at a similar rate for any species of aquatic macrophyte and that aquatic vegetation releases large quantities of nutrients during the first few days of decomposition. Heal and French (1974) stated that the potentially leachable fraction of most plants is in the range of 5–30% of dry weight, with highest values in soft herbaceous leaves and lowest in woody tissue. Mason and Bryant (1975) found that Na, K, Mg and P leached out of *Phragmites* and *Typha* confined in litterbags during the 1st month, after which a constant residual level remained. Calcium levels changed less, due to precipitation out of the water. Nitrogen levels increased perhaps due to microbial activity. The elemental

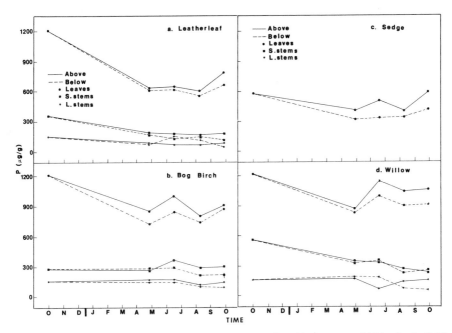

Fig. 3. Changes in P concentrations in the leaves, small and large stems of (a) leatherleaf, (b) bog birch, (c) sedge, and (d) willow placed in aboveground and belowground positions (September 1973). Points equal the mean of 8 samples. Standard errors about the means never varied more than 6% (leaves of woody species), 12% (willow stems), 14% (sedge leaves), and 20% (leatherleaf and bog birch stems).

concentrations changes for the Michigan site for the 10 plant groups are given in the following sections.

Nitrogen   There were significant differences ($P < .10$) in N concentrations for all groups over time except leatherleaf small stems which remained virtually unchanged. Only half of the 10 groups showed significant differences ($P < .10$) between aboveground and belowground samples of vegetation (Fig. 2). These were sedge, willow and bog birch leaves, leatherleaf large stems, and bog birch small stems. In all other tissues, concentrations of N in aboveground and belowground positions were not significantly different.

Sedge leaves had an initial concentration of 1.35%, which then decreased to 0.90% at 7 months followed by a steady rise to 1.32%. The leaves of all woody species had initial concentrations of about 2.0–2.3% and all showed an increase after 370 days. Small (1972) reported N concentration values of 1.7 to 2.3% for new leatherleaf leaves. Nitrogen concentrations in willow leaves in our study showed an increase of 39.4% after 1 year. Bacteria and other microorganisms' growth on the samples may explain these rapid increases in N percentages.

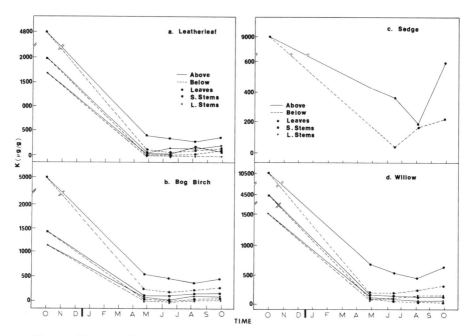

Fig. 4.   Changes in K concentrations in leaves, small and large stems of (a) leatherleaf, (b) bog birch, (c) sedge, and (d) willow placed in aboveground and belowground positions (September 1973). Points equal the mean of 8 samples. Standard errors about the means never varied more than 10% (willow), 22% (sedge), and 25% (leatherleaf and bog birch).

The initial N levels of the stems of all species fell into a rather small range, 0.37 to 0.95%. In general, small and large stems of all species showed decreased levels of N at 215 days except bog birch small stems which showed a slight increase (3.5%) over initial concentrations.

Phosphorus   There were significant differences ($P < .10$) in P concentrations over time for all groups, except for bog birch stems and leatherleaf small stems. Initial P concentrations of about 1,200 μg/g were determined for all leaves of woody species, with sedge having half this concentration. Small (1972) reported new leatherleaf leaves values at 1,100–1,500 μg/g. Concentrations of P in aboveground leaf samples were always higher than concentrations in belowground positions (Fig. 3). At 7 months all aboveground groups showed decreases, except bog birch and willow stems with 7.8% and 14.8% increases, respectively. After 1 year all groups remained relatively unchanged except for increases of 16%, 18%, and 29% for bog birch, leatherleaf and sedge leaves, respectively (Fig. 3a–d).

Potassium   There were significant differences ($P < .0001$) between K concentrations in aboveground and belowground positions. All groups lost

>88.9% of their initial K concentrations within 7 months. Remezov (1961) reported that 80–90% of the K can be lost in 1 year. As K is not a structural part of plant tissue, most K will be leached out very quickly. All stem concentrations were initially 1,200 μg/g or greater and reached near-zero levels after 7 months. Changes <4.3% occurred during the next 5 months. All belowground groups except willow large stems lost greater amounts of K than their aboveground counterparts (Fig. 4a–d).

Calcium Initial Ca concentrations varied more than any other element for the 10 groups of vegetation and ranged between 0.13–1.22%. Fluctuations occurred at 7 months and at 1 year which may be regulated by high Ca concentration changes in interstitial water. The only aboveground groups that did not show increases after 7 months were leatherleaf large stems and all bog birch parts. Only litterbags containing leaves had aboveground concentrations significantly higher ($P < .10$) than belowground concentrations. The general range of concentrations for the combined stems by species after 7 months was: willow (0.58–0.92%) > bog birch (0.48–0.71%) > leatherleaf (0.14–0.26%).

Magnesium At 7 months, Mg, similar to K, showed decreases in every group except willow large stems (+6.6%). Initial Mg concentrations of leaves of bog birch, willow and leatherleaf were 0.30%, 0.21% and 0.13%, respectively. In contrast the initial Mg concentrations of the stems from all species ranged between 0.035–0.077% and varied only negligibly after 7 months. At 1 year losses occurred in all vegetation groups ranging between −29.3% to −67.9%. All leaves showed significant differences ($P < .10$) with location in the first 7 months with the belowground samples always showing greater losses than the aboveground samples.

At 215 and 370 days there were significant differences ($P < .0001$) in the concentrations of each element between the leaves, small stems and large stems of woody species placed aboveground. Similar trends for vegetation placed belowground and at the other aboveground dates were obtained. Exceptions were all elemental concentrations in leatherleaf and Mg in willow ($P < .01$). A posteriori tests indicated that significant differences ($P < .0001$) always occurred between leaves and stems sizes of woody species, except for some elements (K in leatherleaf and K + Ca + Mg in bog birch) where no significant differences occurred between small and large stems.

Elemental Mobility

The rate of leaching of the elements in the Michigan study both by position and at different dates usually followed the order: K > Mg > P > N > Ca. However, in large woody stems the order of P, N and Ca changed slightly. Elemental mobility in leatherleaf stems showed nearly the same order of elemental loss as *Juncus* leaves (K > P > Mg > Ca) determined by Latter and Cragg (1967).

The previous lack of interest in wetlands along with the problems associated with field and laboratory decomposition research on waterlogged peat has resulted in a paucity of data. The available data are restricted to a limited number of plant species, studied generally for relatively short periods of time using sample methods that estimate decomposition rates (Chamie, 1976; Hodkinson, 1975; Latter and Cragg, 1967; Mason and Bryant, 1975; and Reader and Stewart, 1972). Much longer studies using more accurate methods must be developed before northern wetland decomposition processes can be more fully understood.

ACKNOWLEDGMENTS

Research was supported by the National Science Foundation, Grant GI-34812X, Research Applied to National Needs. Additional grant support was received from the Schools of Natural Resources and Rackham in The University of Michigan. Research was conducted in the Schools of Public Health and Natural Resources and also on lands maintained by the Michigan Department of Natural Resources. Appreciation is expressed to V. Chamie, W. A. Wentz, E. Kasischke, M. Quade and others for assistance in gathering and analyzing portions of the data.

LITERATURE CITED

Aomine, S. (1962). A review of research on redox potentials of paddy soils in Japan. *Soil Sci.* **94**, 6–13.

Ausmus, B. S., and Edwards, N. T. (1972). The relationship between two microbial metabolic activity indices: ATP Concentration and CO_2 evolution rate. Eastern Deciduous Forest Biome, Memo Report No. 72-93 Oak Ridge Nat. Lab., Oak Ridge, Tennessee.

Behan, M. J., and Kinraide, T. (1973). Rapid wet ash digestion of coniferous foliage for analysis of K, P, Ca, Mg. Botany Department, The Univ. of Montana, Missoula, Montana (mimeo).

Black, C. A. (1965). Methods of Soil Analysis, Part 2: Chemical and Microbiological Properties. American Society of Agronomy Inc., Madison, Wisconsin.

Boyd, C. E. (1970). Losses of mineral nutrients during decomposition of *Typha latifolia*. *Arch. Hydrobiol.* **66**, 511–517.

Brock, T. D. (1966). Principles of Microbial Ecology. Prentice-Hall Inc., Engelwood Cliffs, New Jersey.

Burkholder, R. R., and Bornside, G. H. (1957). Decomposition of marsh grass by aerobic marine bacteria. *Bull. Torrey Botan. Club* **84**, 366–383.

Chamie, J. P. M. (1976). "The Effects of Simulated Sewage Effluent upon Decomposition, Nutrient Status and Litter Fall in a Central Michigan Peatland." Ph.D. Dissertation, The Univ. of Michigan Press, Ann Arbor, MI.

Chapman, S. B. (1971). A simple conductimetric soil respirometer for field use. *Oikos* **22**, 348–353.

Clark, F. E. (1970). "The Microbial Component of the Ecosystem." Technical Report No. 52. U.S. International Biological Program, Colorado State University, Las Cruces, New Mexico.

Dawson, J. E. (1956). Organic soils. *Advan. Agron.* **8**, 377–401.

Dickinson, C. H. (1974). Decomposition of litter in soil. *In* "Biology of Plant Litter Decomposition." (C. H. Dickinson, and G. J. F. Pugh, eds.), pp. 633–658. Academic Press, New York.

Epstein, E. (1972). "Mineral Nutrition of Plants: Principles and Perspectives." John Wiley and Sons Inc., New York.

Fox, D. J., and Guire, K. E. (1974). "Documentation for Michigan Interactive Data Analysis System (MIDAS)." The Univ. of Michigan Press, Ann Arbor, 86 pp.

Gosz, J. R., Liken, G. E., and Bormann, F. H. (1973). Nutrient release from decomposing leaf and branch litter in the Hubbard Brook Forest, New Hampshire. *Ecol. Monog.* **43**, 173–191.

Heal, O. W., and French, D. D. (1974). Decomposition of organic matter in tundra. *In* "Soil Organisms and Decomposition in Tundra" (A. J. Holding, O. W. Heal, S. F. Maclean, and P. W. Flanagan, eds.), pp. 279–304. The University of Alaska Press, College, Alaska.

Hodkinson, I. D. (1975). Dry weight loss and chemical changes in vascular plant litter of terrestrial origin, occurring in a beaver pond ecosystem. *J. Ecol.* **63**, 131–142.

Ivarson, K. C. (1977). Changes in decomposition rate, microbial population and carbohydrate content of an acid peat bog after liming and reclamation. *Can. J. Soil. Sci.* **57**, 129–137.

Jenny, H., Gessel, S. P., and Bingham, F. T. (1949). Comparative study of decomposition of organic matter in temperate and tropical regions. *Soil Sci.* **68**, 419–432.

Kozlovskaya, L. S. (1963). Role of soil organisms in decomposition of organic remains in swamped forest soils. *In* "The Increase of Productivity of Swamped Forests" (L. S. Kozlovskaya, ed.), pp. 27–39. Israel Program for Scientific Translations, Jerusalem.

Küster, E. (1968). Influence of peat and peat extracts on the growth and metabolism of microorganisms. *International Peat Congress* **2**, 945–952.

Lähde, E. (1969). Biological activity in some natural and drained peat soils with special reference to oxidation-reduction conditions. *Acta For. Fenn.* **94**, 1–69.

Latter, P. M., and Cragg, J. B. (1967). The decomposition of *Juncus squarrosus* leaves and microbiological changes in the profile of *Juncus* moor. *J. Ecol.* **55**, 465–482.

Macfadyen, A. (1970). Soil metabolism in relation to ecosystem energy flow and to primary and secondary production. *In* "Methods of Study in Soil Ecology" (Unesco, ed.), pp. 167–172. Unesco, Paris.

Mason, C. F., and Bryant, R. J. (1975). Production, nutrient content and decomposition of *Phragmites communis* Trin. and *Typha angustifolia* L. *J. Ecol.* **63**, 71–95.

Miller, R. H. (1973). The soil as a biological filter. *In* "Recycling Treated Municipal Wastewater and Sludge through Forest and Cropland" (W. E. Sopper, and L. T. Kardos, eds.), pp. 71–93. The Pennsylvania State Univ. Press. University Park, Pa.

Mina, V. N. (1962). Comparison of methods for determining the intensity of soil respiration. *Soviet Soil Sci.* **10**, 1188–1192.

Nykvist, N. (1959). Leaching and decomposition of litter: I. Experiments on leaf litter of *Fraxinus excelsior*. *Oikos* **10**, 190–211.

Olson, J. S. (1963). Energy storage and the balance of producers and decomposers in ecological systems. *Ecology* **44**, 322–331.

Parkinson, D., Gray, T. R. C., and Williams, S. T. (1971). "Methods for Studying the Ecology of Soil Microorganisms." Blackwell Scientific Publications, Oxford, England.

Perkin-Elmer. (1973). "Analytical methods for atomic absorption spectrophotometry." Perkin-Elmer Corporation, Norwalk, Connecticut.

Reader, R. J., and Stewart, J. M. (1972). The relationship between net primary production and accumulation for a peatland in southeastern Manitoba. *Ecology* **53**, 1024–1037.

Remezov, N. P. (1961). Decomposition of forest litter and the cycle of elements in an oak forest. *Soviet Soil Sci.* **7**, 703–711.

Richardson, C. J., Kadlec, J. A., Wentz, W. A., Chamie, J. P. M., and Kadlec, R. H. (1976). Background ecology and the effects of nutrient additions on a central Michigan wetland. *In* "Proceedings: Third Wetlands Conference" (M. W. Lefor, W. C. Kennard, and T. B. Helfgott, eds.), pp. 34–72. The Univ. of Connecticut Press, Storrs, Conn.

Richardson, C. J., Tilton, D. L., Kadlec, J. A., Chamie, J. P. M., and Wentz, W. A. (1978). Nutrient dynamics of northern wetland ecosystems. *In* "Freshwater Wetlands: Ecological

Processes and Management Potential" (R. E. Good, D. F. Whigham, and R. L. Simpson, eds.), pp. 217–241. Academic Press, New York.

Romarov, V. V. (1968). "Hydrophysics of Bogs." Israel Program for Scientific Translations, Jerusalem.

Rosswall, T. (1973). Plant litter decomposition studies at the Swedish Tundra Site. In "IBP-Progress Report 1972" (M. Sorenson, ed.), pp. 124–133. Swedish Natural Science Research, Stockholm.

Shanks, R. E., and Olson, J. S. (1961). First-year breakdown of leaf litter in Southern Appalachian forest. *Science* 134, 194–195.

Skujins, J. (1973). Dehydrogenase: an indicator of biological activities in arid soils. *Bull. Ecol. Res. Comm.* 17, 235–241.

Small, E. (1972). Ecological significance of four critical elements in plants of raised *Sphagnum* peat bogs, *Ecology* 53, 498–503.

Sorenson, L. H. (1974). Rate of decomposition of organic matter in soils as influenced by repeated air drying-rewetting and repeated additions of organic material. *Soil Biol. Biochem.* 6, 287–292.

Sorokin, Y. I., and Kadota, H. (1972). "Techniques for the Assessment of Microbial Production and Decomposition in Fresh Waters." Blackwell Scientific Publications, Oxford, England.

Tusneem, M. E., and Patrick, W. H. (1972). "Nitrogen Transformations in Waterlogged Soils." Agricultural Experiment Station, Bulletin No. 6, Louisiana State University, Baton Rouge, La.

van Cleve, K. (1971). Energy- and weight-loss functions for decomposing foliage in birch and aspen forests in interior Alaska. *Ecology* 57, 720–723.

Wentz, W. A. (1976). "The Effects of Sewage Effluents on the Growth and Productivity of Peatland Plants." Ph.D. Dissertation. The Univ. of Michigan Press, Ann Arbor, MI.

DECOMPOSITION IN THE LITTORAL ZONE OF LAKES

Gordon L. Godshalk and Robert G. Wetzel

W. K. Kellogg Biological Station
Michigan State University
Hickory Corners, Michigan 49060

Abstract The decomposition of 5 species of aquatic angiosperms representing submersed, floating-leaved, and emergent growth forms was studied in laboratory and field experiments. Laboratory investigations involved the decay of lyophilized senescent plant material in synthetic lake water under aerobic and anaerobic conditions at both 10°C and 25°C. Plant tissue was also placed at sites in the littoral zones of 2 lakes at 2 times of the year in order to monitor in situ decomposition processes.

After incubation periods of up to 180 days, particulate material was dried and analyzed for weight loss and contents of ash, hemicellulose, cellulose, lignin, nonstructural carbohydrate, and total C and N. Media of the laboratory experiments were filtered, dissolved material fractionated by membrane ultrafiltration into molecular weight categories, and assayed for total dissolved organic C, UV absorbance, and fluorescence activity; C and N analyses and IR spectrophotometry were performed on unfractionated filtrates.

Results show that temperature was the most important factor influencing the rates of decomposition and the conversion of particulate matter to dissolved organic matter (DOM) while O_2 concentration controlled the efficiency of decomposition and the conversion of dissolved organic matter to CO_2. The interaction of these 2 environmental parameters produced a continuum of decomposition rates from very slow during cold, anaerobic conditions to rapid under warm, aerated conditions. Decomposition processes were also affected by the structure and morphology of the plants involved, the floating-leaved water lily decaying fastest, with the truly submersed plants at an intermediate rate and the species of emergent bulrushes most slowly. Decay rates were not correlated with the concentration of any particular fiber component but rather to the total amount of all fiber constituents present in the tissue. In situ decomposition was not greatly different qualitatively from that observed in the laboratory.

Microbial metabolism associated with the decomposing plant tissue was measured by assaying the content of ATP and the activity of dehydrogenase. The data from these 2 assays did not compare favorably with each other, and severe inadequacies were encountered in both methods. Of possible use in estimating microbial colonization of decomposing plant tissue is the total C and N contents and the resultant C:N ratio. As found in many other studies, N generally accumulates through time, with a consequent lowering of the C:N ratio.

Key words   *Aerobic, anaerobic aquatic plants, decomposition, dissolved organic matter, lakes, limnology, marshes, Michigan, oxygen, sediments, temperature.*

INTRODUCTION

Production of biomass by the aquatic macrophytes that grow in freshwater marshes and the littoral zones often constitutes at least half of the total organic C inputs to lake systems. Very little of this photosynthetically synthesized C remains in reduced form over geological time in the permanent sediments of the marsh or lake; a majority continues to be important in the metabolism of the aquatic system. Much of the organic C of the plants is respired and released as extracellular dissolved organic matter. Utilization of the particulate tissue directly in consumption by animals is quantitatively insignificant in most systems as compared to degradation by microbes. A large majority of the particulate and dissolved organic C of the net production of the system is processed during detrital decomposition by the microflora (Wetzel, 1975).

The process of decomposition has received much attention in terrestrial conditions, particularly in relation to agriculturally important species. The processes and environmental factors influencing decomposition of organic materials in streams has been investigated recently in some detail. Very few systematic inquiries, however, have been made on the influence of environmental conditions or the effects of morphology and phylogenetic relationships of aquatic angiosperms in regard to decomposition in freshwater marshes or lacustrine littoral areas (see review in Wetzel, 1975). In the analyses discussed here, a systematic appraisal of temperature and O_2 conditions was made on decomposition of freshwater angiosperms exhibiting a variety of growth forms and phylogenetic ancestry to provide an informational base upon which more detailed mechanistic studies could be made. The results presented here are summary in nature; more detailed results and discussion will be found elsewhere (Godshalk, 1977).

METHODS

Controlled experiments were conducted to evaluate the rates of decomposition of littoral macrophytes and the ultimate fate of these plants and their dissolved and particulate organic fractions in the lake ecosystem. Senescent plants of 5 freshwater species of aquatic angiosperms were collected at the end of the growing season as the plants began to senesce, washed free of sediments, and lyophilized. Replicated samples of dried plant material were put in 1-litre flasks of synthetic lake water (Wetzel Medium 5), inoculated with natural sediment (from the littoral zone of Lawrence Lake, Michigan) containing no coarse (>1 mm) organic matter, and incubated up to 180 days under anaerobic and aerated conditions at 10°C and 25°C. The various conditions of O_2 concentration were selected to simulate 3 situations under which naturally occurring decomposition proceeds: (1) completely aerobic conditions, as in wave-swept, turbulent littoral regions; (2) completely anaerobic conditions, as in many quiescent littoral regions where plant material compacts at the sediment–water interface under continuously strongly reducing conditions; and (3) aerobic to anaerobic conditions (i.e., synthetic lake water was initially oxidizing but allowed to become anoxic as a result of decomposition).

Samples were taken after 2, 4, 10, 24, 50, 90, and 180 days of decomposition and the contents of the flasks filtered (0.22 μm Millipore® GS membrane filters). Dissolved organic compounds (<0.2 μm) were fractionated by Amicon® membrane ultrafiltration into several molecular weight categories (<30,000 M; <10,000 M; <1,000 M; and <500 M). Each fraction was examined for total dissolved organic C, ultraviolet absorbance, and fluorescence (Wetzel and Otsuki, 1974); unfractionated filtrate was analyzed for dissolved C and N content, and by infrared spectrophotometry. The particulate material recovered was lyophilized, weighed to determine organic weight loss during decomposition, and analyzed for content of ash, total nonstructural carbohydrates (Smith, 1969), hemicellulose, cellulose, lignin (Goering and Van Soest, 1970), and particulate C and N. In addition, microbial activity of identical fresh detritus at each time interval was assessed by determining dehydrogenase activity (Zimmerman, 1975) and ATP content (Suberkropp and Klug, 1976).

In conjunction with the laboratory experiments we conducted a series of experiments involving the decomposition of the same plant species contained in mesh bags placed in the littoral zones of 2 lakes. These experiments were designed to provide data on decomposition in the natural environment under the conditions that were simulated in the laboratory experiments. Replicate litterbags, sacrificed periodically during the course of incubation, were placed in the moderately developed littoral zone (2 m) of Lawrence Lake (a hard water mesotrophic lake) during spring and summer seasons. Additional sets of litterbags were incubated during fall and winter at the same location, the pelagial zone (7 m) in Lawrence Lake, and in the littoral zone (1.5 m) of Wintergreen Lake (a hypereutrophic hard water lake with a well-developed littoral zone). Assay procedures were identical to those used in the laboratory experiments with the exception that no data could be obtained concerning DOM associated with the decomposing macrophyte material.

RESULTS AND DISCUSSION

Laboratory Experiments

Consistent with the knowledge of basic principles of decomposition in general, temperature would likely affect primarily the rates of decay whereas

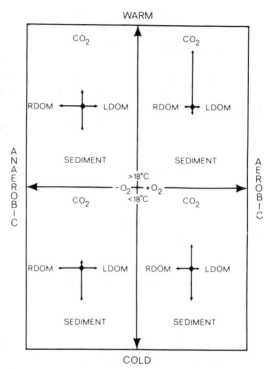

Fig. 1. Relative importance, with respect to rates and total accumulation, of the 4 possible fates of macrophytic tissue observed during laboratory decomposition under various conditions of temperature and O_2 concentration. LDOM = labile dissolved organic matter, RDOM = resistant dissolved organic matter. (From Godshalk and Wetzel, 1977.)

O_2 would influence the efficiency and completeness of decomposition. The relationships between decomposition of aquatic angiosperms and temperature and O_2 concentrations were not found to be that direct but to interact significantly.

The response of decomposing macrophytes to temperature and O_2 concentration at various levels is diagrammed in Fig. 1. Because anaerobic conditions (E_h <100 mV) were established rapidly in the aerobic-to-anaerobic cultures for all species (i.e., ≤10 days at 10°C, ≤2 days at 25°C), the results of these experiments are not differentiated in this report from those of the strict anaerobic experiments. The interaction of temperature and O_2 appears to be as important as either parameter by itself, causing a gradient of decompositional rates from slow in cold anaerobic conditions, to fast in warm aerobic situations. Most notable with respect to environmental control of decomposition are the apparent dependence on temperature of the process of converting particulate organic matter to dissolved organic matter and the constraints put on the conversion of dissolved organic matter to CO_2 by

Table 1. Parameters Describing Equations of Best Fit to Data for Weight Loss during Decomposition of 5 Macrophyte Species under 4 Laboratory Conditions[a]

Species	Anaerobic (25°C)		Aerobic (25°C)	
	a	b	a	b
Nuphar variegatum	0.1114	0.0692	0.1439	0.0930
Myriophyllum heterophyllum	0.0831	0.0625	0.0619	0.0335
Najas flexilis	0.0626	0.0487	0.0444	0.0244
Scirpus subterminalis	0.0167	0.0145	0.0280	0.0182
Scirpus acutus	0.0138	0.0248	0.0170	0.0103
	Anaerobic, 10°C		Aerobic, 10°C	
Nuphar variegatum	0.0436	0.0350	0.0465	0.0451
Myriophyllum heterophyllum	0.0284	0.0328	0.0264	0.0417
Najas flexilis	0.0101	0.0080	0.0148	0.0168
Scirpus subterminalis	0.0003	−0.0193	0.0057	0.0040
Scirpus acutus	0.0791	0.5080	0.0104	0.0219

[a] Equations are exponentials of the form $dW/dt = -kW$, where the decay coefficient, k, decreases exponentially with time ($k = ae^{-bt}$). W = percent of initial ash-free dry weight remaining, t = days of decomposition, a = initial rate of weight loss which declines exponentially at a rate of b.

the presence or absence of O_2. The seasonal and morphometric occurrence of the various conditions of temperature and O_2 in a temperate lake, and the ecological ramifications of environmental constraints on decomposition are presented elsewhere (Godshalk and Wetzel, 1977).

It was further hypothesized that the simple rates of decay of plant material, i.e., the loss in weight over time, would vary in any one environmental condition according to the structure of the plants involved. Emergent plants requiring more structural tissue (hemicellulose, cellulose, lignin) because of their erect growth form and lack of water support for most of the upper portions of the plant, would be more resistant to decomposition and therefore persist in the system for relatively long periods of time. By this same argument, submersed plants, because of their lack of rigid supportive tissue, would contain less of the resistant structural tissue and would decompose fairly rapidly. These general relationships between growth form and structural tissue content are supported by the survey of 21 species of aquatic macrophytes by Polisini and Boyd (1972) who found 53.4 ± 4.2, 66.6 ± 3.1, and 33.7 ± 2.6 g non-cell wall material ±2 SE/100 g plant for submersed, floating-leaved, and emergent macrophytes, respectively.

In our experiments, least-squares regression analysis of organic weight loss through time was used to determine the equation of best fit to the data. These equations for all species decomposing under the various conditions tested are presented in Table 1. It will be noted that under all conditions the floating-leaved water lily *Nuphar variegatum* Engelm. (the plant with the

Table 2.   Content (as Percent of Total Ash-Free Dry Weight) of Fiber Components of Senescent Tissue of 5 Macrophyte Species before Decomposition

Species	Hemicellulose	Cellulose	Lignin	Total
Nuphar variegatum	12.33	17.58	4.79	34.70
Myriophyllum heterophyllum	10.31	17.34	4.69	32.34
Najas flexilis	19.99	31.55	8.44	59.98
Scirpus subterminalis	34.55	26.43	2.59	63.57
Scirpus acutus	32.27	33.67	3.76	69.70

least amount of structural tissue) lost weight faster than all other species. The next most rapid to be decomposed were the submersed species *Najas flexilis* (Willd.) Rostk. & Schmidt, and *Myriophyllum heterophyllum* Michx., followed by *Scirpus subterminalis* Torr. which lost weight at a rate similar to that of the emergent *Scirpus acutus* Muhl. It is important to note the significance of phylogenetic ancestry here. *Scirpus subterminalis* grows vegetatively totally submersed through its entire life cycle and in Lawrence Lake, Michigan, constitutes >50% of the total annual primary production of the lake (Rich, et al., 1971). This locally important plant does not, however, decompose as would be predicted from its growth habit; *S. subterminalis* is the only submersed species in this bulrush genus, but it retains the structural and decompositional characteristics of the genus.

Under anaerobic conditions and 10°C, the proportions of hemicellulose, cellulose, and lignin in the recovered plant material of all species, except *N. flexilis*, remained approximately constant; in *Najas* both hemicellulose and cellulose declined while lignin increased proportionally. Under aerobic conditions, the percentage of lignin increased in all species while hemicellulose and cellulose either remained constant or decreased slightly. At 25°C, changes in proportions of fiber constituents become greater. During anaerobic decomposition, the proportion of lignin increased in all other species but was constant in *S. acutus*, that of cellulose decreased among all species except *Najas* where it increased, and that of hemicellulose increased in *Myriophyllum* and *Nuphar* but declined in *S. subterminalis* and *N. flexilis*. The greatest increases in the proportion of lignin in the recovered plant tissue of all species occurred during aerobic decomposition at 25°C. Hemicellulose content remained approximately constant while cellulose content declined in all species. The rate of lignin ''accumulation'' was most rapid in *Myriophyllum*, indicating that the other components (hemicellulose, cellulose) were being decomposed more rapidly in this species.

The structural characteristics of the plant which control its decomposition are not necessarily the proportions of hemicellulose, cellulose, and lignin in each species but rather the total content of these structural materials. The initial composition of the fiber fractions of each species used in our experiments is shown in Table 2. It can be seen that the most rapidly decomposing plants, *Myriophyllum heterophyllum* and *Nuphar variegatum*, have the lowest total concentrations of fiber. *Najas*, which decomposed about as fast as these other 2 species, had a very high fiber content but this is probably the result of a high density in the plant material of nearly ripe seeds whose protective coats are resistant to breakdown during fiber analysis procedures and would cause an overestimate of the cellulose fraction. The 2 species of bulrush were the slowest decomposers and had the highest total fiber content of all plants studied. As seen in Table 2, no single fiber constituent seems to be correlated with potential decomposition rate. The constant or only slightly shifting proportions of all structural components throughout the experimental period during even precipitous declines in organic weight demonstrate that macrophyte tissue decayed uniformly with respect to its various individual components. For example, substantial amounts of all fiber fractions were present in *Najas* and *Myriophyllum* samples after 90 days of decomposition, aerobically at 25°C, and yet after another 90 days not enough tissue could be recovered to allow fiber assays to be performed.

Total nonstructural carbohydrate (TNC), present only in low concentrations even initially in these senescent plants, was lost rapidly from particulate material, as would be predicted. The TNC values remained approximately constant in all species during anaerobic decomposition at 10°C, and TNC concentrations decreased in all species at 25°C, the decrease in *Najas*, *Myriophyllum*, and *Nuphar* being faster than those in *S. subterminalis* and *S. acutus*. Under aerobic conditions, loss of initial content of TNC was much more rapid than under anaerobic conditions; TNC levels fell to ≈ 0 in all species and remained constant for the duration of sampling (180 days) at 10°C. At the higher temperatures, decomposition of TNC of all species, except *Najas*, showed rapid initial decline in the first 10 to 25 days, no measurable change until after 50 to 90 days, and then a slight increase during the last 90 days of decomposition. *Najas flexilis* initially exhibited a high content of TNC, probably resulting from many nearly mature seeds in the axils of leaf material, which declined steadily throughout the decomposition period. These results imply that losses of simple carbohydrates from decomposing plant matter may not be simply from leaching. However, such results must be interpreted with caution because of probable interferences of humic compounds present during the enzyme extraction of the TNC.

Concentrations of dissolved organic carbon (DOC) exhibited marked contrasts in rates of change through time among molecular weight fractions and under differing conditions of temperature and O_2. Concentrations of DOC $> 30,000\ M$ and $> 10,000\ M$ were consistently 2–3× those of $M < 1,000$. Decomposition of DOC in all fractions was much more rapid under aerobic conditions than anoxic and much more complete at 25°C than at 10°C (an example is given in Fig. 2). Higher molecular weight compounds of the DOC were degraded less, proportionately, than low molecular weight fractions at the colder temperatures. Under anaerobic conditions the levels of high

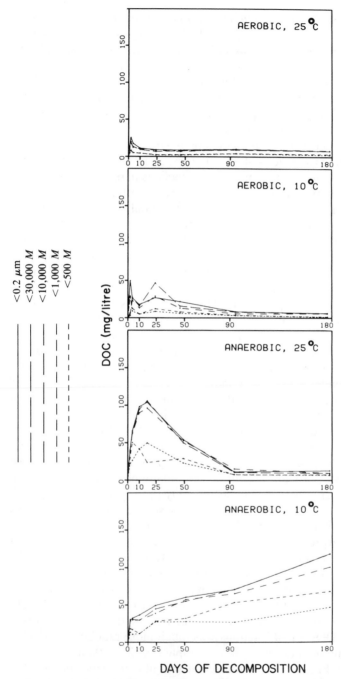

Fig. 2. Concentration (mg/litre) of dissolved organic carbon in various molecular weight fractions of media during laboratory decomposition of *Myriophyllum heterophyllum* under 4 conditions. (From Godshalk and Wetzel, 1977.)

molecular weight DOC were much higher and were degraded very much more slowly than under aerobic conditions at 25°C. At 10°C and when anoxic, all molecular weight fractions increased progressively throughout the half-year of experimentation. Analyses of ultraviolet light absorption and fluorescence indicated a progressive accumulation of yellow organic acids of polyphenolic type under these latter conditions (Wetzel and Otsuki, 1974).

Microbial Metabolism

Measurement of ATP content provides an approximate estimate of living biomass because ATP is not stored appreciably by organisms and degrades extremely rapidly upon death of an organism. The ATP content of microorganisms associated with decomposing macrophytic tissue of *Najas, Nuphar*, and *Myriophyllum* decaying anaerobically at 10°C reached maximum values after ≈25 days and then slowly declined during the rest of the experimental period. The ATP content of both species of *Scirpus* remained very low throughout the entire 180-day incubation period. The ATP levels were somewhat greater at 25°C than at 10°C among all species.

The ATP content was higher during aerobic decomposition than during anaerobic at both temperatures. Content associated with *Najas, Nuphar* and *Myriophyllum* also reached maximum levels early in the incubation period and then declined, whereas ATP associated with both species of *Scirpus* increased continually through time. Highest content of ATP among communities associated with all macrophytic species was found during aerobic decomposition at 25°C. Content of ATP associated with *Najas, Myriophyllum*, and *S. subterminalis* reached maximum values within ≈10 days, *S. acutus* after 50 days, and *Nuphar*, although highly erratic, attained maximum values after 180 days. Only under the latter conditions (aerobic, 25°C) were the differences in ATP content associated with the various species appreciable. The maximum values for microflora associated with decomposing *Najas* and *Myriophyllum* were ≈4× those of *S. subterminalis* and *Nuphar*, but averaged over the entire incubation period, the overall ATP contents of the 4 species were probably not different. The emergent *S. acutus*, on the other hand, had a maximum ATP content about 5× that of *S. subterminalis* and *Nuphar* and at least 2× the content for nearly all of the decomposition period, thus implying a significantly greater microbial biomass associated with this species.

Observed concentrations of ATP were difficult to interpret because of the subtlety of the differences among the conditions and the large numbers of samples precluded replication. Further complications arose from chemical interferences by humic compounds and $CaCO_3$, both of which were abundant in these samples (Cunningham and Wetzel, 1977). We believe that the values found are useful within this study but should not be used in comparisons of ATP values obtained by other investigators because of the disparities caused by organic interferences, variations in sample material, and differ-

Table 3. Initial Nitrogen and Carbon Content (as Percent of Total Dry Weight) and the
C:N Ratio of Senescent Macrophytes

Species	N	C	C:N
Nuphar variegatum	2.4	39.3	16.4
Myriophyllum heterophyllum	2.0	24.7	12.3
Najas flexilis	1.8	31.2	17.3
Scirpus subterminalis	1.2	30.4	25.3
Scirpus acutus	1.5	43.6	29.1

ences in assay efficiencies (G. L. Godshalk and R. G. Wetzel, *personal observation*).

The other method of assaying microbial activity directly during our investigations involved the reduction of a tetrazolium salt by the microbial electron transport system (ETS), specifically a measure of dehydrogenase activity. Anaerobic decomposition at 10°C caused moderate and about equal increases in dehydrogenase activity in all species during the entire experimental period, except in *S. acutus* which peaked after only 50 days then declined. This increase through time was not seen during cold aerobic decomposition as ETS values for *S. subterminalis*, *Najas*, and *Myriophyllum* consistently remained low. However, values for *Nuphar* and *S. acutus* approached the highest values of all species studied under all conditions by day 50, and then declined to values about the same as for the other species by the end of the experiments. At the higher temperature, anaerobic decomposition produced very high ETS values, especially in *Nuphar*, but exhibited no discernible trends over time. Aerobically, at 25°C all species had ETS values approximately equal to those obtained during 10°C aerobic decomposition, and all species shared the trend of higher initial values in the first 25 days preceding a gradual but steady decline through the remainder of the sampling period.

In all species, and in particular for parallel experiments on a marine species decomposing in synthetic sea water, all anaerobic conditions produced ETS values substantially higher than those of aerobic conditions, thus implying a chemical reduction of the tetrazolium salt that was not dependent on enzymatic activity (cf. Schindler et al., 1976). Further, data for ATP and ETS did not correlate, and values of each through time varied independently.

Analysis of the C and N content of decaying macrophyte tissue and the calculated C:N ratio may prove to be a useful parameter to describe microbial colonization. During anaerobic decomposition at 10°C, particulate N remained constant over the entire 180 days of sampling in both *Scirpus* species, the slowly decomposing plants. The more rapidly decomposing plants, *Najas, Myriophyllum* and *Nuphar*, showed an increase in N, which was especially rapid in *Nuphar*, and a resultant drop in the C:N ratio. Aerobically, N content increased continuously in *S. subterminalis* and *Nuphar*; a similar increase occurred in *Najas* and *Myriophyllum*, but only to day 50 and then a decline occurred. The N content of *S. acutus* remained about constant. The C:N ratios for all species under these anaerobic conditions gradually and continually decreased.

At 25°C under anaerobic conditions, N content increased initially in all species then declined gradually for the remainder of the experiments in *Najas, Myriophyllum*, and *Nuphar*. These changes were in contrast to a rapid initial increase and decrease during the first 10 days in *S. subterminalis* and 25 days in *S. acutus*, followed by continual gradual increase in N. Consequently, C:N remained constant for 180 days in *Myriophyllum* and *Nuphar*, declined gradually in *S. subterminalis* and *Najas*, and increased for the first 50 days then declined in *S. acutus*. All species showed an increase in N content during aerobic decay during the first 4 to 10 days and then a decline, except in *S. acutus* which increased through the entire decomposition period. As a result C:N of *S. acutus* slowly decreased for the entire 180 days. In all other species C:N declined during the first 4 to 10 days and then remained constant. Increases in the N content of decomposing macrophyte tissue have previously been observed (e.g., de la Cruz and Gabriel, 1974; Mason and Bryant, 1975). This increase typically has been attributed to accumulation of microbial protein.

Rates of weight loss of the plant tissue were related to the initial N content of the senescent plant material (Table 3). The faster decomposing plants had greater initial N concentrations and low C:N ratios while the slower decomposing plants had low initial N and high C:N ratios.

In situ Studies

Overall, the rates of decay observed in the lake samples did not differ very much qualitatively from those observed under laboratory experiments. In situ decomposition during the spring—summer experiments was somewhat faster than 25°C aerobic decomposition in the laboratory, but the fall— winter series of samples decayed at rates slightly less than those seen in the controlled experiments conducted at 10°C. In all cases, the various species decomposed at the same rates with respect to each other as in the laboratory experiments, i.e., *Nuphar* > *Myriophyllum* > *Najas* > *S. subterminalis* > *S. acutus*. No discernible differences were found between decomposition occurring at the ecologically different incubation sites.

CONCLUSIONS

Based on the data obtained in these experiments, the needs for future work on the decomposition of macrophytes are clear to us. Attention must be paid to the variety of plant species colonizing the habitat of interest because different species do decompose at different rates. This differential decomposition, taken along with the production of each species, affects the total annual metabolism of C in the system and even within zones of ecosystems. The nutritive value of detrital vegetation to consumers is influenced by the qualitative species-specific characteristics of the decaying macrophytes. In addition, there will surely be long-term effects on sedimentation and system geomorphology as a result of accumulation and deposition of undecomposed material.

The single most important fate of reduced C in aquatic ecosystems (viz, the ultimate conversion to CO_2 in the detrital food chain) has received attention far short of its ecological significance. Further progress depends on the perfection and application of methods to assay microbial products and to study processes which are associated in marshes and littoral waters with interfering complex organic compounds and inorganic salts. Only with careful attempts to answer questions of ecological importance concerning microbial metabolism of reduced C compounds of macrophytic origin will insight into the metabolism of aquatic ecosystems be achieved.

ACKNOWLEDGMENTS

The technical assistance of J. S. Sonnad, S. Morrison, and A. J. Johnson is greatly appreciated. These investigations were supported, in part, by subventions from the National Science Foundation (GB-40311, BMS-75-20322, and OEC 74-24356-A01) and the U.S. Energy Research and Development Administration (EY-76-S-02-1599, COO-1599-118). This is Contribution No. 318, W. K. Kellogg Biological Station of Michigan State University.

LITERATURE CITED

Cunningham, H. W., and Wetzel, R. G. (1977). Fulvic acid interferences on ATP determinations in sediments. *Limnol. Oceanogr. (In press)*.

de la Cruz, A. A., and Gabriel, B. C. (1974). Caloric, elemental, and nutritive changes in decomposing *Juncus roemerianus* leaves. *Ecology* 55, 882–886.

Godshalk, G. L. (1977). "The Decomposition of Aquatic Plants in Lakes." Ph.D. Thesis, Mich. St. Univ. 309 pp.

Godshalk, G. L., and Wetzel, R. G. (1977). Decomposition of macrophytes and the metabolism of organic matter in sediments. *In* "Interactions between Sediments and Fresh Water" (H. Golterman, ed.), pp. 258–264. Dr. W. Junk. N. V. Publishers, The Hague.

Goering, H. K., and Van Soest, P. J. (1970). "Forage Fiber Analysis. Apparatus, Reagents, Procedures, and Some Applications." U.S. Dept. Agric., Agric. Res. Serv., Agric. Handbook No. 379. 20 pp.

Mason, C. F., and Bryant, R. J. (1975). Production, nutrient content and decomposition of *Phragmites communis* Trin. and *Typha angustifolia* L. *J. Ecol.* 63, 71–95.

Polisini, J. M., and Boyd, C. E. (1972). Relationships between cell-wall fractions, nitrogen, and standing crop in aquatic macrophytes. *Ecology* 53, 484–488.

Rich, P. H., Wetzel, R. G., and Van Thuy, N. (1971). Distribution, production, and role of aquatic macrophytes in a southern Michigan marl lake. *Freshwater Biol.* 1, 3–21.

Schindler, J. E., Williams, D. J., and Zimmerman, A. P. (1976). Investigation of extracellular electron transport by humic acids. *In* "Environmental Biogeochemistry" (J. O. Nriagu, ed.), vol. 1, pp. 109–115. Ann Arbor Science Publishers Inc., Ann Arbor, Michigan.

Smith, D. (1969). "Removing and Analyzing Total Nonstructural Carbohydrates from Plant Tissue." Univ. Wisconsin, Madison, Coll. Agric. Life Sci., Res. Rep. 41. 11 pp.

Suberkropp, K., and Klug, M. J. (1976). Fungi and bacteria associated with leaves during processing in a woodland stream. *Ecology* **57**, 707–719.

Wetzel, R. G. (1975). "Limnology." W. B. Saunders Company, Philadelphia, Pennsylvania.

Wetzel, R. G., and Otsuki, A. (1974). Allochthonous organic carbon of a marl lake. *Arch. Hydrobiol.* **73**, 31–56.

Zimmerman, A. P. (1975). Electron transport analysis as an indicator of biological oxidations in freshwater sediments. *Verh. Internat. Verein. Limnol.* **19**, 1518–1523.

DECOMPOSITION PROCESSES: SUMMARY AND RECOMMENDATIONS

John L. Gallagher[1]

Marine Institute
The University of Georgia
Sapelo Island, Georgia 31327

In the preceding chapters one has seen that our knowledge of decomposition in freshwater wetlands is far from complete especially our knowledge of belowground materials. Most studies in freshwater wetlands have also included numerous investigations of weight loss and composition changes of dead material confined in litterbags. Some bags have been hung from poles to simulate decay of the attached dead plant community, others placed on the ground to monitor decomposition in the litter layer and in at least one instance buried to resemble the situation in the peat layer. Although much criticism has been directed at the use of litterbags, they do provide a reproducible approximation of decomposition rates which integrate the effects of daily or weekly environmental variation. The curves consistently have a negative exponential shape although values on the axes have a wide range. Curves showing the decomposition rates range from 100% in weeks to <50% after several years. Those plotted by Chamie and Richardson (1978) for northern peatlands indicated ≈40% of the material disappeared the 1st year with only slight losses occurring in the subsequent 2-year period. On the other end of the spectrum are the rates occurring in freshwater tidal marshes. Simpson et al. (1978) indicated 60 to 80% of the plant material disappeared in 90 days. Odum and Heywood (1978) collected similar data on 3 freshwater species in which all of the material disappeared in <150 to 300 days. The differences in rates are due to both environmental and species-specific characteristics. As would be expected, those materials exposed to

---

[1] Present address: U.S. Environmental Protection Agency, Corvallis Environmental Research Laboratory, 200 S.W. 35th Street, Corvallis, Oregon 97330.

warm, moist aerobic environments decay more rapidly than those subjected to conditions which are cold, anaerobic or dry. I have measured O_2 consumption and CO_2 evolution rates of the community associated with dead freshwater marsh plants and found Q_{10} values between 2.0–3.0. In another study we soaked the dead plants in water and then followed respiratory rates as the plants dried (J. L. Gallagher and P. Hawkins, *personal observations*) and found a linear relationship with the highest rates associated with the highest moisture contents. These results help explain why Odum and Heywood (1978) found the most rapid loss from litterbags placed in tidal marsh areas which were regularly flooded. In addition to differences which occur due to environment, much of the observed variation in decomposition rates can be attributed to species-specific differences.

Woody plants decompose more slowly than herbaceous and the more succulent herbaceous species faster than the more fibrous. Godshalk and Wetzel (1978) found decay rates were correlated with the total fiber content rather than with a particular fiber component. Gallagher and Pfeiffer (1977) observed that the respiratory rates of the microbial community associated with the attached dead plants are highest in species having large surface:volume ratios. These large surfaces provide more sites for initial microbial colonization, and facilitate exchange between the tissue and the environment (Seliskar et al., 1977). The differences in decomposition rates of the leaves, small stems and large stems of bog birch observed by Chamie and Richardson (1978) were due no doubt to differences in surface phenomena as well as chemical composition differences. Although the initial geometry of the plant material influences the decomposition rate, fragmentation of whole plant parts greatly increases surface area. In most cases reducing particle size increased the decomposition rates measured by Davis and van der Valk (1978).

Some of the weight losses noted with litterbags are due to losses of mineral rather than C compounds. Both Chamie and Richardson (1978) developed budgets for losses or gains of many mineral elements during decomposition in northern peatlands and prairie glacial marshes, respectively. Only N data were available for the freshwater tidal system (Odum and Heywood, 1978). Losses result from leaching or mineral release while gains may come from sorptive processes, deposition of minerals in fine fissures or assimilation by attached microbes. Mineral dynamics during decomposition is an important process but the major portion of that information is discussed in Part III of this volume.

All of the above data focus on decomposition of aerial plant parts. Chamie and Richardson (1978) described the fate of aerial parts which become buried, but there are virtually no data on decomposition rates for the root systems of freshwater wetland plants. It could be presumed that rates generally would be relatively low because physical conditions are often not those considered ideal for decay in view of the fact that peat often accumulates in

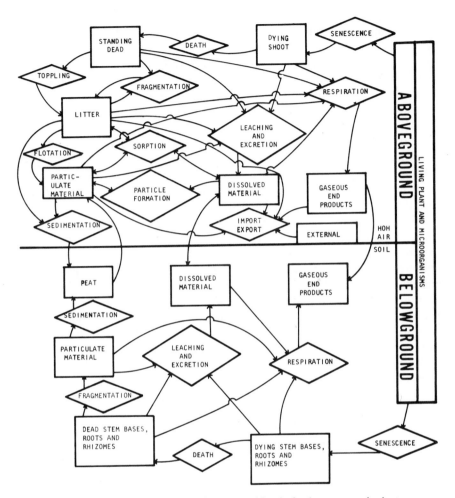

Fig. 1.   Conceptual model of decomposition in freshwater marsh plants.

these soils. Chamie and Richardson (1978) did now however find that buried
aerial parts consistently decayed at a slower rate than similar tissue exposed
at the surface. Probably there is a wide range of root decay rates depending
on the nature of the soil–root system, the coarseness of the roots and the
chemical composition of the particular system under study. Root–rhizome
systems that are woody and coarse in texture would be expected to decay at
a slower rate than more succulent fine-textured systems. In freshwater
marshes and other systems the C:N ratio of the decaying material has been
reported to be important in regulatory decomposition rates. Although soil
aeration and nutrient characteristics are also probably important in determining the decay rates of underground material they are largely uninvesti-
gated in freshwater wetlands.

It is apparent that much more descriptive work must be done before we have a clear picture of what happens in the decomposition aspects of freshwater marsh dynamics. There is, however, enough knowledge available from the work done by the authors of the various chapters on decomposition when coupled with information transferable from saltwater marsh systems to construct a conceptual model of decomposition in freshwater marshes. Such a model is shown in Fig. 1.

The rectangular boxes represent pools of materials and the diamonds, processes acting on these pools. The living aboveground portion of a plant becomes a dying plant through the senescence process. Losses from the dying plant occur due to respiration, as well as leaching and secretion. At these initial decomposition stages respiration is the sum of macrophyte and invading microbial processes. Factors regulating senescence of aerial and underground plant parts, as well as factors determining the relative importance of leaching *versus* respiration, are not well understood and are significant questions for future research. As membranes lose their integrity during senescence, it is suspected the freshwater marsh plants would release large quantities of dissolved material at that time. Scanning electron photomicrographs of living, senescing and dead marsh plants indicate microbial populations are much higher on the dead plants than on the senescing tissue. Populations are similarly higher on the senescing than on the living leaves (D. M. Seliskar, *personal communication*). Because microbial populations are relatively sparse on the senescing leaf surface much of the dissolved material should escape their filtering actions and reach the aquatic phase of the system.

Following death it would be expected that respiration by microbes increases as the microbial colonization of the tissue progressed. Total respiration is probably lower initially as it is in salt marsh plants (Gallagher and Seliskar, 1976) because macrophyte metabolism has stopped. At this point excretion by the microbes becomes a significant portion of the dissolved organic matter (DOM) loss. Godshalk and Wetzel (1978) demonstrated that the proportion of respired material going into gaseous end products or to dissolved material depended on the O_2 level in the incubation water. High levels of O_2 favored the more complete decomposition process.

The standing dead plants become part of the litter either by rotting at the base and toppling (Davis and van der Valk, 1978) or by fragmentation of leaf and stem tissue due to biological or physical action. The litter pool can undergo the same respiration and leaching–excretion losses. Probably more effort has gone into studying this compartment than any other. In view of the connectivity of the litter compartment with others and the number of processes acting on that pool, it should be a focus of attention. The conceptual model shows 8 processes are associated with the litter compartment. The major input is from the standing dead, but dissolved material may be sorbed from the water. As fragmentation within the pool reduces particle size thus increasing litter surface area, sorptive processes are enhanced. New parti-cles may also form by condensation of dissolved material in the water.

Figure 1 shows that litter may be exported directly from the marsh surface either (1) in rafts as is shown by a long direct line or (2) indirectly through flotation of small particles as is shown through an indirect pathway. In systems where decomposition rates are slow or where litter removal by flushing tides (Odum and Heywood, 1978) or wind (Davis and van der Valk, 1978) is minimal, substantial quantities of litter may be incorporated into the sediment (Chamie and Richardson, 1978).

Sedimentation and resuspension of material from the soil is but one of several exchange processes which can occur at either the soil—water or soil—air interface. The exchange of gaseous end products of metabolism or dissolved material obviously takes place but little is known about the na-ture of the products, the quantities exchanged or the factor regulating either the types of products or the fluxes.

The decomposition of the underground material probably follows path-ways similar to those identified for the aerial portions although few of these have been documented for freshwater wetland systems. The model for un-derground decomposition appears less complex than that for the aerial de-composition simply because we know less about it. As our knowledge about this part of the system increases, new pathways will probably need to be added.

Most of the major differences between the aerial and underground systems can be traced directly or indirectly to the differences between the milieu (fluid or gas *versus* solid) of the plant material. In the aboveground portion of the system the milieu consists of a gas and (or) water phase where exchange of material between the dead plants and their environment is by convection as well as diffusion. End products which could inhibit decomposition are readily carried away. Whole plants, fragments or particles are moved with relative ease to environments where conditions may favor their decomposi-tion.

Conversely, conditions belowground make diffusion processes relatively more important. Movement of dead underground material from one place to another is very limited. The major environmental changes to which the material is exposed result from the evolution of the microenvironment brought about by the decomposition of the plant material itself. One such change probably encountered is a change in the redox potential. An oxidized zone is often maintained around the living roots. As the roots or shoots die this zone may be expected to become smaller and less intense before disap-pearing when the aerenchymous tissue in the stem bases, rhizomes and roots collapses. The rhizosphere which is normally the microenvironment of high-est microbial density in the soil, would change from community aerobic to anaerobic metabolism and produce a number of new products.

In summary, our knowledge of the decomposition of underground material as outlined in Fig. 1 is primitive in all of the systems reported in this volume. In contrast, much progress is being made with the aerial research. In some systems, such as, northern peatlands (Chamie and Richardson, 1978), prairie glacial marshes (Davis and van der Valk, 1978) freshwater tidal marshes (Odum and Heywood, 1978) descriptive work is progressing rapidly and a few studies with regulatory factors have been undertaken. The study of decomposition in the littoral zone of lakes is farther along (Godshalk and Wetzel, 1978) and efforts are now focusing on examining particular pathways and the factors which regulate them. As more of the descriptive work becomes complete in the other wetland systems, research in those systems will also naturally move in that direction.

In view of the past rapid rate of the commercial exploitation and destruction of freshwater wetlands, future research plans assume an added urgency because answers are needed for protection and management purposes. It seems that the most rapid progress in research in decomposition in these systems can be made by focusing on understanding the processes at work. This understanding will enable the scientist to give the resource manager reasonable answers to as yet unconceived questions about impacts on freshwater wetlands.

ACKNOWLEDGMENTS

The author is grateful to the authors of the chapters in this section who contributed much time and numerous ideas to this summary. The work was supported by Grant OCE 75-20842-A01 from the National Science Foundation and by a grant from the Sapelo Island Research Foundation; contribution No. 338 from the University of Georgia Marine Institute.

LITERATURE CITED

Chamie, J. P. M., and Richardson, C. J. (1978). Decomposition in Northern wetlands. In "Freshwater Wetlands: Ecological Processes and Management Potential" (R. E. Good, D. F. Whigham and R. L. Simpson, eds.) pp. 115–130. Academic Press, New York.

Davis, C. B., and van der Valk, A. G. (1978). Litter decomposition in prairie glacial marshes. In "Freshwater Wetlands: Ecological Processes and Management Potential" (R. E. Good, D. F. Whigham and R. L. Simpson, eds.) pp. 99–113. Academic Press, New York.

Gallagher, J. L., and Pfeiffer, W. J. (1977). Aquatic metabolism of the communities associated with attached dead shoots of marsh plants. Limnol. Oceanogr. 22, 562–565.

Gallagher, J. S., and Seliskar, D. M. (1976). The metabolism of senescing Spartina alterniflora leaves. Annual Meeting American Institute of Biological Sciences. Tulane Univ. New Orleans, USA.

Godshalk, G. L., and Wetzel, R. G. (1978). Decomposition in the littoral zone of lakes. In "Freshwater Wetlands: Ecological Processes and Management Potential" (R. E. Good, D. F. Whigham, and R. L. Simpson, eds.), pp. 131–143. Academic Press, New York.

Odum, W. E., and Heywood, M. A. (1978). Decomposition of intertidal freshwater marsh plants. In "Freshwater Wetlands: Ecological Processes and Management Potential" (R. E. Good, D. F. Whigham, and R. L. Simpson, eds.), pp. 89–97. Academic Press, New York.

Seliskar, D. M., Gallagher, J. L., and Pearson, T. C. (1977). Microbial colonization of leaves entering the detrital food webs in swamps and marshes. Annual Meeting American Society of Limnology and Oceanography, Mich. State Univ. East Lansing, USA.

Simpson, R. L., Whigham, D. F., and Walker, R. (1978). Seasonal patterns of nutrient movement in a freshwater tidal marsh. *In* "Freshwater Wetlands: Ecological Processes and Management Potential" (R. E. Good, D. F. Whigham and R. L. Simpson, eds.) pp. 243–257. Academic Press, New York.

PART III

Nutrient Dynamics

CHEMICAL COMPOSITION OF WETLAND PLANTS

Claude E. Boyd

Department of Fisheries and Allied Aquacultures
Auburn University Agricultural Experiment Station
Auburn, Alabama 36830

Abstract Examination of data on the chemical composition of wetland plants reveals that variation in composition increases in the following order: within-site intraspecific variation, between-site intraspecific variation, and interspecific variation. Furthermore, chemical composition changes as some species mature and variation in chemical composition occurs between different plant structures. No average composition for an individual species or ecological grouping of species based on data from the literature would be reliable. Therefore, actual measurements should be made in studies requiring data on chemical composition. Energy content values are more constant than chemical parameters.

Key words *Aquatic macrophytes, aquatic weeds, biogeochemical cycling, elemental analysis, marshes, phytochemistry, wetland plants, wetlands.*

INTRODUCTION

Although data on the chemical composition of wetland plants are of intrinsic interest, such information is also useful in studies of food webs, biogeochemical cycling, and physiological ecology. Furthermore, recent attempts to use aquatic plants for animal feeds (Boyd, 1968; Easley and Shirley, 1974; Linn et al., 1975a, 1975b) and for nutrient removal from polluted water (Rogers and Davis, 1972; Boyd, 1976) require data on chemical composition. The intention of this paper is to present a concise description of the ecologically pertinent chemical characteristics of wetland plants exclusive of phytoplankton. Generalizations are emphasized, and where possible, factors regulating chemical composition are discussed. I relied primarily on data from my own research and made little or no use of data from a number of important reports including Mayer and Gorham (1951), Gorham (1953), Riemer and Toth (1968), Stake (1967, 1968), Straškraba (1968), Cowgill (1973a, 1973b, 1974), Gaudet (1973), and Lawrence (1971). The excellent but lengthy review by Hutchinson (1975) may be consulted for detailed information of the chemical composition of aquatic plants.

ASH AND ELEMENTAL CONSTITUENTS[1]

Intraspecific Variation

Before considering the comparative elemental composition of wetland plants, some knowledge of the variation in composition of different samples of the same species is helpful. Both within-site variation (location, plant part, and maturity) and between-site variation are considered.

Plant Part The limited data available suggest that the elemental composition of different organs is markedly different. Riemer and Toth (1970) reported differences in concentrations of several elements in leaf blades and petioles of Nymphaeaceae. For example, leaf blades contained >2× the levels of N found in petioles but petioles had much higher concentrations of Na than leaf blades. Cowgill (1973b, 1974) also noted differences in elemental composition of leaf petiole, leaf, flower stalk, and flower of Nymphaeaceae. Denton (1966) demonstrated that leaves of *Jussiaea repens* and *Eichhornia crassipes* contained higher percentages of C and N than roots of these species. Roots of both species had higher levels of ash and Na than leaves. Roots of *Jussiaea* had greater concentrations of K than leaves while the opposite was true for *Eichhornia*. Leaves and roots of both species had similar percentages of P, Ca, and Mg. Musil and Breen (1977) also found significant differences in the concentrations of several elements between various parts of *Eichhornia crassipes*. For example, K concentrations in roots, petioles, and pseudolaminae were 1.42%, 5.07%, and 2.82%, respectively. Boyd (1969a) reported that foliage of *Justicia americana* contained much greater concentrations of several elements than did roots. Percentages of N and P in foliage were 2.9 and 4.1× greater, respectively, than those of roots. Marked differences in concentrations of different elements between culms, rhizomes, roots and umbels occur in *Cyperus papyrus* (Gaudet, 1975). For example, average N concentrations were: culm, 0.86%, rhizome, 2.76%, root, 1.73%, and umbel 2.62%.

Location within Stand Findings of several studies suggested that variation in elemental composition of samples of shoots or of whole emergent aquatic species from single stands was less than variation in samples of the same

---

[1] Unless otherwise indicated, all data are presented on a dry weight basis.

Table 1. Variation in Chemical Composition of 5 Replicate Samples of Shoots *Typha* and *Juncus* Collected from Single Stands of These 2 Plants

Sample	Percentage of dry weight				
	C	N	P	Ca	K
Typha latifolia					
1	44.6	1.03	0.18	0.95	1.84
2	46.1	0.96	0.17	0.85	2.00
3	45.2	1.03	0.17	1.02	2.00
4	44.5	0.90	0.16	0.97	2.28
5	44.8	1.13	0.18	0.86	2.12
$\bar{x}$	45.0	1.01	0.17	0.93	2.05
SE	0.3	0.04	0.004	0.03	0.07
C.V. (%)	1.45	8.55	4.92	7.86	7.97
Juncus effusus					
1	47.5	1.28	0.17	0.18	0.79
2	48.6	1.47	0.17	0.16	0.89
3	48.1	1.41	0.16	0.17	0.65
4	47.9	1.48	0.18	0.14	0.74
5	47.8	1.18	0.15	0.14	0.59
$\bar{x}$	48.0	1.36	0.17	0.16	0.73
SE	0.2	0.06	0.005	0.01	0.05
C.V. (%)	0.85	9.57	6.71	11.18	16.10

species collected from different sites (Boyd, 1969a, Boyd and Hess, 1970, Boyd and Walley, 1972a). Variation between samples taken from different locations within a site is illustrated in Table 1 for shoots from stands of *Typha latifolia* and *Juncus effusus*. The composition of the 5 samples of each species was similar and all coefficients of variation were <20% and most were <10%. These are lower coefficients of variation than are usually encountered for between-site variation (Table 2). Gaudet (1975) working with *Cyperus papyrus* also demonstrated that within-site variation in elemental composition was negligible when compared to between-site variation. Gaudet emphasized that the findings applied only to comparisons of specified plant structures of similar age. Conversely, Musil and Breen (1977) reported that the chemical composition of whole *Eichhornia crassipes* plants differed between plants collected from the margin and the center of the community. For example, plants from the margin contained 2.32% K and 0.39% P but plants from the center contained 3.88% K and 0.47% P. Further studies of within-site variation would be useful to workers deciding on the number of samples to take from a given plant stand to yield reliable estimates of elemental composition.

Maturity   Boyd (1969a, 1970a) demonstrated that concentrations of several elements declined drastically as *Typha latifolia* and *Justicia americana*

Table 2. Summary of Elemental and Ash Concentrations in 2 Species of Plants Collected from Different Sites[a]

Con-stit-uent	Pithophora kewensis			Typha latifolia		
	Range of values (all sites)	$\bar{x} \pm SE$	C.V. (%)	Range of values (all sites)	$\bar{x} \pm SE$	C.V. (%)
Ash	16.42–44.76	26.18 ± 1.00	23.2	3.96–10.25	6.75 ± 0.32	25.6
C	28.60–39.53	35.43 ± 0.65	8.0	43.33–47.16	45.91 ± 0.19	2.3
N	2.02–5.19	3.04 ± 0.19	26.9	0.86–2.12	1.37 ± 0.05	21.8
P	0.14–0.64	0.32 ± 0.03	44.2	0.08–0.41	0.21 ± 0.01	33.9
S	0.69–2.34	1.47 ± 0.10	29.7	0.05–0.53	0.13 ± 0.02	65.2
Ca	0.67–10.90	2.46 ± 0.98	55.3	0.35–1.62	0.89 ± 0.05	33.8
Mg	0.13–0.45	0.24 ± 0.02	39.2	0.06–0.31	0.16 ± 0.01	35.8
K	0.98–6.24	3.43 ± 0.35	43.9	0.91–4.39	2.38 ± 0.15	35.2
Na	0.04–0.32	0.08 ± 0.01	81.5	0.05–1.09	0.38 ± 0.05	83.5

[a] *Pithophora* was collected from 19 sites (Boyd, 1966) and *Typha* from 30 sites (Boyd and Hess, 1970). All data are percentage of dry weight basis.

aged. Nitrogen concentrations in *Justicia* decreased from 2.83% in May to 1.63% in August. In *Typha*, N levels dropped from 2.40% in April to 0.51% in July. Changes of large proportions were also measured for P, K, Mn, and Zn in *Justicia* and for ash, P, S, Ca, Mg, and K in *Typha*. Seasonal changes in nutrient content have also been demonstrated for *Phragmites communis* and *Typha angustifolia* (Mason and Bryant, 1975), *Carex lacustris* (Bernard and Solsky, 1977) and wetland vegetation of the arctic tundra (Chapin et al., 1975). Boyd and Scarsbrook (1975) failed to demonstrate changes in N and P concentrations as the floating plant, *Eichhornia crassipes*, matured. Nichols and Kenney (1976) demonstrated that old shoots of *Myriophyllum spicatum* absorbed less N than new shoots of this species. The N content of *Chara* sp. decreased as the growing season progressed (Boyd, 1966).

Between-Site Variation   Data presented in Table 2 illustrate the magnitude of variation in levels of certain elements and ash in samples of *Pithophora kewensis*, a large filamentous alga, and *Typha latifolia* (shoots). Samples were from 19 and 30 sites, respectively, which represented a wide range of environmental characteristics (Boyd, 1966; Boyd and Hess, 1970). Intraspecific comparisons represented plants at comparable stages of maturity. Except for C, percentages of all elements and ash covered a wide range with the highest value for a particular measurement often exceeding the lowest value by several times. Coefficients of variation for ash and elements other than C usually exceeded 25% and some were >50%. Similar series of data are available for *Cyperus papyrus* (Gaudet, 1975), *Phragmites communis* (Květ, 1973), *Eichhornia crassipes* (Boyd and Vickers, 1971), *Saururus cernuus* (Boyd and Walley, 1972a), and *Lemna* sp. (Culley and Epps, 1973). Adams et al. (1973) also analyzed a number of species, each from several sites. All findings confirm that the elemental composition of individual wetland species may vary greatly between sites.

Boyd and Hess (1970) demonstrated positive correlations between concentrations of P, Ca, Mg, K, and Na in waters of collecting sites and percentages of these elements in shoots of *Typha latifolia*. Magnesium and Na concentrations in mud were also positively correlated with concentrations in *Typha* shoots. None of the correlations accounted for >32% of the variability in tissue concentrations. Positive correlations were also reported for concentrations of P, Ca, Mg, K and Na in muds and concentrations of these elements in *Saururus cernuus* shoots (Boyd and Walley, 1972a). None of the correlations accounted for >36% of the variation in tissue concentrations of a given element. Steward and Ornes (1975) reported that accumulation of K by sawgrass (*Cladium jamaicense*) was positively correlated with reserve supplies of K in marsh soil. Gosset and Norris (1971) demonstrated a positive correlation between the N and P contents of the tissues of *Eichhornia crassipes* and those of the environment. However, Boyd and Vickers (1971) were unable to correlate elemental concentrations in the environment with the elemental content of whole plant samples of *Eichhornia*.

Cultures of aquatic plants absorb nutrients in relation to external concentration (Gerloff and Skoog, 1954, 1957; Gerloff and Krombholz, 1966). However, in nutrient uptake experiments the only variable is usually the concentration of a single element. Many interrelationships exist between concentrations and ratios of ions in the environment and the uptake of the ions by plants (Sutcliffe, 1962). Sites for the studies mentioned above represented a wide range of nutrients and nutrient ratios. Environmental factors other than inorganic nutrients (e.g., pH, light, temperature, water depth, substrate type) also varied greatly between sites and no doubt influenced nutrient uptake. Furthermore, genetic differences between plants on different sites may influence elemental composition.

Data used in preparing Table 2 were analyzed by regression analysis to determine if relationships existed between concentrations of different elements. In both species, carbon concentrations (Y) were negatively correlated $(P < .01)$ with ash values (X); $r = -.95$ and $r = -.83$ for *Pithophora* and *Typha*, respectively. These strong correlations were expected since ash represents the inorganic components and C the organic constituents of plant samples. Calcium concentrations increased with ash values in *Pithophora* $(r = .57; P < .05)$ suggesting that Ca was a major contributor to ash. Much of the Ca was likely encrusted on the external surfaces of *Pithophora* as marl rather than being a component of the cellular constituents. Potassium concentrations were negatively correlated with ash in *Pithophora* $(r = -.62; P < .01)$. For *Typha*, Ca concentrations did not increase with ash values, but there was a strong positive correlation between K concentration and ash $(r = .85; P < .01)$ suggesting that variations in ash values resulted from differences in K concentration. Neither Ca and Mg nor K

Table 3. Summary of Elemental and Ash Concentrations in Different Species of Wetland Plants[a]

Constituent	Species (n)	Range of values (all species)	$\bar{x} \pm SE$	C.V. (%)
Ash (%)	40	6.1–40.6	14.03 ± 1.21	53.4
C (%)	28	29.3–48.8	41.06 ± 0.71	9.2
N (%)	27	1.46–3.95	2.26 ± 0.14	32.7
P (%)	35	0.08–0.63	0.25 ± 0.02	58.0
S (%)	25	0.11–1.58	0.41 ± 0.08	98.3
Ca (%)	35	0.20–8.03	1.34 ± 0.24	105.5
Mg (%)	35	0.08–0.95	0.29 ± 0.03	143.6
K (%)	35	0.42–4.56	2.61 ± 0.22	51.0
Na (%)	35	0.07–1.52	0.51 ± 0.08	93.1
Fe (ppm)	33	133–3866	1420 ± 169	68.0
Mn (ppm)	33	120–5390	1143 ± 227	113.9
Zn (ppm)	33	20–267	80 ± 10	71.2
Cu (ppm)	33	1–190	40 ± 6	93.8
B (ppm)	31	1.2–112	14.3 ± 4.5	176.2

[a] Data from Denton (1966), Boyd (1966, 1968, 1970b), Boyd and Vickers (1971), Boyd and Walley (1972b).

and Na concentrations were correlated in either species. Phosphorus concentrations increased with increasing N levels in *Typha* ($r = .67$; $P < .01$) but not in *Pithophora*.

Interspecific Variation

Elemental Composition of Wetland Species   Data for the ash and elemental composition of individual wetland species were obtained from 6 sources (Denton, 1966; Boyd, 1966, 1968, 1970b; Boyd and Vickers, 1971; Boyd and Walley, 1972b) and averaged to arrive at the "average elemental composition" of typical wetland species (Table 3). Data for each constituent did not always involve the same number of species; however, not less than 27 or more than 40 species were included per constituent. Macrophytic algae and submersed, floating-leafed, floating, and emergent vascular plants were represented. Data for a given species were often based on samples from several sites. Samples of all species were of lush, green plants hopefully of comparable maturity. The average values are similar to elemental values reported for mesic species (Gerloff et al., 1964). However, an "average elemental composition" for wetland species is rather meaningless because different species often differ drastically in concentrations of ash or of 1 or more elements. For example, N concentrations ranged from 1.46% to 3.95% and percentages of Ca varied from 0.20 to 8.03. In all instances, coefficients of variation for comparisons between species (Table 3) were greater than coef-

Table 4. Chemical Composition of Plants from Par Pond, Savannah River Plant, Aiken, South Carolina[a]

Species	Percentage of dry weight						
	N	P	S	Ca	Mg	K	Na
Submersed plants							
Myriophyllum heterophyllum	2.35	0.16	0.24	1.47	0.26	1.25	1.87
Ceratophyllum demersum	2.66	0.26	0.30	0.77	0.42	4.01	1.16
Najas guadalupensis	...	0.15	0.28	0.98	0.47	3.49	0.61
Eleocharis acicularis	...	0.24	0.28	0.53	0.33	2.86	0.54
Utricularia inflata	2.40	0.12	0.26	0.67	0.21	1.98	1.52
Potamogeton diversifolius	2.86	0.27	0.50	1.14	0.19	3.08	0.44
$\bar{x}$	2.56	0.20	0.31	0.93	0.31	2.78	1.02
Floating-leafed plants							
Nymphaea odorata	1.86	0.18	0.14	1.06	0.14	1.28	1.35
Nuphar advena	3.79	0.40	0.32	1.08	0.27	1.88	1.47
Nelumbo lutea	2.62	0.19	0.16	1.56	0.23	2.27	0.28
Brasenia schreberi	1.95	0.14	0.11	1.79	0.26	0.99	0.66
$\bar{x}$	2.56	0.23	0.18	1.37	0.22	1.60	0.94
Emergent plants							
Typha latifolia	0.93	0.14	0.15	0.76	0.15	2.65	0.28
Hydrocotyle umbellata	2.56	0.18	0.16	1.85	0.47	1.73	0.98
Scirpus americanus	1.22	0.18	0.59	0.50	0.22	2.83	0.09
Juncus effusus	1.24	0.27	0.26	0.38	0.11	0.89	0.40
Panicum hemitomon	1.50	0.14	0.23	0.38	0.25	1.06	0.19
Eleocharis quadrangulata	1.10	0.10	0.15	0.20	0.10	1.81	0.12
Sagittaria latifolia	...	0.30	0.15	0.55	0.18	4.04	0.14
Pontederia cordata	1.40	0.24	0.22	0.96	0.15	2.58	0.83
$\bar{x}$	1.42	0.19	0.24	0.70	0.20	2.20	0.38

[a] Data from Boyd (1970*b*) and Polisini and Boyd (1972).

ficients of variation for comparisons between samples of a given species from different sites (Table 2).

More specific information on the elemental composition of wetland plants from Par Pond near Aiken, South Carolina (Boyd, 1970*b*) is presented in Tables 4 and 5. Different species growing in close proximity in the same lake and during the same season varied widely in composition. Furthermore, certain patterns may be seen if species with similar growth habits are compared. Nitrogen and Na levels were higher in submersed and floating-leafed plants than in emergent plants. Calcium concentrations tended to be higher in floating-leafed species than in the other 2 groups. Magnesium levels were greatest in submersed species. Iron concentrations were higher in submersed species than in other groups. Floating-leafed species contained greatest levels of Zn and B. Phosphorus, Mn and Cu concentrations were similar for all 3 groups. Ash values were not available for Par Pond plants,

Table 5. Chemical Composition of Plants from Par Pond, Savannah River Plant, Aiken, South Carolina[a]

Species	Dry weight (ppm)				
	Fe	Mn	Zn	Cu	B
Submersed plants					
Myriophyllum heterophyllum	2000	473	54	44	10.6
Ceratophyllum demersum	1053	486	100	30	4.3
Najas guadalupensis	712	201	48	48	...
Eleocharis acicularis	2920	192	68	42	...
Utricularia inflata	2112	480	108	47	7.6
Potamogeton diversifolius	1240	160	60	36	5.3
$\bar{x}$	1673	332	73	41	7.0
Floating-leafed plants					
Nymphaea odorata	600	128	32	36	11.3
Nuphar advena	740	300	50	35	8.2
Nelumbo lutea	126	607	50	40	10.9
Brasenia schreberi	500	265	267	32	10.4
$\bar{x}$	492	325	100	36	10.2
Emergent plants					
Typha latifolia	120	412	30	37	5.2
Hydrocotyle umbellata	1245	196	53	53	7.7
Panicum hemitomon	133	292	31	26	2.3
Eleocharis quadrangulata	560	120	45	20	3.7
Sagittaria latifolia	460	355	46	57	...
Pontederia cordata	200	970	67	60	7.9
$\bar{x}$	453	391	45	42	5.4

[a] Data from Boyd (1970b) and Boyd and Walley (1972b).

but data presented by Boyd (1968) were averaged to give the following ash values: macrophytic algae (19.3%), submersed vascular plants (18.4%), floating-leafed vascular plants (8.0%), and emergent vascular plants (9.7%). The high percentages of ash in macrophytic algae and submersed vascular plants may not be entirely representive of uptake of elements because large quantities of inorganic debris which accumulate on surfaces of submersed plants cannot be completely removed by washing. The elemental composition of submersed plants may also be influenced by epiphytic diatoms which cannot be completely removed from external surfaces.

Some aquatic plants appear to accumulate especially high concentrations of 1 or more elements. A few submersed species, especially macrophytic algae of the genus *Chara*, precipitate marl on external surfaces and thus show high Ca concentrations (Wetzel, 1960; Boyd, 1966). In a study of macrophytic algae, Boyd (1966) demonstrated that B was accumulated by *Pithophora kewensis*, *Cladophora* sp., and *Lyngbya* sp.; S by *Pithophora kewensis, Cladophora* sp., and *Hydrodictyon reticulatum*; Na by species of *Spirogyra*; and Ca by *Chara*. *Lyngbya* sp., a blue-green alga, had a greater percentage of N than other species. Examination of Tables 4 and 5 reveals that certain species of vascular plants accumulate higher concentrations of some elements than other species. Conversely, certain species also tend to have lower concentrations of some elements than are usually encountered. *Oedogonium* sp. had an especially low P concentration (Boyd, 1966) as compared to other macrophytic algae. The selective accumulation or exclusion of a particular element is apparently a characteristic of the species and is influenced only to a degree by external concentration of the element (Boyd, 1966).

Variation in N concentrations between species is probably influenced strongly by the inherent morphology of species being compared. Species which are large and require extensive supporting tissue, especially reed swamp species, contain more cell-wall material and less protoplasmic constituents than smaller species which need less supporting tissue. Nitrogen is contained primarily in protoplasmic material and N concentrations in aquatic plants are negatively correlated with percentages of cell-wall material (Polisini and Boyd, 1972). It follows that species which produce high standing crops usually have lower percentages of N than species which produce low standing crops.

Series of analyses of different species of aquatic plants similar to those reported in Tables 4 and 5 may be found in Chapin et al. (1975) and Adams et al. (1973). Both of these studies indicate that different species may differ markedly in elemental composition. For example, N contents of 13 species ranged from 1.00% to 2.55% while P concentrations in these same species varied from 0.09% to 0.24% (Chapin et al., 1975).

Correlations between Elemental Concentrations   Data used in preparing Table 4 were analyzed by regression analysis to determine if patterns existed between concentrations of elements in different species. As expected, percentage of C decreased as ash levels increased ($r = -.82$; $P < .01$). Furthermore, Ca concentrations increased with ash values ($r = .81$; $P < .01$) suggesting that differences in Ca concentrations had an appreciable influence on ash. Species which contained more P usually contained more K than other species ($r = .53$; $P < .01$). No other significant correlations were obtained.

Organic Constituents

Proximate Analyses   Most analyses of wetland plants for organic constituents were conducted to determine their potential value as feedstuffs. Boyd (1968) analyzed a variety of species and found great differences in estimated crude protein (N × 6.25), ether extract, and cellulose levels. A summary of the data is presented in Table 6. In general, macrophytic algae had higher concentrations of crude protein than vascular plants. Differences

Table 6. Summary of Proximate Analyses for a Variety of Species of Wetland Plants[a]

Parameter	Species (n)	Range of values (all species)	$\bar{x} \pm SE$	C.V. (%)
Crude protein (%)	40	8.5–31.3	15.5 ± 0.83	33.9
Ether extract (%)	39	1.63–8.11	4.2 ± 0.28	42.3
Cellulose (%)	40	10.0–40.9	25.9 ± 0.97	23.7
Energy content (kilojoules/g)	36	10.33–17.99	15.48 ± 2.80	11.3

[a] Data from Boyd (1968).

in crude protein levels among the major ecological groups of vascular plants were similar to those discussed earlier for N as expected because crude protein content is estimated from N data. No obvious differences in cellulose or ether extract could be attributed to ecological groupings.

Percentages of cell-wall material in species of plants from a single lake ranged from 19.8 to 72.1 (Polisini and Boyd, 1972). Plants with highest percentages of cell-wall material had the highest standing crops, but the lowest concentrations of crude protein.

Samples of *Pistia stratioides, Eichhornia crassipes,* and *Hydrilla verticillata* were obtained from different sites in Florida (Boyd, 1969b). Wide variation in crude protein, cellulose, ether extract, and total available carbohydrate concentrations were noted among the samples of each of the 3 species. For example, the highest protein concentration for a particular species was 3 or more times greater than the lowest concentration for that species.

Actual Protein and Amino Acids   Although different species of aquatic plants varied greatly in amino acid concentrations and total protein levels, there was a remarkable consistency between the amino acid composition of the protein (Boyd, 1970c, 1973). This suggests that the percentage of actual protein can be estimated with some reliability from crude protein content. Even though crude protein usually overestimated actual protein (sum of amino acids) in aquatic plants by 10 to 20%, actual protein increased with increasing crude protein. Therefore, estimated crude protein (N × 6.25) is an adequate estimate of actual protein in aquatic plants for most ecological purposes.

Miscellaneous Compounds   Amounts of other constituents including tannins (Boyd, 1968), alkaloids (Gibbs, 1974), xanthophylls (Creger et al., 1963), chlorophylls (Boyd, 1970b), and vitamins (Nelson et al., 1938) have been determined for samples of wetland plants. Hutchinson (1975) should be consulted for further information.

ENERGY CONTENT

Boyd (1968) reported that the energy content values for 36 species of aquatic plants ranged from 10.33 to 17.99 kJ/g. Most species (27) had values

>14.6 kJ/g. The mean for all species was 15.48 ± 0.29 (SE) kJ/g with a coefficient of variation of 11.3%. There was a high negative correlation between percentages of ash (X) and energy contents (Y) $(r = -.89;$ $P < .01)$. The regression equation was $\hat{Y} = 4.367 - 0.047X$. When energy values are expressed on an ash-free dry weight basis, values for the 36 species ranged from 16.57 to 18.33 kJ/g. The mean was 18.03 ± 0.17 (SE) with a coefficient of variation of 5.2%. Species with higher percentages of ether extract tended to have higher energy content values on both a dry weight $(r = .45;$ $P < .05)$ and an ash-free dry weight basis $(r = .46;$ $P < .05)$. Correlation coefficients between crude protein and energy values were not significant. Good and Good (1975) reported that the energy content of shoot samples collected from 6 aquatic plant communities ranged from 16.32 to 19.87 kJ/g ash-free dry weight. Root samples often had slightly lower energy values than shoot samples. Dykyjová and Přibil (1975) also observed that variation in the energy content of higher aquatic plants is relatively low. These findings reveal that interspecific variation in energetic density is much less than variation in chemical constituents other than C.

Energy content values for *Typha latifolia* increased from 17.41 to 19.04 kJ/g between April and July (Boyd, 1970c). Good and Good (1975) reported only small seasonal changes in the energy content of aquatic plants. A range of 18.07 to 18.99 kJ/g was obtained for samples of live *Juncus effusus* shoots collected during a 12-month period (Boyd, 1971). These changes are not nearly as great as those reported for most elements. Energy content values for samples of *Typha latifolia* from 11 sites ranged from 17.15 to 18.41 kJ/g dry weight with a coefficient of variation of 2% (Boyd, 1970c).

Available data suggest that variation in energy content values is much less than variation in chemical composition. Boyd (1970c) suggested that energy content information was probably given too much significance in ecological studies.

LITERATURE CITED

Adams, F. S., Cole, H., Jr., and Massie, L. B. (1973). Elemental constitution of selected aquatic vascular plants from Pennsylvania: submersed and floating leaved species and rooted emergent species. *Environ. Pollut.* **5**, 117–147.

Bernard, J. M. and Solsky, B. A. (1977). Nutrient cycling in a *Carex lacustris* wetland. *Can. J. Bot.* **55**, 630–638.

Boyd, C. E. (1966). The elemental composition of several freshwater algae. Ph.D. Dissertation, Auburn University, Auburn, Alabama.

Boyd, C. E. (1968). Fresh-water plants: a potential source of protein. *Econ. Bot.* **22**, 359–368.

Boyd, C. E. (1969a). Production, mineral nutrient absorption, and biochemical assimilation by *Justicia americana* and *Alternanthera philoxeroides*. *Arch. Hydrobiol.* **66**, 139–160.

Boyd, C. E. (1969b). The nutritive value of three species of water weeds. *Econ. Bot.* **23**, 123–127.

Boyd, C. E. (1970a). Production, mineral accumulation and pigment concentrations in *Typha latifolia* and *Scirpus americanus*. *Ecology* **51**, 285–290.

Boyd, C. E. (1970b). Chemical analyses of some vascular aquatic plants. *Arch. Hydrobiol.* **67**, 78–85.

Boyd, C. E. (1970c). Amino acid, protein, and caloric content of vascular aquatic macrophytes. *Ecology* **51**, 902–906.

Boyd, C. E. (1971). The dynamics of dry matter and chemical substances in a *Juncus effusus* population. *Am. Midl. Nat.* **86**, 28–45.

Boyd, C. E. (1973). Amino acid composition of freshwater algae. *Arch. Hydrobiol.* **72**, 1–9.

Boyd, C. E. (1976). Accumulation of dry matter, nitrogen and phosphorus by cultivated water hyacinths. *Econ. Bot.* **30**, 51–56.

Boyd, C. E. and Hess, L. W. (1970). Factors influencing shoot production and mineral nutrient levels in *Typha latifolia*. *Ecology* **51**, 296–300.

Boyd, C. E. and Scarsbrook, E. (1975). Influence of nutrient additions and initial density of plants on production of waterhyacinth *Eichhornia crassipes*. *Aquat. Bot.* **1**, 253–261.

Boyd, C. E. and Vickers, D. H. (1971). Variation in the elemental content of *Eichhornia crassipes*. *Hydrobiologia* **38**, 409–414.

Boyd, C. E. and Walley, W. W. (1972a). Production and chemical composition of *Saururus cernuus* L. at sites of different fertility. *Ecology* **53**, 927–932.

Boyd, C. E. and Walley, W. W. (1972b). Studies of the biogeochemistry of boron. 1. Concentrations in surface waters, rainfall, and aquatic plants. *Am. Midl. Nat.* **88**, 1–14.

Chapin, F. S. III, Van Cleve, K., and Tieszen, L. L. (1975). Seasonal nutrient dynamics of tundra vegetation at Barrow, Alaska. *Arctic and Alpine Res.* **7**, 209–226.

Cowgill, U. M. (1973a). Biogeochemistry of rare-earth elements in aquatic macrophytes of Linsley Pond, North Branford, Connecticut. *Geochim. Cosmochim. Acta* **37**, 2329–2345.

Cowgill, U. M. (1973b). Biogeochemical cycles for the chemical elements in *Nymphaea odorata* Ait. and the aphid *Rhopalosiphum numphaeae* (L.) living in Linsley Pond. *Sci. Total Environm.* **2**, 259–303.

Cowgill, U. M. (1974). The hydrogeochemistry of Linsley Pond. II. The chemical composition of the aquatic macrophytes. *Arch. Hydrobiol. Suppl.* **45**, 1–119.

Creger, C. R., Farr, F. M., Castro, E., and Couch, J. R. (1963). The pigmenting value of aquatic flowering plants. *Poultry Sci.* **42**, 1262.

Culley, D. D., Jr. and Epps, E. A. (1973). Use of duckweed for waste treatment and animal feed. *J. Water Pollution Contr. Fed.* **45**, 337–347.

Denton, J. B. (1966). Relationships between the chemical composition of aquatic plants and water quality. M. S. Thesis, Auburn University, Auburn, Alabama.

Dykyjová, D. and Přibil, S. (1975). Energy content in the biomass of emergent macrophytes and their ecological efficiency. *Arch. Hydrobiol.* **75**, 90–108.

Easley, J. R. and Shirley, R. L. (1974). Nutrient elements for livestock in aquatic plants. *Hyacinth Contr. J.* **12**, 82–84.

Gaudet, J. J. (1973). Growth of a floating aquatic weed, *Salvinia*, under standard conditions. *Hydrobiologia* **41**, 77–106.

Gaudet, J. J. (1975). Mineral concentrations in papyrus in various African swamps. *J. Ecol.* **63**, 483–491.

Gerloff, G. C. and Krombholz, P. H. (1966). Tissue analysis as a measure of nutrient availability for the growth of angiosperm aquatic plants. *Limnol. Oceanogr.* **11**, 529–537.

Gerloff, G. C. and Skoog, F. (1954). Cell contents of nitrogen and phosphorus as a measure of their availability for the growth of *Microcystis aeruginosa*. *Ecology* **35**, 348–353.

Gerloff, G. C. and Skoog, F. (1957). Availability of iron and manganese in southern Wisconsin lakes and the growth of *Microcystis aeruginosa*. *Ecology* **38**, 551–556.

Gerloff, G. C., Moore, D. G., and Curtis, J. T. (1964). The mineral content of native plants of Wisconsin. *Wisconsin Agr. Exp. Sta. Res. Rep. No. 14*, Madison, Wisconsin.

Gibbs, R. D. (1974). "Chemotaxonomy of flowering plants." McGill and Queen's University Press, Montreal, Canada.

Good, R. E. and Good, N. F. (1975). Vegetation and production of the Woodbury Creek-Hessian Run freshwater tidal marshes, *Bartonia* **43**, 38–45.

Gosset, D. R. and Norris, W. E. (1971). Relationship between nutrient availability and content of nitrogen and phosphorus in tissues of the aquatic macrophyte *Eichhornia crassipes* (Mart.) Solms. *Hydrobiologia* **38**, 15–28.

Gorham, E. (1953). Chemical studies on the soils and vegetation of water-logged habitats in the English Lake District. *J. Ecol.* **41**, 345–360.

Hutchinson, G. E. (1975). "A treatise on limnology volume III–Limnological botany." Wiley-Interscience, New York.

Květ, J. (1973). Mineral nutrients in shoots of reed *Phragmites communis* (Trin.). *Pol. Arch. Hydrobiol.* **20**, 137–146.

Lawrence, J. M. (1971). Dynamics of chemical and physical characteristics of water, bottom muds, and aquatic life in a large impoundment on a river. Phase II. Final Rep. on OWRR Project B-010-ALA. Auburn Univ. Agr. Sta., Auburn, Alabama.

Linn, J. G., Staba, E. J., Goodrich, R. D., Meiske, J. C., and Otterby, D. E. (1975*a*). Nutritive value of dried or ensiled aquatic plants. I. Chemical composition. *J. Anim. Sci.* **41**, 601–609.

Linn, J. G., Goodrich, R. D., Otterby, D. E., Meiske, J. E., and Staba, E. J. (1975*b*). Nutritive value of dried or ensiled aquatic plants: II. Digestibility by sheep. *J. Anim. Sci.* **41**, 610–615.

Mason, C. F. and Bryant, R. J. (1975). Production, nutrient content and decomposition of *Phragmites communis* Trin. and *Typha angustifolia* L. *J. Ecol.* **63**, 71–95.

Mayer, A. M. and Gorham, E. (1951). The iron and manganese content of plants present in the natural vegetation of the English Lake District. *Ann. Bot.* **15**, 247–263.

Musil, C. F. and Breen, C. M. (1977). The influence of site and position in the plant community on the nutrient distribution in, and content of *Eichhornia crassipes* (Mart.) Solms. *Hydrobiologia* **53**, 67–72.

Nelson, J. W., Palmer, L. S., Wick, A. N., Sandstrom, W. M., and Lindstrom, H. V. (1939). Nutritive value and chemical composition of certain fresh-water plants of Minnesota. Univ. Minnesota Agr. Exp. Sta., Tech. Bull. 136.

Nichols, D. S. and Kenney, D. R. (1976). Nitrogen nutrition of *Myriophyllum spicatum*: uptake and translocation of ^{15}N by shoots and roots. *Freshwater Biol.* **6**, 145–154.

Polisini, J. M. and Boyd, C. E. (1972). Relationships between cell–wall fractions, nitrogen, and standing crop in aquatic macrophytes. *Ecology* **53**, 484–488.

Riemer, D. N. and Toth, S. J. (1968). A survey of the chemical composition of aquatic plants in New Jersey. New Jersey Agr. Exp. Sta. Bull. 820.

Riemer, D. N. and Toth, S. J. (1970). Chemical composition of five species of *Nymphaeaceae*. *Weed Sci.* **18**, 4–6.

Rogers, H. H. and Davis, D. E. (1972). Nutrient removal by water hyacinth. *Weed Sci.* **20**, 423–427.

Stake, E. (1967). Higher vegetation and nitrogen in a rivulet in central Sweden. *Schweiz. Z. Hydrol.* **29**, 107–124.

Stake, E. (1968). Higher vegetation and phosphorus in a small stream in Sweden. *Schweiz. Z. Hydrol.* **30**, 353–373.

Steward, K. K. and Ornes, W. H. (1975). The autecology of sawgrass in the Florida Everglades. *Ecology* **56**, 162–171.

Straškraba, M. (1968). Der Anteil der höheren Pflanzen an der Produktion der stehen den Gewasser. *Mitt. Int. Ver. Limnol.* **14**, 212–230.

Sutcliffe, J. F. (1962). "Mineral salts absorption in plants." Pergamon Press, New York.

Wetzel, R. G. (1960). Marl encrustations on hydrophytes in several Michigan lakes. *Oikos* **11**, 223–228.

NUTRIENT MOVEMENTS IN LAKESHORE MARSHES

R. T. Prentki, T. D. Gustafson[1], and M. S. Adams

Institute for Environmental Studies and Botany Department
University of Wisconsin–Madison, Madison, Wisconsin 53706

Abstract   Seasonal phosphorus allocation in *Typha latifolia* L. was investigated by analysis of above- and belowground parts in a lakeshore marsh at Lake Mendota, Wisconsin. Maximum total P stocks of 4.3 and 2.2–2.5 g P/m^2 were found in summer and winter, respectively. A maximum upward translocation of 140 mg P/m^2, ⅓ due to mobilization of belowground reserves, was observed the first 2 weeks of June. Over an entire summer, 40% of the 3.2 g P/m^2 accumulated aboveground was reallocated from belowground plant parts. However, only 23% of this amount returned to belowground in the fall; 2.5 g P/m^2 was left on the marsh surface as litter plus leachate. *Typha* roots continued to absorb P past aboveground senescence and into early winter, resulting in post-shoot-death uptake of up to 9% of the next season's P requirement.

Analysis of lakeshore marsh literature via a input–output model suggests that macrophytic translocation of nutrients is an important source of internal loading. In those marshes without major surface water inputs, seasonal accumulation and abandonment of nutrients above the soil interface is likely to be the dominant term in nutrient budgets.

Key words   *Aquatic plants; emergent macrophytes; Lake Mendota; marshes; nitrogen; nutrient budgets; nutrient cycling; phosphorus; translocation;* Typha latifolia; *Wisconsin.*

INTRODUCTION

Intuitively one expects that lakeside marshes may have some, even a profound, effect upon the natural flow of nutrients from the terrestrial land-

[1] Present Address: Department of Biology, Juniata College, Huntingdon, Pennsylvania 16652.

scape to open lake. The high annual production and the robust emergent plants which characterize many of these marshes have been visualized as a biological depository for plant nutrients, especially for eutrophying nitrogen and phosphorus (e.g., Tóth, 1972; Sloey et al., 1978). Effectiveness of a marsh intercepting nutrients from throughflow water will ultimately depend upon subsequent processing such as denitrification or incorporation into the sediments followed by eventual burial below rooting depth. The above-ground shoot biomass is an obvious sink but its nutrient content also represents a potential source of export from the marsh if nutrients are left at the surface either in dissolved or particulate form after shoot death.

In this chapter we address the question of nutrient movements through lakeshore marshes. The literature's paucity of data on all other aspects of the nutrient cycle forces us to focus on the activities of the plants. The initial section if therefore not an inclusive literature review of relevant aspects of nutrient relations in marsh plants. In the second section we report our specific study of P uptake and allocation in *Typha latifolia* L. This study attempts to lay the groundwork for evaluation of the impact of a cattail stand on the marsh nutrient budget by looking in detail at life history and seasonal progression in size of plant nutrient pools. As the final section we take the liberty of offering a mass balance model for nutrient transport through a lakeside marsh.

NUTRIENT LEVELS IN LAKESHORE
MARSH PLANTS

Nutrient Concentrations

Tissue nutrient levels are the most studied nutrient parameters in lakeshore marsh literature (for reviews see Riemer and Tóth, 1968; Keefe, 1972; Dykyjová, 1973a; Hutchinson, 1975). Differences among observed nutrient levels have been attributed to additive effects of seasonal trends (e.g., Boyd, 1969, 1971; Dykyjová, 1973b), site fertility (Boyd and Hess, 1970), site exposure to wave action (Dykyjová and Hradecká, 1976), species (Mason and Bryant, 1975; Boyd, 1970a), and ecotype (Björk, 1967).

Seasonal changes in tissue levels are often parallel for different nutrients and these trends show consistent patterns across both site and species. Nitrogen, P, K, and S concentrations normally parallel one another, typically declining from spring to fall (Boyd, 1969, 1970a, 1971; Dykyjová, 1973b; Květ, 1973). One important latitudinal exception to continuous seasonal decline of tissue concentration for this nutrient grouping occurs in the arctic where the bulk of amphibious plant biomass and nutrients are always belowground (see Hobbie, 1972; Prentki, 1976; Shaver and Billings, 1975; Chapin et al., 1975). Initial arctic summer plant growth is dependent upon meager nutrient reserves held in limited, overwintering green tissue and the few millimetres or centimetres of thawed soil. Only as the unfrozen layer increases by several times through the first half of the 2-month growing

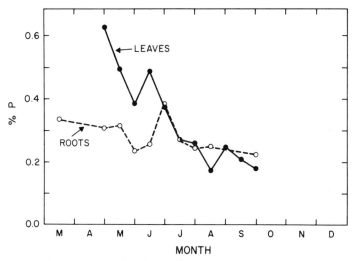

Fig. 1. Phosphorus concentrations in *Typha latifolia* leaves and roots, as percentages of dry weight, in University Bay Marsh, Lake Mendota, Wisconsin, 1974.

season (Miller and Prentki, *In press*), freeing the bulk of plant nutrient reserves and activating the uptake system, do aboveground tissue concentrations of N, P, and K peak (McRoy and Alexander, 1975; Chapin et al., 1975).

Of the other aboveground macronutrients, Ca and Mg do not exhibit consistent behavior. Sodium is also irregular except for a pronounced autumnal peak in *Phragmites communis* Trin. Björk (1967) and Mason and Bryant (1975) suggest that this peak is related to senescence, but Kvĕt (1973) attributes it to new shoot growth.

Belowground tissue nutrient data are more limited, but appear consistent for lakeshore marsh plants. Dykyjová and Hradecká (1976) found that *Phragmites* roots decrease in N, P, K, Ca, and Mg continuously through the growing season. We have observed that *T. latifolia* roots similarly decrease in at least P. Rhizome concentrations of both species deplete in summer but recover to high levels in fall. Initial spring P concentrations in overwintering shoot bases (important storage organs in *T. latifolia*) are 2× those of roots or rhizomes (Figs. 1 and 2). Shoot bases become depleted of P by early summer, die, and in the year studied, were visibly decomposed by 1 July (Gustafson, 1976). New shoot bases and rhizomes form in midsummer and are initially very high in P, having 2× the maximum percent P content of adult organs. These high P levels rapidly drop as shoot bases and rhizomes mature.

Interpretation of tissue nutrient dynamics in many lakeshore marsh studies is ambiguous because of the absence of parallel biomass data. Real changes in plant-part nutrient contents caused by translocation and/or up-

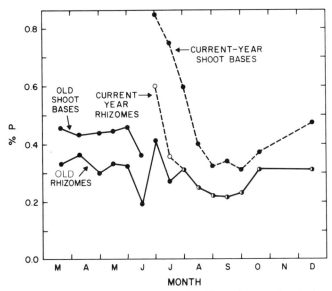

Fig. 2. Phosphorus concentrations in *Typha latifolia* old (overwintering) and current-year (1974) shoot bases and rhizomes, as percentages of dry weight, in University Bay Marsh, Lake Mendota, Wisconsin, 1974. Although old and current-year rhizomes can still be distinguished in the field, their tissue P concentrations are very similar after July and have therefore been averaged after this date.

take are difficult to distinguish from the similar tissue level changes due to growth dilution or carbohydrate transfers. Tissue levels are demonstrated poor predictors of nutrient standing stocks in *T. latifolia* and *Scirpus americanus* Pers. stands; areal biomass alone was found to be a much better predictor (Boyd and Hess, 1970; Boyd, 1970a).

Critical nutrient concentrations (the internal levels above which plant yield is no longer nutrient limited) have not been established for emergent marsh species. However, Gerloff (1969, 1975) and Wong and Clark (1976) have reported critical concentrations for several submerged angiosperms on either whole plant or index segment basis. Critical concentrations, as percentage of dry weight, range from 0.75%–1.6% for N (6 species), 0.07%–0.16% for P (8 species), 0.35%–1.7% for K (3 species), 0.22% and 0.28% for Ca (2 species), 0.10% and 0.18% for Mg (2 species) and 0.08% for S (1 species). These can be compared with tissue levels of marsh plants given in parentheses in Table 1 for the time of peak nutrient standing stock. Nitrogen, P, K, Ca, and Mg concentrations in the emergents all frequently fall within ranges found growth-limiting for submersed plants. Concentrations much lower than those in Table 1 occur for plants in lakeshore marshes: 0.51% N (Boyd, 1970a), 0.39% K (Bernatowicz, 1969), and 0.08% S (Boyd, 1970a) in *Typha latifolia*, and 0.01% P (Bernatowicz, 1969), 0.014% Ca (Mason and Bryant, 1975) and 0.048% Mg (Květ, 1973), in *Phragmites communis*. These

Table 1. Tissue Nutrient Concentrations in Lakeshore Marsh Plants

Species	Location	Peak nutrient standing stock in g/m² (% dry wt concentration)							Reference
		N	P	K	Ca	Na	Mg	S	
Phragmites communis	Opatovický Fishpond (Czechoslovakia)	41. (1.9)	5.3 (0.18)	38. (1.7)	7.4 (0.27)	...	2.7 (0.09)	...	Dykyjová and Hradecká (1976)
	Nesyt Fishpond (Czechoslovakia)	28. (1.6)	2.9 (0.16)	31. (1.7)	6.3 (0.36)	0.87 (0.083)	1.2 (0.068)	...	Květ (1973)
	Alderfen Broad (England)	43. (4.2)	2.0 (0.20)	11. (1.0)	2.3 (0.21)	2.5 (0.26)	1.1 (0.10)	...	Mason and Bryant (1975)
Typha angustifolia	Alderfen Broad	13. (1.6)	3.2 (0.40)	20. (2.0)	7.2 (0.60)	9.1 (0.82)	1.6 (0.16)	...	Mason and Bryant (1975)
	Opatovický Fishpond	33. (1.2)	4.6 (0.15)	47. (1.7)	21. (0.73)	...	5.2 (0.18)	...	Dykyjová (1973b)
Typha latifolia	Lake Kobyli (Czechoslovakia)	25. (1.6)	1.6 (0.10)	19. (1.7)	11. (0.65)	5.3 (0.31)	2.4 (0.23)	...	Květ (1975)
	Par Pond (South Carolina)	5.3 (0.87)	0.77 (0.11)	14. (2.1)	5.5 (0.89)	1.6 (0.24)	1.1 (0.18)	0.65 (0.18)	Boyd (1970a)
	Pond A (South Carolina)	12. (1.1)	1.7 (0.15)	14. (1.2)	...	...	...	...	Boyd (1971)
	[a]Pond B (South Carolina)	5.6 (1.1)	0.68 (0.13)	10. (1.9)	...	...	...	...	Boyd (1971)
	[a]Steel Creek Bay (South Carolina)	9.0 (1.2)	0.95 (0.12)	10. (1.3)	...	...	...	...	Boyd (1971)
	Lake Mendota (Wisconsin)	31.[b] (2.2)	3.2 (0.49)	...	...	...	...	...	This study
Eleocharis quadrangulata	Par Pond	8.3 (0.94)	1.3 (0.13)	6.5 (0.61)	1.8 (0.20)	2.4 (0.27)	0.61 (0.07)	...	Boyd and Vickers (1971)

Table 1. Continued

Species	Location	Peak nutrient standing stock in g/m² (% dry wt concentration)							Reference
		N	P	K	Ca	Na	Mg	S	
Scripus americanus	[a]Par Pond	1.7 (1.2)	0.23 (0.15)	2.7 (1.9)	0.97 (0.64)	0.31 (0.20)	0.45 (0.29)	0.99 (0.64)	Boyd (1970*a*)
Schoenoplectus lacustris	Opatovický Fishpond	53. (1.3)	11. (0.26)	57. (1.4)	10. (0.25)	...	6.3 (0.15)	...	Dykyjová (1973*b*)
	[a]Opatovický Fishpond	30. (1.1)	7.2 (0.25)	39. (1.4)	9.1 (0.38)	...	3.7 (0.13)	...	Dykyjová (1973*b*)
Glyceria maxima	Opatovický Fishpond	44. (1.6)	7.1 (0.26)	62. (2.3)	3.5 (0.12)	...	3.4 (0.11)	...	Dykyjová (1973*b*)
Acorus calamus	Opatovický Fishpond	21. (2.9)	2.5 (0.35)	27. (3.7)	10. (0.85)	...	1.8 (0.18)	...	Dykyjová (1973*b*)
Carex aquatilis	[c]Pond J (Alaska)	2.5 (2.7)	...	...	...	...	...	...	Alexander et al. (*In press b*)
Arctophila fulva	[c]Pond J	2.7 (2.5)	...	...	...	...	...	0.65	Alexander et al. (*In press b*)
x̄ (excluding stands marked[a] or [c])		28.	3.6	29.	7.8	3.6	2.5		

[a] Denotes erosional or exposed stand.
[b] *Personal communication* (C. S. Smith, 1977).
[c] Arctic marsh.

latter tissue percents with the exception of K and S values are well below likely critical concentrations and may be indicative of serious nutrient deficiencies.

Nutrient Standing Stocks

Seasonal peak standing stocks of nutrients in several emergent lakeshore stands are compiled from the literature in Table 1. In each case N and K represent the greatest nutrient pools, usually present in tens of grams per square metre; relative rankings of other nutrients are variable. The range for each element is about an order of magnitude and is mainly due to the contrast between oligotrophic South Carolinian lakes (Boyd, 1970a, 1971; Boyd and Vickers, 1971) and fertilized Czechoslovakian fishponds (Květ, 1973, 1975; Dykyjová, 1973b; Dykyjová and Hradecká, 1976): the 2 locations which dominate the literature and possibly represent the extremes of temperature lakeshore marshes. The low arctic N standing stocks are due to low biomass; the 2.7% and 2.5% N contents (Alexander et al., *In press b*) are actually fairly high.

Peaking of different nutrients at any one of the Table 1 study sites clusters within a couple of weeks; however, the slopes of accumulation curves vary considerably between sites. Apparently site differences are the controlling factor affecting timing of nutrient peaks. Calcium is the only element considered which accumulates more slowly than biomass. The ambiguity in timing of nutrient peaks in the same species on different lakes is analogous to other aspects of plant development such as biomass. For example, various European studies of *Phragmites* list biomass maxima as occurring in mid-July (Květ, 1973), mid-August (Mason and Bryant, 1975), and mid-September (Dykyjová and Hradecká, 1976). Corresponding nutrient maxima occur at time of biomass peak in Květ (1973), afterwards in Mason and Bryant (1975), and before in Dykyjová and Hradecká (1976). Whigham et al. (1978) have reported similar production and biomass variability in freshwater tidal marshes.

In general, however, initial aboveground accumulation curves are much steeper for nutrients than for biomass. Thus Dykyjová and Hradecká (1976) observed that *Phragmites* shoots accrued 60% of N, 73% of K, 92% of P, 20% of Ca, and 56% of Mg peak standing stocks during the first 45 days of the growing season while completing only 30% of biomass production. In the same fashion, Boyd (1971) found that relative accumulations of N, P, and K in *Typha latifolia* were 2× that of biomass through early May. Similar results are reported for several other temperate latitude lakeside emergents (Boyd, 1969, 1970a; Boyd and Vickers, 1971; Dykyjová, 1973b), but the arctic species *Carex aquatilis* Wahlenb. and *Arctophila fulva* (Trin.) Anderss. behave differently, gaining shoot N and biomass at the same rate (Alexander et al., *In press b*). This again suggests that slow thawing of tundra soils retards plant use of belowground nutrients.

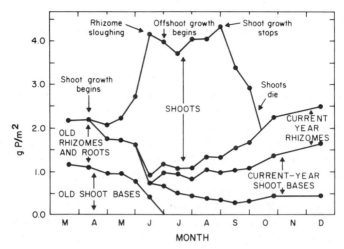

Fig. 3. Cumulative seasonal stocks of P in *Typha latifolia* plant parts (as cumulative g P/m²) in University Bay Marsh, Lake Mendota, Wisconsin, 1974. Roots are a very minor pool (≤0.13 g P/m²) and have been combined with old rhizomes for visual simplification.

The magnitude and seasonal importance of belowground stores are nowhere near as well known as those for aboveground stores. Emergent species with longer-lived storage organs, such as *Phragmites* with rhizome life of at least 3 years (Fiala, 1973), and *C. aquatilis* with root life of 6 to 8 years (Shaver and Billings, 1975), might be expected to retain a greater fraction of plant nutrients belowground than species with shorter-lived organs such as the *Typha* genus with rhizome life of only 1.5 to 2 years (Fiala, 1973; Gustafson, 1976). Dykyjová and Hradecká (1976) found that maximum accumulations of N, P, K, and Mg in *Phragmites* were 1.4×–1.8× greater below- than aboveground. In tundra pond marshes *C. aquatilis* belowground P is ≈0.4 g P/m² (Barsdate and Prentki, 1973) and peak aboveground standing stock ≈0.08 g P/m² (Prentki, 1976) or 5× higher belowground than aboveground. Our own studies on Lake Mendota *T. latifolia* found winter belowground storage of N to be 40% of peak summer aboveground stock (C. S. Smith, 1977, *personal communication*), and P to be equivalent to 70% (Fig. 3). The importance of belowground nutrient storage, as indexed by the ratio of maximum belowground storage to peak standing stock, appears to be roughly proportional to the life-span of the storage organ. If this relationship holds true for additional marsh species, then it may be a useful indicator of plant dependence upon belowground reserves for the vast majority of marsh species for which belowground nutrient data are lacking.

Nutrient Uptake, Translocation, and Release

Estimates of nutrient uptake and translocation by marsh plants are few. Such data, in ecological context as opposed to laboratory physiological studies, are derived either indirectly from seasonal changes in standing stocks or directly with isotopic tracers. In situ tracer uptake measurements are hindered by the impossibility of getting uniform tracer distribution in soil and plant tissue and by cumulative errors inherent in extrapolating from short-term uptake to seasonal accumulation.

Given the present state of the art, we feel monitoring standing stocks provides the best alternative. For many species, spring and summer increases in aboveground standing stock furnish reasonable estimates of gross transfers from belowground. A primary fault in this procedure is the effect of sampling frequency on apparent rates of nutrient movement. Perennial marsh plants typically have short, intense periods of nutrient uptake and aboveground accumulation (Dykyjová, 1973b; Dykyjová and Hradecká, 1976; Boyd, 1969, 1970a, 1971; Boyd and Vickers, 1971; Mason and Bryant, 1975). Any lack of synchrony in sampling during this active period will cause underestimation of the real translocation rate. Furthermore, any process which depletes aboveground plant stores: leaching, sloughing, shoot death, reallocation belowground, or grazing, will also cause turnover to be underestimated. During the period of aboveground nutrient accumulation we believe these processes to be minor in many marsh species. Here it is sufficient to state this; our reasoning will be discussed later in this chapter.

Upward translocation rates can be estimated from works referenced in Table 1 (as noted) by dividing increases in standing stocks between successive samplings by the time period between samplings. Considering only the highest rate calculable for each macrophyte stand, we find for the more studied elements among-site ranges of 34–750 mg N, 59–970 mg K, 5–190 mg P, 37–550 mg Ca, and 18–80 mg Mg $\cdot$ m$^{-2}\cdot$ day^{-1}. Note that these translocation rates are averaged over 2 to 9 weeks; shorter-term measurements would likely have resulted in higher calculated rates.

Whether the bulk of the nutrients being translocated upward are from *de novo* root uptake or from reallocation of belowground reserves is not known for most species. The belowground samplings of Dykyjová and Hradecká (1976) were too infrequent to establish this for *Phragmites*. Boyd's (1969) study of the nutrient accumulation by *Justicia americana* Pers. (a plant with a very small perennial root stock and thus no substantial reallocation) gives base-line data for accumulation rates due to uptake alone of 540 mg N, 34 mg P, 960 mg K, 320 mg Ca, and 230 mg Mg $\cdot$ m$^{-2}\cdot$ day^{-1}. These data clearly indicate that root uptake can be high enough to account for observed rates of nutrient accumulation in other stands. In our study of *T. latifolia* (reported later in this chapter) we show that *de novo* root uptake accounts for 95 mg P $\cdot$ m$^{-2}\cdot$ day^{-1} or ⅔ of a 140 mg P $\cdot$ m$^{-2}\cdot$ day^{-1} translocation. Conversely, in plants with greater belowground storage, such as *Phragmites* or *C. aquatilis*, it is quite possible that reallocation is a proportionally greater nutrient pump.

There are 3 plant processes: leaching, death of leaves and shoots, and reallocation to belowground, which are of particular consequence both for the role they may play in cycling nutrients within the marsh and as sources of error in estimating uptake or translocation from standing stock.

The first process, leaching, may be an important summer pathway returning soil nutrients to marsh surfaces, but is less likely to be of sufficient magnitude to influence greatly the plant itself. Szczepańska and Szczepański (1973) were able to leach measurable (but not vital) amounts of NH_3, NO_3, dissolved organic nitrogen, orthophosphate, K, Ca, and Na from various lakeshore emergents. Potassium was the most leachable, at a rate of 6 mg K/m^2 of reeds per "storm," but 50 such storms would only decrease "average" K peak standing stock (29 g K/m^2, Table 1) by 1%. Leaching losses of N and P also occur for *C. aquatilis* (R. T. Prentki, *personal observation*) and guttation may release additional nutrients from leaf tips of this plant (Chapin and Bloom, 1976). Total release of P for *C. aquatilis* from arctic tundra pond marshes under near in situ conditions is ≈ 1.1 μg P/g dry wt leaf per day in summer (Barsdate and Prentki, 1972). This amounts to a growing season loss of 0.4 mg P/m^2 in the *Carex* marshes, or $\approx 0.5\%$ of the actual aboveground standing stock. In contrast, loss of P from *Sparganium* in Shagawa Lake, Minnesota, is 1.2 g P/m^2 over 102 days (Lie, 1975), or equivalent to 33% of an "average" peak standing stock, and is thus a more major loss to the plant than that found in the other citations. Thus for *Sparganium*, standing stocks would undervalue translocation by ⅓, but for the other 2 marsh types probably by <1%.

Any summer shoot or leaf death prior to peaking of nutrient standing stocks will also cause underestimation of to-shoot translocation. For marsh plants with infrequent summer shoot or leaf death, such as *Phragmites* and *Typha*, nutrient accumulation should reflect movement with little error. Mason and Bryant (1975) estimate that peak biomass underestimates *Phragmites* production by 5 to 15% and *T. angustifolia* production by 23% due to shoot and leaf death. Gustafson (1976) estimates shoot death equivalent to < 5% of production in *T. latifolia*. Parallel underestimation would occur in nutrient translocation calculations based on standing stocks; however, errors of this magnitude fall within natural between-site and within-site variation for C and nutrients (Whigham et al., 1978; Boyd and Hess, 1970; Gustafson, 1976). For plants such as *Carex lacustris* which has more continuous production and death of shoots in summer than the aforementioned species, peak live biomass can greatly underestimate summer shoot production (in the case of *C. lacustris* by 100%; Bernard and Gorham, 1978) and also nutrient flow. In species of this or similar growth form standing stocks should not be used as accurate estimators of either nutrient or carbon flow.

The fates of aboveground nutrients in senescent plants or plant parts have not been quantitatively established. Some substantial amounts of nutrients may have already been incorporated into fruiting bodies (Boyd, 1970*a*;

Dykyjová and Hradecká, 1976). Nutrients are also leached more easily from older shoots than from healthy vegetation (Tukey, 1970), but this loss cannot be distinguished from translocation belowground without belowground measurement.

Leaching is also especially rapid after plant tissue dies. Boyd (1970b) observed that dead *Typha latifolia* shoots submerged in December in Par Pond, South Carolina, lost almost all K and Na, $\approx 50\%$ of P and Ca, and 75% of Mg within 3 weeks. Nitrogen, however, either increases or remains constant in the plant litter for many months. Terrestrial decomposition studies (e.g., Enwezor, 1976) have shown N is not mineralized in organic matter unless the C:N ratio is less than about 20:1 by weight; at higher ratios N is immobilized in decomposer biomass. Boyd's *Typha* litter (assuming C as 45% of dry weight) had a 52:1 ratio, well above the immobilization limit.

In cold climates, low decomposer activity may result in immobilization of additional elements as well. Barsdate and Prentki (1973) submerged green *Carex aquatilis* leaves in Alaskan tundra ponds for 33 days in open-mesh bags and sealed, $HgCl_2$-poisoned, leaching controls. Dry weight losses in open-mesh bags were $2\times$ those of controls, 46% *versus* 24%. However, for all macronutrients followed, losses as percent of original amount present, were greater in controls than open-mesh bags: 74% *versus* 64% for P, 25% *versus* 9% for Ca, 41% *versus* 33% for Mg, and 63% loss *versus* a net 84% increase for Na. Thus initial nutrient release occurred through leaching, and decomposers were a net sink over the duration of the experiment, incorporating more nutrients into biomass than they mineralized.

Other investigators (Mason and Bryant, 1975; Planter, 1970a; Davis and van der Valk, 1978) have documented rapid, initial release of most nutrients from dead lakeshore marsh vegetation. These plants, however, appear normally to have C:N ratios over the N immobilization limit of 20:1 (see Keefe, 1972), and therefore their initial decomposition should tend to remove rather than release N. Studies of Chamie and Richardson (1978), Godshalk and Wetzel (1978), and Odum and Heywood (1978) in this volume, and of Brinson (1977), in addition to that of Boyd, support this premise. After this initial leaching–immobilization phase is over, plant-litter nutrients are probably slowly lost, limited by the rate of processing of the bulk litter.

The combination of rapid leaching of nutrients from dead or dying plants and the ability of litter microflora to scavenge nutrients from outside the litter system also make estimation of downward translocation by difference between peak standing stocks and fall litter content very suspect. Unfortunately very seldom are underground data collected to calculate this transfer directly. This is not necessarily only a problem in fall, however, for those plants with frequent shoot death in summer (e.g., the genus *Carex*; Bernard and Gorham, 1978) are likely simultaneously translocating nutrients and carbohydrates down one shoot and up another.

PHOSPHORUS TRANSLOCATION AND
ALLOCATION IN TYPHA LATIFOLIA

Marsh plants are typically perennial and heavily rhizomatous, possibly allowing for a strategy of strong nutrient conservation. On the other hand, these plants commonly form robust stands suggesting that these are very fertile habitats where there is little need to save nutrients from one growing season to the next. Because these plants strongly dominate the marsh, the question of plant conservation has major implications for the marsh as a whole. The concept of the marsh operating as a nutrient trap would be threatened if the emergent vegetation is, in fact, quarrying buried nutrients and abandoning them, in quantity, at the surface.

With existing data it is seldom possible to calculate internal conservation of nutrients by the marsh plants, that is, to distinguish spring uptake from reallocation of belowground stores or to measure the amount of autumnal translocation from senescent shoots. Björk (1967) measured nutrient concentrations in *Phragmites* "sap" from various plant parts, and concluded that appreciable fall translocation probably occurs, but could not give a quantitative estimate. Chapin et al. (1975) inferred (based on aboveground measurements only) that nearly half of aboveground N, P, and K in several wet tundra macrophyte species must be translocated belowground at season end. Klopatek (1975), on the other hand, estimated belowground translocation at 13% of aboveground P and N standing stocks in *Scirpus fluviatilis*, by assuming that half of the nutrients that disappeared aboveground during senescence were translocated and that half were leached. Although the latter 2 studies give quantitative estimates of reallocation to underground plant organs, neither confirms by belowground measurement that such a reallocation does in fact occur.

Two problems in translocation and reallocation are especially pertinent to the issue of the efficiency of internal nutrient conservation in marsh plants:

1) Emergent marsh plants accumulate nutrients aboveground in spring and early summer at a much higher rate than biomass. How much of this accumulation is due to reallocation of belowground reserves?

2) By autumn, huge quantities of nutrients which were stored in summer standing crops quickly disappear. How much of this apparent plant loss is translocated belowground?

We have addressed these 2 questions here in a study of the role of internal P cycling in *T. latifolia*, by means of seasonal above- and belowground harvests. Shoot and leaf death prior to stand senescence are minimal in *Typha* species (Mason and Bryant, 1975; Gustafson, 1976) and a plant-part accumulation budget should therefore closely reflect internal plant reallocation.

Materials and Methods

The *Typha latifolia* population used in this study was a 0.1 ha, robust, urban, monospecific stand in a marsh which receives appreciable phosphorus-rich runoff directly from road culverts or indirectly through a small upstream marsh (Ahern, 1976). Most runoff is the result of snowmelt or thunderstorms.

Typha standing stocks in 5 or more random quadrats were sampled at 2-week intervals during the 1974 growing season and less frequently in winter by the partial harvest technique aboveground (Ondok and Dykyjová, 1973) and excavation belowground (Westlake, 1968). Biomass estimates were made for all roots, current-year shoots, rhizomes, shoot bases (the corm-like thickening at the junction of shoot and rhizome), and overwintering shoot bases and rhizomes. Rhizomes were aged by noting their position in relation to shoot bases. A detailed description of sampling procedures, statistics, seasonal biomass, and stand life history are reported in Gustafson (1976). Percent standard errors of shoot biomass estimates fall within an arbitrary $\pm$ 10% of the $\overline{X}$ criterion; unfortunately belowground estimates exceed this criterion (>7 to <20% of $\overline{X}$ for shoot base and <13 to <26% of $\overline{X}$ for rhizome biomasses).

Biomass samples were subsampled for carbohydrates and several nutrients in addition to P. Carbohydrates are reported in Gustafson (1976), N and other nutrient data reduction are not completed and will be reported elsewhere.

Tissue P measurements were made on replicate digests of 0.2 g material in a nitric–perchloric acid mixture (Sommers and Nelson, 1972) by a phosphomolybdate procedure (John, 1970). Standard errors of percent tissue P stem base and rhizome means exceed the 10% of $\overline{X}$ criterion by 3% on 3 occasions, 30 April and 1 August for stem bases and 1 July for current-year rhizomes. Shoots, roots, and leaves (when separated from stems) were pooled across quadrats prior to chemical analysis, allowing no standard error estimates. Standing stocks of P were calculated individually for each quadrat and then averaged.

In December 1975, January 1976, and June 1977, *T. latifolia* roots were collected for radiophosphorus uptake measurements. Plants were excavated in blocks and January (frozen) roots were thawed and acclimatized at 5°C for 24 hours. Other roots were immediately excised and tested upon return to laboratory. Phosphate uptake was measured at 150 mg P/litre, pH 7, and 5°C (winter roots) or 20°C (summer roots) as in Chapin (1974) using the desorption procedure B of Epstein et al. (1963). The PO_4 concentration and pH utilized reflect about median observed values in the marsh soil, and temperatures used were those of the root zone in June and those below the frost line in December. Radiophosphorus activity and uptake were assayed by dry counting in a liquid scintillation counter, with counting efficiency calibrated against subsequent Cherenkov activity of a sample subset following nitric–perchloric acid (HNO_3–$HClO_4$) digestion. The correlation coefficient between dry and digest counts was $r = .99$, ($n = 20$). Means and variances of uptake were not significantly different for December and thawed January roots and results on these 2 groups were therefore pooled as winter roots.

In the following discussion, a single asterisk signifies statistical significance at $\alpha = .05$, and two asterisks at $\alpha = .01$.

Tissue Phosphorus Concentrations

Phosphorus concentrations for different plant parts show distinct and different seasonal patterns. Phosphorus in initial aboveground growth is very high, 0.63% P, but declines sharply (slope = -0.019 tissue % P per week, $r = -.92**$) through September (Fig. 1). Root concentrations vary much less seasonally, but from July onward are similar to leaf concentrations.

Both overwintering (old) shoot bases and rhizomes show a sharp, significant decrease (**, Student's t-test) in P concentration from earlier rather constant percentages during the first 2 weeks of June (Fig. 2). This is the time of maximal shoot base and rhizome translocation of reserve carbohydrate to shoots and period of highest rate of canopy growth. In late June, the current season's rhizomes and shoot bases begin to develop. In a fashion parallel to that of the shoots, these current-year organs have a very high (0.85%) phosphorus concentration when very young and when their biomass is low; as the juvenile rhizomes and shoot bases enlarge and mature, these high tissue concentrations fall off sharply. It is of interest that, until autumnal shoot senescence, tissue percent P is inversely correlated with P standing stock in these current-year organs. Regression equations are:

Current-year shoot base P (g/m²) = $-0.72 \times \%$ tissue P + 0.92, $r = .90*$, and

current-year rhizome P (g/m²) = $-0.57 \times \%$ tissue P + 0.47, $r = .60$ (NS).

Although with but 5 data pairs, only the shoot base correlation is significant, it is obvious that changes in tissue nutrient concentration should not be assumed to parallel loss or accumulation.

Phosphorus Standing Stock

The calculated standing stocks of P for whole plant and plant parts are much more valuable in setting forth seasonal phenomena (Fig. 3). Plant P stocks in the *T. latifolia* stand are lowest at 2.2 g P/m² to 2.5 g P/m² in winter when aboveground parts of the plant have senesced. Summer plant P content is almost 2× higher, reaching a 4.3 g P/m² peak on 1 September, the date of maximum shoot biomass.

The shoot P pool attains the highest levels of any plant compartment

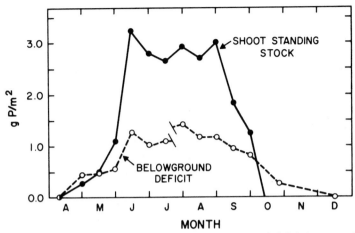

Fig. 4. Phosphorus standing stock in shoots and calculated P deficit belowground (in g P/m²) in University Bay Marsh, Lake Mendota, Wisconsin, 1974. The belowground deficit is calculated as: winter, 1973, belowground stock (2.2 g P/m²) minus belowground stock for each date through mid-July, and then as: winter, 1974, stock (2.5 g P/m²) minus belowground stock for each date afterwards.

during the year, reaching 3.2 g P/m² or 78% of the total plant P in mid-June. Shoot P content then remains fairly constant until shoot growth ceases in September (Figs. 3 and 4). During aboveground senescence in September and early October, a net (live) shoot loss of 70 mg $P \cdot m^{-2} \cdot day^{-1}$ occurs; however, the maximal shoot-to-belowground translocation (based on increase in belowground pool) can account for only a small fraction of this loss, ≈ 20 mg $P \cdot m^{-2} \cdot day^{-1}$. Less than 0.05 g P/m² is incorporated into seed heads, and thus the bulk of nontranslocated phosphorus, 50 mg $P \cdot m^{-2} \cdot day^{-1}$, must either be leached or lost in dead tissue. The small allotment of P in seed heads is in agreement with observations of McNaughton (1966) that *Typha* dependence on sexual reproduction lessens with increasing latitude.

Rhizomes and shoot bases are the important winter storage organs for P (Fig. 3). Roots are only a very minor P pool (≤ 0.13 g P/m²) because of their low biomass and need not be considered. (In Fig. 3, root P content is included with that of rhizomes in order to simplify and in order to maintain mass balance.) Initially in spring the overwintering 2.2 g P/m² in the plants is about equally divided between rhizomes and shoot bases; this reserve is rapidly depleted as shoot development accelerates in spring. By mid-June the major portion of the previous season's belowground biomass (all old shoot bases and most old rhizomes) is sloughed, representing a plant loss of ≈ 0.6 g P/m² based on the residual P content of these organs just prior to sloughing. As a result of this and to-shoot translocation, the belowground P pool reaches an annual minimum just as current-year rhizomes and shoot bases start to develop. Total P in the juvenile organs increases throughout the remainder of the summer; however, P accumulation does not occur as quickly as rhizome and shoot base growth, resulting in aforementioned decreasing percent tissue concentrations. A rhizome carbohydrate reserve is established well before a P reserve, with belowground tissues becoming ≈40% starch by time of shoot death in October (Gustafson, 1976). Slightly more P was stored belowground in fall 1974 than had been stored the prior winter; this is due to higher belowground biomass the second winter, rather than to higher percent P composition.

In spite of the death of aboveground parts, P accumulation in shoot bases (the region to which most of the roots are attached) appears to continue into December, as evidenced by increases in both P standing stock and tissue percentage in this organ (Figs. 2 and 3). The bulk of root zone soil in our *T. latifolia* stand does not usually freeze until January, and as late as mid-December soil temperatures below the frost line are 5°C. The standing stock evidence for late fall P uptake is supported by direct measurements on excised roots (Table 2). Phosphorus uptake rates by (mature) roots attached to current-year shoot bases were 201 μg P·g^{-1} dry wt·h^{-1} in June and 36.5 μg P·g^{-1} dry wt·h^{-1} in winter under identical pH and phosphate concentration but seasonal temperature regimes. Attributing the summer vs. winter 5.5× difference in uptake per gram of root to the temperature differential alone would require a Q_{10} of 3.1. This is a reasonable value, and in fact, is near the 2.9 value calculable for (low altitude ecotype) *T. latifolia* from similar but nonseasonal data in McNaughton et al. (1974). Thus there appears to be no drop-off in physiological activity of mature roots between summer and freeze-up. After the June sloughing of old belowground organs is completed, no further root loss is observed, and there is no reason why *T. latifolia* should not continue to supplement its shoot base P store until roots and shoot bases freeze.

Annual Phosphorus Budget

An annual phosphorus budget can easily be constructed from seasonal cumulative standing stocks (Fig. 3), albeit some aspects are more easily visualized by directly comparing shoot P accumulation to belowground deficit (based on reduction from winter-stored levels, Fig. 4).

This latter data transformation illustrates the degree to which belowground stores supply P to developing shoots. All shoot P in the 1st month of growth is derivable from belowground reserves, and the total reallocated from belowground, 1.3 g P/m^2, represents 40% of that required aboveground over the entire growing season. Residual shoot accumulation, 3.2 − 1.3 = 1.9 g P/m^2, therefore must come from root uptake over a very short period (mid-May through mid-June) at a rate of 65 mg P·m^{-2}·day^{-1} (95 mg P·m^{-2}·day^{-1} during 31 May–15 June). After mid-June, translocation to shoots appears to cease.

After this point, belowground P reserves slowly and gradually replenish.

Table 2. Phosphorus Uptake by Excised Summer and Winter Roots ($\bar{x} \pm$ SE [N])

Season	Plant part	Uptake ($\mu g \; P \cdot g^{-1} \cdot h^{-1}$)	Temp (°C)
June 1977			
	Roots attached to: current-year shoot base	201 ± 9.8 [59]	≈ 20
Winter 1975–1976			
	Roots attached to: current-year shoot base	36.5 ± 6.9 [7]	≈ 5
	budding shoot	3.4 ± 0.78 [26]	≈ 5

A portion of this recovery occurs prior to definite aboveground P loss and prior to the major carbohydrate translocation (Gustafson, 1976); this we have attributed to residual root activity. Regression against time of the net increase in belowground storage from 15 June through 1 September produces a daily uptake rate (slope) of 5.8 mg $P \cdot m^{-2} \cdot day^{-1}$ (SE = 1.2, df = 5) and a correlation coefficient (r) of .95**. The linearity of the apparent root uptake rate is consistent with: (1) little variation in summer 1974 soil temperature (Gustafson, 1976); (2) lack of change in physiological activity of mature roots through fall; (3) conservation of mature root biomass after mid-June sloughing; and (4) the very low phosphate uptake capacity of those new roots which develop (Table 2).

To calculate reallocation of shoot P to belowground during senescence (1 September–25 October) requires correction of belowground standing stock increase for uptake by the root system. This can be done under 3 sets of assumptions, leading to moderately varying results. A lower-limit estimate of translocation is obtained by subtracting the product of the pre-September root uptake rate × time period from the total belowground increase of 0.90 g P/m² (from Fig. 4), or 0.90 g P/m² − (0.0058 g $P \cdot m^{-2} \cdot day^{-1}$ × 54 days) = 0.59 g P/m². This would be a reallocation of only 18% of peak aboveground standing stock. This value is supported by the closeness of the assumed root uptake rate to the measured 0.27 g P/m² or 0.0053 g $P \cdot m^{-2} \cdot day^{-1}$ increase in belowground standing stock that occurs after shoots are dead, from 25 October to 15 December (Fig. 4). A more conservative estimate of root uptake would involve the 3.1 Q_{10} correction for the observed decreasing soil temperatures September through October (Gustafson, 1976). This would reduce estimated 1 September through 25 October root uptake by half to 0.15 g P/m². Downward translocation would then be 0.75 g P/m², or 23% of peak aboveground standing stock. If fall root uptake is ignored altogether, translocation would then be assumed to account for the entire 0.90 g P/m² belowground increase, or 28% of peak standing stock. Thus our 3 estimates range from 0.59 to 0.90 g P/m², from 18 to 28% of peak standing stock. We believe that the median of 0.75 g P/m² is the most reasonable

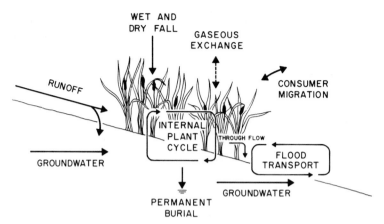

Fig. 5.   A conceptual input–output model for a lakeshore marsh.

compromise and we accept this value as the best estimate; however, neither of the 2 other choices would greatly affect the following conclusions.

To summarize, *T. latifolia* requires annual uptake of 3.1 g P/m², 2.5 g P/m² to counter net aboveground loss by shoots plus 0.6 g P/m² to balance that sloughed belowground. Root uptake in late fall may account for up to 9% of this total P requirement. Much more rapid root uptake in spring directly supplies 60% of the 3.2 g P/m² summer shoot needs. The residual 40% of shoot needs are reallocated from rhizome and shoot base stores. In the fall only a lesser amount, 23% of the peak shoot standing stock is returned to belowground storage.

The annual retention of P by *T. latifolia* is only 1.3 g P/m² (peak plant content plus seed head content minus aboveground and belowground losses) or 30% of the annual peak 4.3 g P/m². This 30% retention of P corresponds well to the 25% recovery of net production as stored carbohydrates by the same cattail stand (Gustafson, 1976). Apparently the ability of the plant to conserve P is only slightly greater than its ability to conserve carbon. The 2.5 g P/m² left annually by *T. latifolia* above or at the soil surface should make the macrophyte a major source of P in the lakeshore marsh ecosystem; this aspect will be discussed below.

NUTRIENT PATHWAYS IN LAKESHORE MARSHES

The importance of macrophytic nutrient pumps to lakeshore marshes is difficult to assess in the presence of only meager literature on other, non-macrophytic components of the nutrient cycle. Some appreciation, however, can be gained by comparison with other loadings from an input–output budget. Our perception of the likely important rates in such a budget includes wet and dry fall, gaseous exchanges, runoff and throughflow, groundwater seepage, flood or current transport, and consumer migration

(Fig. 5). Loss of nutrients by permanent burial below rooting depth is not a real transport, but rather a case of the ecosystem receding from the nutrients. The net upward translocation of nutrients by marsh plants impedes this process but may accelerate nutrient losses through surface pathways.

In the following text we limit discussion to N and P alone; these are the 2 nutrients of greatest importance in eutrophication of lakes and marshes, and are the only nutrients with an appreciable literature data base. We have also expressed annual P and N flows as percentages of respective peak aboveground standing stocks for study sites if possible, or otherwise as percentages of the "average" peak standing stocks (28. g N/m² and 3.6 g P/m²) calculated at the end of Table 1. The reasons for this transformation are twofold: (1) it normalizes rates for N and P into the same units, and (2) it allows easier comparison of different pathways with internal cycling by marsh vegetation.

The first inputs to our budget to consider in this context are wet and dry fall. In the city of Madison, Wisconsin, (the location of our *T. latifolia* study), rain and snow contribute 0.8 g $N \cdot m^{-2} \cdot yr^{-1}$ and 0.02 g $P \cdot m^{-2} \cdot yr^{-1}$ and dry fall somewhat greater amounts, 1.6 g $N \cdot m^{-2} \cdot yr^{-1}$ and 0.08 g $P \cdot m^{-2} \cdot yr^{-1}$ (Kluesener and Lee, 1974). The magnitude of this addition to our *T. latifolia* stand is equivalent to 8% per year of N peak standing stock and 3% per year of P. In similar fashion, Neyst Fishpond, Czechoslovakia, receives 0.8 g N and 0.06 g $P \cdot m^{-2} \cdot yr^{-1}$ or 3% per year of N and 2% per year of P "average" peak standing stocks as rainfall (Úlehlová et al., 1973). Tropical rainfall in Kawaga Swamp on Lake Victoria contributes 4% of N and "negligible" P per year on a [live] papyrus standing stock basis (Gaudet, 1976). In arctic Alaskan tundra marshes, precipitation contributions are much less than found for temperate or tropical marshes at only 0.001 g P/m² in annual wet fall plus winter dry fall (Prentki, 1976). However, in terms of arctic (*C. aquatilis*) standing stock equivalence, this is 1.5% per year, a percentage identical to that of warmer climates.

Taken together, these precipitation data indicate that wet fall can be expected to meet 3% of annual marsh plant aboveground N requirements, and dry fall, 5% more. Phosphorus input is consistently low, with particularly little range, 0.4 to 2% per year in wet fall when expressed as percentage of peak standing stocks, even though the actual amounts precipitated decrease 1,000× going from a Czechoslovakian fishpond to Alaskan tundra. Thus, neither N nor P appear to be supplied to lakeshore marshes to any appreciable extent by precipitation.

Gaseous exchanges are a second aerial pathway of potential importance for N. Nitrogen can be gained by the marsh through absorption of NH_3 by water or plants (Hoeft et al., 1972) and through N_2-fixation or else lost through denitrification.

These 3 processes are only poorly understood in freshwater marshes. The magnitude of NH_3 absorption is unevaluated, but the other 2 processes have been studied in at least 1 marsh, Lake Wingra Marsh, Madison, Wisconsin.

Here, Isirimah and Kenney (1973) searched for but could not detect any N_2-fixation, but did detect very slow rates of denitrification. In experiments on marsh soil at initial NO_3 concentrations of 90–120 μg N/g moist soil, only 3 μg N·g^{-1}·day^{-1} was denitrified at 25°C, and 0.7 μg N·g^{-1}·day^{-1} at 10°C. Extrapolating the 10°C rate to 365 days and to the top 25 cm of marsh soil, gives an order of magnitude denitrification estimate of 0.2 g N·m^{-2}·yr^{-1}, a loss balancing <0.1 of the precipitation input and equivalent to only 0.6% per year of "average" peak standing stock. Thus neither denitrification nor nitrogen-fixation appear to be major pathways in Lake Wingra Marsh. It should be noted, however, that this conclusion appears to be in disagreement with the importance claimed for these 2 processes in other Wisconsin, nonlakeshore marshes by Lee et al. (1975).

Runoff or other nonchannelized surface flows entering lakeshore marshes can provide much higher amounts of nutrients than either precipitation or gas-phase capture, depending upon watershed and runoff-source characteristics. Gaudet (1976) provides 1 set of estimates for runoff loading of Kawaga Swamp. In this African tropical swamp, runoff supplies the equivalent of 0% per year of the N, but 350% per year of the P contained in papyrus biomass. The swamp loses N to throughflowing waters, but retains 97% of the incoming P load. The amounts lost to the ecosystem by throughflow are equivalent to 9% per year of papyrus standing stocks for both N and P.

A somewhat different picture of the importance and of runoff loading and throughflow in lakeshore marshes can be derived for Lake Wingra Marsh. A storm drain entering the head of the marsh delivers annual loads of 8 m^3 of water, 43 g N, and 5.6 g P per square metre of marsh (Prentki et al., 1977); or 150% per year of N and 160% per year of P "average" peak standing stocks. Because annual retention of runoff P in the marsh is only $\approx 10\%$ (Loucks et al., 1977), almost all P which enters from the storm drain leaves during the annual cycle, and the loss to the ecosystem is a 140% per year equivalency to "average" peak marsh plant standing stock. Unfortunately, there is no estimate for marsh retention of N.

The few comparisons possible from these 2 runoff examples are limited and are not sufficient in this case to demonstrate any trends. We can conclude only that runoff loading of N and P and throughflow of P can be high, exceeding the 100% per year equivalency to peak standing stocks.

One further hydrologic pathway requiring evaluation is groundwater movement. Unfortunately the diffuse boundaries of lakeshore marshes make estimation of this parameter difficult and groundwater flow in lakeshore marshes is therefore seldom measured. We have been able to collate sufficient data to construct groundwater nutrient loadings and exports for only 1 system, the Lake Wingra Marsh.

In this marsh, deep recharge is 0.2 m^3·m^{-2}·yr^{-1} (Loucks et al., 1977). Assuming recharge concentrations similar to those in nearby wells and springs (Oakes et al., 1975; Prentki et al., 1977), groundwater loadings are

0.80 g N and 0.0086 g $P \cdot m^{-2} \cdot yr^{-1}$ or 2.9% per year of N and 0.2% per year of P "average" peak standing stocks. Shallow groundwater discharge out of the lower end of the marsh is 3.4-fold recharge, or 0.7 $m^3 \cdot m^{-2} \cdot yr^{-1}$ (Loucks et al., 1977). Interstitial dissolved N in the lake end of the marsh averages of 0.8 g N/m^3 (Isirimah and Kenney, 1973). This concentration multiplied by the discharge rate yields a groundwater export of 0.6 $g \cdot m^{-2} \cdot yr^{-1}$ or 2.1% per year of an "average" peak standing stock. We have measured interstitial total dissolved P concentrations of 64 to 250 mg P/m^3 in the uppermost 20 cm of soil at the same end of the marsh, with the lowest concentration occurring below the water table, i.e., within the zone of groundwater movement. Assuming a P concentration of 64 mg P/m^3 in groundwater flow from the marsh, export is 0.045 g $P \cdot m^{-2} \cdot yr^{-1}$, or 1.2% per year of a standing stock. When groundwater input and export of N and P are compared, groundwater flow appears to result in a net accumulation of N of 0.2 g $N \cdot m^{-2} \cdot yr^{-1}$ but net loss of P of 0.036 g $P \cdot m^{-2} \cdot yr^{-1}$. Rates in terms of standing stock equivalence indicate that for this marsh, groundwater transport is much less than internal nutrient transport by plants or runoff loads.

Flooding is the remaining hydrologic mechanism in our conceptual lakeshore marsh (Fig. 5), and moves nutrients both into and out of the marsh. When the eulittoral is water-covered, 2 subprocesses occur: deposition of wind-driven and waterborne limnological debris (Klötzli and Züst, 1973; Pieczynská, 1975) and export of dissolved nutrients (Planter, 1970b, 1973; Schröder, 1973; Banoub, 1975; Lie, 1975; Pieczynská et al., 1975). Unfortunately we know of no estimate for the former and investigations of the latter seldom have the juxtaposition of nutrient concentration and water flow data necessary to quantify actual transport.

In 1 study for which transport can be estimated, brief spring flooding in Alaskan tundra ponds leached P from *Carex aquatilis* in pond margins and exported a portion, ≈4.6 mg P/m^2 [6% per year equivalence to standing stock] (Alexander et al., *In press a*). This loss represents only 15% of the total P leached during this period, and thus flood transport was only a similar 15% effective (Prentki, 1976).

In a second study, Mason and Bryant (1975) measured N and P standing stocks of *Phragmites communis* and *Typha angustifolia* before and during winter-long flooding of an English marsh and calculated the corresponding plant stand losses, assuming negligible belowground translocation. In fall, 1972, 7.9 g N and 0.63 g P per square metre were lost, and in 1973, 18.3 g N and 1.3 g P per square metre. The authors allowing for increases in lake volume and in total lakewater N and P concentrations, could account for only 5–13% of N and 0–5% of P loss from the plants. In terms of peak standing stocks, the losses to the lake averaged 3% per year for N and 0.8% per year for P.

Based on these 2 examples, flood exports of N and P are both ≈3% per year of respective standing stocks. Exports in both were equivalent to only a very small increment of that brought to the marsh surface and then lost by the vegetation.

Consumer migration is the one pathway in our input–output model which involves higher trophic levels. Excluding human activities, the greatest effect of consumers is in grazing and manuring. Grazing may be very intense on a local basis and can completely denude and destroy small stands (Dykyjová, 1973b). The net amount of nutrients physically transported over marsh boundaries by consumers, however, is much less than that cycling through the animals. Thus while Dobrowolski (1973) estimates that herbivorous birds annually graze ≈2% of macrophyte production in several Polish lakes, transport of grazed nutrients out of the lakes by birds is likely much less than the 2% per year equivalence of nutrient standing stocks. As for manuring effects, nesting gulls were found to add 0.075 mg $P \cdot m^{-2} \cdot yr^{-1}$ to a cattail pool in Agassiz National Wildlife Refuge in Minnesota (McColl and Burger, 1976). This is ≈43% of the external P load to that pool, but again equivalent to only 2% per year of an "average" peak standing stock of marsh vegetation. Thus neither of the latter 2 studies suggests consumer impact is anywhere near that of primary producers.

Generalizations about transport along aerial, hydrologic, and consumer pathways (based on the very limited number of systems that we have followed in this discussion) are admittedly speculative, but do allow us to place the internal plant nutrient cycle and permanent burial into the perspective of our input–output model (Fig. 5). For the plant, upward translocation through the soil interface can be taken as equivalent to ≥100% per year of standing stock with exception of plants with appreciable nutrient storage in overwintering shoots (e.g., *Carex lacustris*; Bernard and Solsky, 1977). Estimates of annual aboveground plant losses of N and P range from 87% per year to 50% per year of peak shoot stocks (Klopatek, 1975; Chapin et al., 1975; this study). In all of the above examples only N and P in runoff ever approximate even the lower limit of this plant loading to the marsh. A parallel situation exists in relation to permanent burial. The maximum rate of permanent burial of nutrients in the marsh can only approach the sum of external inputs to the marsh. If runoff inputs are low, loading by the plants will be severalfold gross input to the marsh and thus severalfold burial. In such cases it is doubtful whether burial could be an efficient process.

This point has important consequences for concepts of marsh nutrient conservation. If conservation is defined such that permanent burial is a loss from the ecosystem, then the net upward translocation by marsh vegetation can be hypothesized to increase conservativeness of the system. That is, the plants increase system conservation by being leaky themselves; the less efficiently plants reutilize their own nutrients, the more they must deplete root zone soils, and the fewer nutrients are left susceptible to permanent burial. However, a more common construct for marsh conservation involves the role of marshes in nutrient interception (Tóth, 1972; Sloey et al., 1978).

Unless continued harvesting of plants or other nutrient pools is envisioned, burial or denitrification (for N) must be the ultimate sinks for intercepted nutrients. If permanent burial is an inefficient process in lakeshore marshes due to the magnitude of upward nutrient translocation by marsh vegetation, then this suggests that lakeshore marshes may also be ineffective nutrient traps.

ACKNOWLEDGMENTS

The authors are indebted to C. S. Smith and C. T. Carnes, for their technical assistance. Part of this work was supported by NSF-Ecosystem Analysis, Grant No. BMS-75-19777.

LITERATURE CITED

Ahern, J. (1976). "Impact and Management of Urban Stormwater Runoff." M. S. Thesis. University of Wisconsin-Madison, Wisconsin.

Alexander, V. A., Barsdate, R. J., Hobbie, J. E., Miller, M. C., and Prentki, R. T. (In press a). Chemistry of ponds. In "Limnology of Arctic Ponds" (J. E. Hobbie, ed.). Dowden, Hutchinson, and Ross, Stroudsburg, Pennsylvania.

Alexander, V. A., Bierle, D. A., Daley, R. J., Fenchel, T. M., Hobbie, J. E., Miller, M. C., Reed, J. P., Rublee, P. A., Stanley, D. W., Stross, R. G., and Traaen, T. (In press b). Biology of ponds. In "Limnology of Arctic Ponds" (J. E. Hobbie, ed.). Dowden, Hutchinson, and Ross, Stroudsburg, Pennsylvania.

Banoub, M. W. (1975). The effect of reeds on the water chemistry of Gnadensee (Bodensee). Arch. Hydrobiol. 75, 500–521.

Barsdate, R. J., and Prentki, R. T. (1972). Nutrient dynamics in tundra ponds. In "Proceedings 1972 Tundra Biome Symposium" (S. Bowen, ed.), pp. 192–199. Lake Wilderness Center, University of Washington, Washington. U.S. Tundra Biome, Fairbanks, Alaska.

Barsdate, R. J., and Prentki, R. T. (1973). Nutrient metabolism and water chemistry in lakes and ponds of the Arctic coastal tundra. U.S. Tundra Biome Data Report 73–27.

Bernard, J. M., and Gorham, E. (1978). Life history aspects of primary production in sedge wetlands. In "Freshwater Wetlands: Ecological Processes and Management Potential" R. E. Good, D. F. Whigham, and R. L. Simpson, eds.), pp. 39–51. Academic Press, New York.

Bernard, J. M., and Solsky, B. A. (1977). Nutrient Cycling in a Carex lacustris wetland. Can. J. Bot. 55, 630–638.

Bernatowicz, S. (1969). Macrophytes in the Lake Warniak and their chemical composition. Ekologia Polska (Seria A) 17, 447–467.

Björk, S. (1967). Ecologic investigations of Phragmites communis. Studies in theoretic and applied limnology. Folia Limnol. Scand. 14, 1–248.

Boyd, C. E. (1969). Production, mineral nutrient absorption, and biochemical assimilation by Justicia americana and Alternanthera philoxeroides. Arch. Hydrobiol. 66, 139–160.

Boyd, C. E. (1970a). Production, mineral accumulation and pigment concentrations in Typha latifolia and Scirpus americanus. Ecology 51, 285–290.

Boyd, C. E. (1970b). Losses of mineral nutrients during decomposition of Typha latifolia. Arch. Hydrobiol. 66, 511–517.

Boyd, C. E. (1971). Further studies on productivity, nutrient and pigment relationships in Typha latifolia populations. Bull. Torrey Bot. Club 98, 144–150.

Boyd, C. E., and Hess, L. W. (1970). Factors influencing shoot production and mineral nutrient levels in Typha latifolia. Ecology 51, 296–300.

Boyd, C. E., and Vickers, D. H. (1971). Relationships between production, nutrient accumulation, and chlorophyll synthesis in an *Eleocharis quadrangulata* population. *Can. J. Bot.* **49**, 883–888.

Brinson, M. M. (1977). Decomposition and nutrient exchange of litter in an alluvial swamp forest. *Ecology* **58**, 601–609.

Chamie, J. P. M., and Richardson, C. J. (1978). Decomposition in northern wetlands. *In* "Freshwater Wetlands: Ecological Processes and Management Potential" (R. E. Good, D. F. Whigham and R. L. Simpson, eds.), pp. 115–130. Academic Press, New York.

Chapin, F. S. III. (1974). Morphological and physiological mechanisms of temperature compensation in phosphate absorption along a latitudinal gradient. *Ecology* **55**, 1180–1198.

Chapin, F. S. III, and Bloom, A. (1976). Phosphate absorption: adaptation of tundra graminoids to a low temperature, low phosphorus environment. *Oikos* **26**, 111–121.

Chapin, F. S. III, Van Cleve, K., and Tieszen, L. L. (1975). Seasonal nutrient dynamics of tundra vegetation at Barrow, Alaska. *Arctic and Alpine Res.* **7**, 209–226.

Davis, C. B., and van der Valk, A. G. (1978). Litter decomposition in prairie glacial marshes. *In* "Freshwater Wetlands: Ecological Processes and Management Potential" (R. E. Good, D. F. Whigham, and R. L. Simpson, eds.), pp. 99–113. Academic Press, New York.

Dobrowolski, K. A. (1973). Role of birds in Polish wetland ecosystems. *Polskie Archiwum Hydrobiologii* **20**, 217–221.

Dykyjová, D. (1973*a*). Accumulation of mineral nutrients in the biomass of reedswamp species. *In* "Ecosystem study on Wetland Biome in Czechoslovakia" (S. Hejný, ed.), pp. 151–161. Czechoslovakian IBP/PT-PP Rep. No. 3, Trebon, Czechoslovakia.

Dykyjová, D. (1973*b*). Content of mineral macronutrients in emergent macrophytes during their seasonal growth and decomposition. *In* "Ecosystem Study on Wetland Biome in Czechoslovakia" (S. Hejný, ed.), pp. 163–172. Czechoslovakian IBP/PT-PP Rep. No. 3, Trebon, Czechoslovakia.

Dykyjová, D., and Hradecká, D. (1976). Production ecology of *Phragmites communis*. 1. Relations of two ecotypes to the microclimate and nutrient conditions of habitat. *Folia Geobot. Phytotaxon.* **11**, 23–61.

Enwezor, W. O. (1976). The mineralization of nitrogen and phosphorus in organic materials of varying C:N and C:P ratios. *Plant Soil* **44**, 237–240.

Epstein, E., Schmid, W. E., and Rains, D. W. (1963). Significance and technique of short-term experiments on solute absorption by plant tissue. *Plant Cell Physiol.* **4**, 79–84.

Fiala, K. (1973). Growth and production of underground organs of *Typha angustifolia* L., *Typha latifolia* L. and *Phragmites communis* Trin. *Polskie Archiwum Hydrobiologii* **20**, 59–66.

Gaudet, J. J. (1976). Nutrient relationships in the detritus of a tropical swamp. *Arch. Hydrobiol.* **78**, 213–239.

Gerloff, G. C. (1969). Evaluating nutrient supplies for the growth of aquatic plants in natural waters. *In* "Eutrophication: Causes, Consequences, Correctives" pp. 537–555. National Academy of Sciences, Washington, D. C.

Gerloff, G. C. (1975). "Nutritional Ecology of Nuisance Aquatic Plants." National Environmental Research Center, Corvallis, Oregon.

Godshalk, G. L., and Wetzel, R. G. (1978). Decomposition in the littoral zone of lakes. *In* "Freshwater Wetlands: Ecological Processes and Management Potential" (R. E. Good, D. F. Whigham, and R. L. Simpson, eds.), pp. 131–143. Academic Press, New York.

Gustafson, T. D. (1976). "Production, Photosynthesis and Storage and Utilization of Reserves in a Natural Stand of *Typha latifolia* L." Ph.D. Thesis. University of Wisconsin-Madison, Wisconsin.

Hobbie, J. E. (1972). Carbon flux through a tundra pond ecosystem at Barrow, Alaska. *In* "Proceedings 1972 Tundra Biome Symposium" (S. Bowen, ed.), pp. 206–208. Lake Wilderness Center, University of Washington, Washington. U. S. Tundra Biome, Fairbanks, Alaska.

Hoeft, R. G., Kenney, D. R., and Walsh, L. M. (1972). Nitrogen and sulfur in precipitation and sulfur dioxide in the atmosphere in Wisconsin. *J. Environ. Qual.* 1, 203–208.

Hutchinson, G. E. (1975). "A Treatise on Limnology. Volume III. Limnological Botany." Wiley-Interscience, New York.

Isirimah, N. O., and Kenney, D. R. (1973). "Contribution of Developed and Natural Marshland Soils to Surface and Subsurface Water Quality," (Water Resources Center Technical Completion Report to U. S. Department of the Interior). University of Wisconsin-Madison, Wisconsin.

John, M. K. (1970). Colorimetric determination of phosphorus in soil and plant materials with ascorbic acid. *Soil Sci.* 109, 214–220.

Keefe, C. W. (1972). Marsh production: A summary of the literature. *Contrib. Mar. Sci.* 16, 163–181.

Klopatek, J. M. (1975). The role of emergent macrophytes in mineral cycling in a freshwater marsh. *In* "Mineral Cycling in Southeastern Ecosystems" (F. G. Howell, J. B. Gentry, and M. H. Smith, eds.), pp. 367–393. ERDA Symposium Series.

Klötzli, F., and Züst, S. (1973). Conservation of reed-beds in Switzerland. *Polskie Archiwum Hydrobiologii* 20, 229–235.

Kluesener, J. W., and Lee, G. F. (1974). Nutrient loading from a separate storm sewer in Madison, Wisconsin. *J. Water Pollution Control Federation* 46, 920–936.

Kvĕt, J. (1973). Mineral nutrients in shoots of reed (*Phragmites communis* Trin.). *Polskie Archiwum Hydrobiologii* 20, 137–147.

Kvĕt, J. (1975). Growth and mineral nutrients in shoots of *Typha latifolia* L. *Symposia Biologica Hungarica* 15, 113–123.

Lee, G. F., Bentley, E., and Amundson, R. (1975). Effects of marshes on water quality. *In* "Coupling of Land and Water Systems" (A. D. Hasler, ed.), pp. 105–127. Springer-Verlag, New York.

Lie, G. B. (1975). "Phosphorus Release from Macrophytes and Macrophyte Ecology in Shagawa Lake—1974." National Environmental Research Center, Corvallis, Oregon.

Loucks, O. L., Prentki, R. T., Watson, V. J., Reynolds, B. J., Weiler, P. R., Bartell, S. M., and D'Alessio, A. B. (1977). "Studies of the Lake Wingra Watershed: An Interim Report," IES Report 78. Center for Biotic Systems, University of Wisconsin-Madison, Wisconsin.

Mason, C. F., and Bryant, R. J. (1975). Production, nutrient content and decomposition of *Phragmites communis* Trin. and *Typha angustifolia* L. *J. Ecol.* 63, 71–95.

McColl, J. G., and Burger, J. (1976). Chemical inputs by a colony of Franklin's Gulls nesting in cattails. *Am. Midl. Nat.* 96, 270–280.

McNaughton, S. J. (1966). Ecotype function in the *Typha* community-type. *Ecol. Monogr.* 36, 297–325.

McNaughton, S. J., Campbell, R. S., Freyer, R. A., Mylroie, and Rodland, K. D. (1974). Photosynthetic properties and root chilling responses of latitudinal ecotypes of *Typha latifolia* L. *Ecology* 55, 168–172.

McRoy, C. P., and Alexander, V. (1975). Nitrogen kinetics in aquatic plants in arctic Alaska. *Aquat. Bot.* 1, 3–10.

Miller, M. C., and Prentki, R. T. (*In press*). Physics of the ponds. *In* "Limnology of Arctic Ponds" (J. E. Hobbie, ed.). Dowden, Hutchinson, and Ross, Stroudsburg, Pennsylvania.

Oakes, E. L., Hendrickson, G. E., and Zuehls, E. E. (1975). "Hydrology of the Lake Wingra Basin, Dane County, Wisconsin." U.S. Geological Survey Water-Resources Investigations 17-75.

Odum, W. E., and Heywood, M. A. (1978). The decomposition of intertidal freshwater marsh plants. *In* "Freshwater Wetlands: Ecological Processes and Management Potential" (R. E. Good, D. F. Whigham, and R. L. Simpson, eds.), pp. 89–97. Academic Press, New York.

Ondok, J. P., and Dykyjová, D. (1973). Assessment of shoot biomass of dominant reed-beds in Trebon Basin, methodical aspects. *In* "Ecosystem Study on Wetland Biome in Czechoslovakia" (S. Hejný, ed.), pp. 79–82. Czechoslovakian IBP/PT-PP Rep. No. 3, Trebon, Czechoslovakia.

Pieczyńska, E. (1975). Ecological interactions between land and the littoral zones of lakes (Masurian Lakeland, Poland). In "Coupling of Land and Water Systems" (A. D. Hasler, ed.), pp. 263–276. Springer-Verlag, New York.

Pieczyńska, E., Sikorska, U., and Ozimek, T. (1975). The influence of domestic sewage on the littoral zone of lakes. Polskie Archiwum Hydrobiologii 22, 141–156.

Planter, M. (1970a). Elution of mineral components out of dead reed Phragmites communis Trin. Polskie Archiwum Hydrobiologii 17, 357–362.

Planter, M. (1970b). Physico-chemical properties of the water of reedbelts in Mikolajskie, Taltowisko, and Sniardwy Lakes. Polskie Archiwum Hydrobiologii 17, 337–356.

Planter, M. (1973). Physical and chemical conditions in the helophytes zone of the lake littoral. Polskie Archiwum Hydrobiologii 20, 1–7.

Prentki, R. T. (1976). "Phosphorus Cycling in Tundra Ponds." Ph.D. Thesis. University of Alaska, Fairbanks, Alaska.

Prentki, R. T., Rogers, D., Watson, V. J., Weiler, P. R., and Loucks, O. L. (1977). "Summary Tables of Lake Wingra Basin Data," IES Report 85. Center for Biotic Systems, University of Wisconsin-Madison, Wisconsin.

Riemer, D. N., and Toth, S. J. (1968). A survey of the chemical composition of aquatic plants in New Jersey. New Jersey Agr. Exp. Sta. Bull. 820.

Schröder, R. (1973). Die Freisetzung von Pflanzennährstoffen im Schilfgebiet und ihr Transport in das Freiwasser am Beispiel des Bodensee-Untersees. Arch. Hydrobiol. 71, 145–158.

Shaver, G. R., and Billings, W. D. (1975). Root production and root turnover in a wet tundra ecosystem, Barrow, Alaska. Ecology 56, 401–409.

Sommers, L. E., and Nelson, D. W. (1972). Determination of total phosphorus in soils: a rapid perchloric acid digestion procedure. Soil Sci. Soc. Am. Proc. 35, 902–904.

Sloey, W., Spangler, F., and Fetter, C. (1978). Management of freshwater marshes for nutrient assimilation. In "Freshwater Wetlands: Ecological Processes and Management Potential" (R. E. Good, D. F. Whigham, and R. L. Simpson, eds.), pp. 321–340. Academic Press, New York.

Szczepańska, W., and Szczepański, A. (1973). Emergent macrophytes and their role in wetland ecosystems. Polskie Archiwum Hydrobiologii 20, 41–50.

Tóth, L. (1972). Reeds control eutrophication of Balaton Lake. Water Res. 6, 1533–1539.

Tukey, H. B., Jr. (1970). The leaching of substances from plants. Ann. Rev. Plant. Physiol. 21, 305–324.

Úlehlová, B., Husák, S., Dvořak, J. (1973). Mineral cycles in reed stands of Nesyt Fishpond in southern Moravia. Polskie Archiwum Hydrobiologii 20, 121–129.

Westlake, D. F. (1968). Methods used to determine the annual production of reed swamp plants with extensive rhizomes. In "Methods of Productivity Studies in Root System and Rhizosphere Organisms," pp. 226–234. NAUKA, Leningrad.

Whigham, D. F., McCormick, J., Good, R. E., and Simpson, R. L. (1978). Biomass and primary production in freshwater tidal wetlands of the Middle Atlantic Coast. In "Freshwater Wetlands: Ecological Processes and Management Potential" (R. E. Good, D. F. Whigham, and R. L. Simpson, eds.), pp. 3–20. Academic Press, New York.

Wong, S. L., and Clark, B. (1976). Field determination of the critical nutrient concentrations for Cladophora in streams. J. Fish. Res. Board Can. 33, 85–92.

NUTRIENT DYNAMICS OF FRESHWATER RIVERINE MARSHES AND THE ROLE OF EMERGENT MACROPHYTES

Jeffrey M. Klopatek

Environmental Sciences Division[1]
Oak Ridge National Laboratory
Oak Ridge, Tennessee 37820

Abstract The open characteristic of freshwater riverine marshes results in a continual subsidy and withdrawal of nutrients, with the specific patterns in a given wetland being dependent on seasonal hydrological fluctuations and biological activity. A key feature of the marsh separating it from terrestrial ecosystems is its inundated and anaerobic soils which show statistically significant ($P < .01$) seasonal variations in available P and K and exchangeable Ca and Mg. Based primarily on investigations in Wisconsin, the role of emergent macrophytes within the marsh nutrient cycle is described. Concentrations of N, P, K, Ca, and Mg are shown to follow predictable trends over the macrophyte growing season. Regression analyses revealed nutrient uptake by the macrophytes was significantly correlated ($r^2 = .98, P < .01$) with total soil N and available P. For the most part, significant correlations were not present for the other elements. This is explained by their mobility and the possibility of luxury uptake. Models of the flow of nutrients in a *Scirpus fluviatilis* stand are depicted. The effects of a marsh drawdown on the nutrient cycle and nutrient dynamics of *Salix interior*, a typical riverine marsh shrub, are also discussed.

Key words Carex, *emergent macrophytes, freshwater marsh, marsh soils, nitrogen, nutrients, phosphorus, potassium*, Salix, Scirpus, Typha, *Wisconsin, wetlands*.

INTRODUCTION

Most freshwater marshes and other wetlands which possess overland hydrologic inputs and outputs are open systems, receiving a constant subsidy

[1] Operated by Union Carbide Corporation for the Energy Research and Development Administration.

of water and nutrients from external sources (Klopatek, 1975). These systems exhibit high biological productivity placing them among the most productive ecosystems of the world (Westlake, 1963; Lieth, 1975). In temperate regions, where a defined growing season exists, these ecosystems capture, store, and release energy in a pulse-like fashion resembling a sine wave. Although biologically productive only part of the year, riverine marshes experience a continual input and output of water, nutrients, and organic matter year-round. These flows have brought marshes to their highly eutrophic state but inhibit their progression to a terrestrial condition.

In addition to being sites of organic matter production, riverine marshes serve in a regulatory capacity. Existing at the land–water interface they are important hydrologic recharge areas and, at times of high rainfall and subsequent floodwater, tend to diminish the stress on downstream waterways by modifying and easing the rate of impending flow (Niering, 1968). Marshes may also act as valves or sinks to regulate or trap the flow of nutrients from surrounding terrestrial systems (Kitchens et al., 1975; Lee et al., 1975). In fact, it has been shown that freshwater wetlands can act as wastewater treatment sites (Kiefer, 1968; Spangler et al., 1976; Mitsch, 1976).

Blum's (1972) botanical definition of a river provides a conceptual view of the aquatic environment of a riverine marsh. The river water flowing through the marsh is a nutrient medium with unidirectional motion. It possesses complex and diverse chemical and physical parameters which are likely to be fluctuating at different frequencies and in different manners without compromising a high level of productivity. Because marsh currents are slow to almost undiscernable, the limiting factor of water velocity plays little or no role except in the main channel. As a result, a much greater diversity and number of species can survive (Granert, 1973; Whitton, 1975) and rooted emergent macrophytes can exist throughout most of the marsh.

The mineral cycles within a freshwater riverine marsh have not been adequately quantified although, in recent years, a number of studies have examined the flux of elements in marshes and characterized various compartments of the marsh system (Klopatek, 1975; Lindsley et al., 1976). This discussion stresses the major concepts concerning nutrient dynamics in riverine freshwater marshes with special emphasis on the role of emergent macrophytes. Many of the data are from Theresa Marsh, an extremely fertile and highly productive marsh in southeastern Wisconsin (Klopatek, 1975; Klopatek and Stearns, 1977).

AQUATIC NUTRIENTS

One must quantify the hydrologic cycle to quantify the nutrient flux through a freshwater riverine marsh. Unfortunately the complexities involved in the hydrological flows within and throughout Theresa Marsh precluded an accurate quantitative balance of nutrient cycles. Quantifying the mineral cycles within or through the marsh would require monitoring stream

Table 1. Range of Various Elements in the Inflow and Outflow Waters of Theresa Marsh, Wisconsin, 1971 and 1972

Nutrients and physico-chemical parameters	Marsh outflow (mg/l)	Inflows (mg/l)		
		Lomira Creek	Kohlsville Creek	Rock River
NO$_2$	<0.01–0.04	<0.01–0.04	<0.01–0.04	<0.01–0.06
NO$_3$	0.10–1.68	0.07–2.03	0.17–2.16	<0.03–1.51
NH$_4$	0.13–1.59	0.10–2.61	0.07–1.69	0.03–5.81
Organic N	0.65–4.80	0.54–1.74	0.33–1.51	0.77–2.23
Total N	1.52–6.77	1.13–5.19	1.72–4.06	1.38–8.32
Dissolved P	0.10–0.50	0.17–2.40	0.02–0.30	0.04–0.69
Total P	0.11–0.69	0.17–2.42	0.05–0.31	0.05–0.78
K	0.9–9.1	3.6–19.6	1.4–4.5	1.4–4.8
Ca	56–168	60–130	55–112	60–137
Mg	23–73	32–71	31–72	31–69
Na	5.5–24.0	9.5–51.4	4.6–19.0	6.0–20.2
Chloride	17–54	27–118	11–34	15–34
Sulfate	15–99	15–81	23–71	19–90
Conductivity (μmho)	519–942	662–1165	549–705	578–731
pH	7.4–8.2	7.5–8.5	7.6–8.4	7.5–8.5
Alkalinity (mg CaCO$_3$/l)	268–420	291–386	288–346	258–402

inflows and outflow measurement of the rates of evapotranspiration, gas exchange, surface runoff, and subsurface flow. Despite the complex hydrology of the marsh it is possible to obtain information about processes occurring within it from comparisons of the nutrient concentrations in inflow and outflow waters. Elemental ranges (Table 1) indicate that the water flowing into and out of the marsh can be considered nutrient-rich.

Nitrogen is perhaps the most complex element in the system to characterize. The processes of mineralization, immobilization, nitrification, and denitrification all occur within the marsh. Figure 1 displays the pattern of the N chemical species in the outflow water of Theresa Marsh over an 18-month period. The high concentrations during summer 1971 resulted from a marsh drawdown for a fish removal program. It is important to note the high concentrations in June and July when the marsh was draining, and the decline, rise, and second decline in late August and early September when the marsh was reflooded, drained, and again reflooded. In the spring N is flushed out of the marsh system during periods of high flow and ice melt. Because of concentration differences between output and input, N appears to be assimilated during most of the growing season (early April for free-floating and submerged macrophytes [McNelly and Klopatek, 1973] and even earlier for algae) and released in the fall. This phenomenon has been reported by others examining marsh outflow waters (Bentley, 1969; Amundson, 1970; Nicholls and MacCrimmon, 1974). Based on inflow–outflow concentrations during summer months, Theresa Marsh appears to

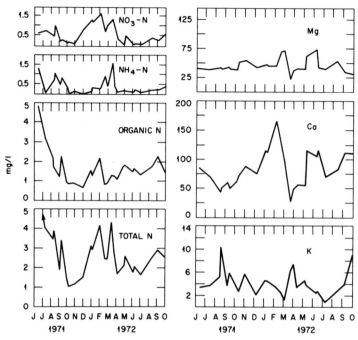

Fig. 1. Concentrations of NO_3, NH_3, organic N, total N, K, Ca, and Mg in the outflow water of Theresa Marsh during 1971 and 1972.

retain N in ways similar to other Wisconsin marshes (Lee et al., 1975) even to the extent of maintaining higher inorganic N concentrations within the marsh confines than in the outflow waters. Brezonik (1968) has shown that NH^+_4 may be the prime inorganic N source for freshwater plants that obtain their nutrients directly from the water and that it may be assimilated very rapidly. Due to the reduced amounts of N in the outflow water when compared to the inflow concentrations, it appears that the marsh may also be a major source of dentrification during this period, a factor extremely important in pollution control (Lee et al., 1975; Keeney, 1973; Patrick and Reddy, 1976). Levels of inorganic N in the inflow waters and in the marsh never appear below limiting levels and organic N is the primary form of N transported out of the marsh, particularly from early spring to late fall.

The level of P in the water of Theresa Marsh was never low enough to become limiting (Table 1). In fact, P concentrations were often many times those stated for eutrophic waters. Phosphorus levels of water flowing out of the marsh were reduced noticeably during the growing season.

Bentley (1969) and Klopatek (1975) have shown that P release from a marsh exhibits a cyclic pattern in spring and fall, with much of the spring P release coming from high P concentrations (1.14 mg/l total-P) locked up in the winter ice covering the marsh (Bozniak and Kennedy, 1968; Klopatek, 1974). In the summer months the marsh is a P sponge, absorbing much of the P flowing through it but never acting as a complete barrier to P. Much of the P retention is due to the primary producers who sorb P from the water. Comparison of the inflow and outflow concentrations of P to the flow of Cl (a conservative element biologically) showed a >50% reduction of total and inorganic P flowing through Theresa Marsh. Suspended matter tends to settle in the slower flowing reaches of the marsh carrying with it absorbed and organically bound P. The sediments, especially the highly organic marsh soils, are known to act as sinks particularly for overflowing waters high in inorganic P (Kramer et al., 1972; Ponnamperuma, 1972; Patrick and Khalid, 1974; Syers et al., 1973).

The seasonal patterns of Ca, Mg, and K in the outflow water at Theresa Marsh (Fig. 1) are similar to those reported by Lee et al. (1975). As discussed by Golterman (1975) the concentration of dissolved compounds in rivers follows 2 broad patterns: (1) the concentration of dissolved salts varies inversely with the river flow rate and results from a more or less constant input, and (2) the concentration of dissolved salts remains constant because the water reaches equilibrium with soils through which it percolates or because the concentration approaches the saturation value. Figure 1 shows that when the river flows through the marsh the patterns described by Golterman (1975) for prevalent cations may be subject to a number of additional changes. (1) In temperate regions with cold winters, the slow moving marsh water can be expected to freeze over with an accompanying precipitation of both cations and anions. For example, at Theresa Marsh the alkalinity (milligrams $CaCO_3$/l) of the outflow water during the winter months was >400 while that of the ice averaged near 50. (2) During periods of intense photosynthesis the decrease in CO_2 can result in precipitation of Ca as $Ca(HCO_3)_2$ or $CaCO_3$ (Ruttner, 1963; Planter, 1970a). (3) Towards the end of the growing season ions will be leached out of the marsh vegetation with a resultant increase in the surrounding water. Sodium and K are among the first cations to be leached with Ca, Mg, and Si (used for structural components) being eluted much more slowly (Planter, 1970b). The increase of K in surface runoff and in river waters (at the end of the growing season, Fig. 1) is due to leaching from plant material (Konenko et al., 1974). Sodium and Cl flows through the marsh were dependent on inflow concentrations related in turn to sewage treatment waste. Sulfate concentrations, consistently higher in the outflow water of the marsh than in the inflow (Table 1), indicate either atmospheric input or a net loss of S from the marsh.

It should be noted that virtually nothing is known about the uptake, storage, and release of K (and also little about Ca and Mg among others) by bacteria, phytoplankton, and zooplankton (Likens, 1975). The assessment of the cycling of these elements within a marsh system is impossible without a much greater understanding of their movement and the processes involved.

Table 2. Monthly Averages ($\bar{x} \pm$ SD) of Chemical Parameters of Soil in Theresa Marsh, Wisconsin

Date	pH	Organic matter (%)	Total N (%)	Available (ppm)		Exchangeable (ppm)	
				P (Bray P-2)	K	Ca	Mg
1971							
Jul	⋯	⋯	1.36 ± 0.44	1.07 ± 52	151 ± 624	⋯	⋯
Sep	6.5 ± 0.7	⋯	1.94 ± 0.57	115 ± 74	118 ± 46	11,740 ± 3,100	2,672 ± 438
Oct	6.5 ± 0.5	⋯	1.78 ± 0.56	118 ± 62	98 ± 32	9,566 ± 2,874	2,080 ± 477
1972							
Feb	6.6 ± 0.7	⋯	1.84 ± 0.72	139 ± 68	148 ± 46	12,160 ± 2,860	2,735 ± 502
Apr	6.5 ± 0.4	⋯	⋯	⋯	140 ± 53	10,833 ± 2,209	2,361 ± 618
May	6.4 ± 0.5	43.2 ± 13.9	1.84 ± 0.62	123 ± 57	133 ± 76	12,060 ± 2,989	2,318 ± 578
Jun	6.5 ± 0.6	43.4 ± 14.6	1.88 ± 0.57	50 ± 24	169 ± 67	5,720 ± 1,566	1,219 ± 238
Jul	6.5 ± 0.6	41.0 ± 14.8	1.77 ± 0.48	68 ± 38	161 ± 58	12,730 ± 3,283	2,300 ± 647
Aug	6.4 ± 0.5	40.4 ± 15.2	1.75 ± 0.60	203 ± 113	230 ± 60	9,542 ± 2,073	2,080 ± 413
Sep	6.4 ± 0.6	42.6 ± 17.1	1.84 ± 0.67	164 ± 93	222 ± 117	9,713 ± 2,417	2,060 ± 527

MARSH SOILS

Marsh soils of alfisol–mollisol landscapes are generally more fertile than the upland soils that surround them (Cook and Powers, 1958; Klopatek, 1974). Marsh soils differ from surrounding soils due primarily to their inundation which results in an anaerobic state throughout the soil column except for a thin aerobic oxidized layer at the sediment–water interface (Ponnamperuma, 1972). The anaerobic state and low redox potentials of marsh soils retard the decomposition and mineralization of organic matter typical in adjacent terrestrial ecosystems. In Theresa Marsh the organic matter content of the soil averaged from 40–44% (Table 2). Soil nutrient values are for samples taken from the 3- to 15-cm depth, the zone where most of the roots and rhizomes were located.

The nutrient dynamics of marsh soils are significantly different from their terrestrial counterparts. The anaerobic soil column maintains the reduced compounds and ions of NH_4^+, H_2S, Mn^{++}, Fe^{++}, and CH_4 instead of their oxidized counterparts NO_3^-, SO_4^{--}, Mn^{++++}, Fe^{+++}, and CO_2 (Harter, 1966). The oxidized surface layer performs an important function in the marsh facilitating a sink phenomenon for P and other elements (Syers et al., 1973; Patrick and Khalid, 1974; Jorgensen et al., 1975) and acting as a release mechanism for N (Kenney, 1973; Patrick and Reddy, 1976).

In inundated soils, NH_4^+ is the principal, if not the only, nitrogen form available to higher plants. Ammonium, if not on the cation exchange complex (Patrick and Mahapatra, 1968) or taken up by rooted macrophytes, may diffuse upward to the thin aerobic layer. Depending on the concentration gradient it either goes through a sequential nitrification–denitrification reaction or is released to the overlying water. Nitrate may also enter the sediment and sequentially be released to the atmosphere through dentrification. These processes have been documented by Isirimah (1972), Keeney (1973), Patrick and Reddy (1976). Denitrification is the principal mechanism for loss of N from flooded soils (Patrick and Tusneem, 1972), whereas the incorporation of N in organic matter is the principal source of N in wetland soils.

Boyd (1970c) and Klopatek (1975) have shown that N content of the marsh soils is positively and highly correlated with organic matter content and is released only through mineralization. Thus, the total N content of the soil remains fairly constant throughout the year (Buckman and Brady, 1969). Nitrogen content of Theresa Marsh soils averaged from 4 to 5% of the organic matter content (Table 2).

As mentioned previously submerged soils, especially those high in organic matter, can be considered as P sinks. However, the release of sediment-bound P to the overlying waters may occur if the water–soil P concentration difference is sufficient (Stumm and Leckie, 1971; Patrick and Khalid, 1974; Syers et al., 1973; Jorgensen et al., 1975). Takahashi (1965), Patrick and Mahapatra (1968), and Klopatek (1975) have indicated that Bray P-2 available P is the best indicator of plant-available P. Levels of available P within the marsh were nearly $2\times$ those found in the surrounding upland soils indicating soil inundation greatly increases the availability of P. Changes of Bray P-2 available P occurred in Theresa Marsh during the growing season (Table 2). Large changes of macrophyte root zone available P have also been found by Lindsley et al. (1976) and Mason and Bryant (1975) reported an increase in interstitial inorganic P from the beginning to the end of the growing season in *Phragmites communis* communities.

The high levels of exchangeable Ca and Mg in the soils of the marsh (Table 2) reflect the dolomitic nature of the drainage basin. Exchangeable Ca and Mg both exhibit a significant decrease then increase in June and July coinciding with peak uptake by the emergent macrophytes. This was inverse to the concentrations found in the outflow water. Increases of Ca^{++}, Mg^{++}, K^+, and Na^+ in submergent soils result primarily from solvent action of CO_2 and cation-exchange reactions (Ponnamperuma, 1965). Typically Ca is correlated with the organic matter content of soils (Buckman and Brady, 1969) although June was the only month that the r^2 value (.35) was $<.70$. Calcium followed a pattern similar to that reported by Bayly and O'Neill (1972). Magnesium did not show the same close correlations with organic matter and may result from the higher solubility of $MgCO_3$ than $CaCO_3$ and also the preferential binding of Ca^{++} on cation exchange sites (Buckman and Brady, 1969).

Little is known about the movement of K in marsh soils. Pringle and Van Ryswyk (1965) suggest that in peat soils, K is often the most deficient plant nutrient. In soils at Theresa Marsh this was not the situation (Table 2). Potassium is a mobile ion and moves readily through the soils (Reitemeier, 1957). It may be lost fairly easily, through leaching, and may also be added from the overlying water (van Schreven, 1970) depending upon the chemical equilibrium. In Theresa Marsh available K content of the soil during the growing season was significantly ($P < .01$) decreased from May–July. This corresponded to the peak uptake period by the vegetation and as indicated by van Schreven (1970) is the primary mode of K loss from the soil.

EMERGENT MACROPHYTES

In comparison to terrestrial vegetation, emergent macrophytes are unique in that they can respire anaerobically and have the ability and need to transport O_2 from the atmosphere to their roots (Ponnamperuma, 1965). The emergents differ from other aquatic macrophytes by obtaining their nutrients almost completely from the soil (Sculthorpe, 1967; Klopatek, 1975). Emergent macrophytes examined at Theresa Marsh included *Typha latifolia*, *Scirpus fluviatilis*, *Carex lacustris*, *Sparganium eurycarpum*, and *Phalaris arundinacea*. Other species found within the marsh are listed in McNelly and Klopatek (1973) and Klopatek (1974). For a description of sampling methodology and nutrient analyses, see Klopatek (1975) and Klopatek and Stearns (1977).

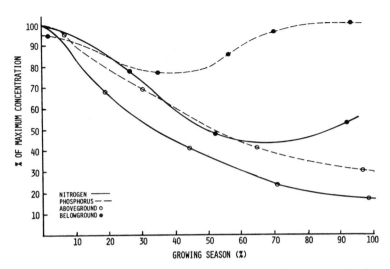

Fig. 2. The relative change in the concentration of N and P in above- and belowground structures of *Typha* sp. during the growing season.

The literature on nutrient concentrations of freshwater macrophytes has greatly expanded in recent years allowing the nutrient regimes of many macrophytes (at least the aboveground structures) to be fairly well categorized (Boyd, 1969, 1970a, 1971a; Boyd and Hess, 1970; Korelyakova, 1970; Maystrenko et al., 1969; Stake, 1967, 1968; Bayly and O'Neill, 1972; van Dyke, 1972; Klopatek, 1975; Lindsley et al., 1976; Bernard and Solsky, 1977; and Boyd, 1978). Garten (1977) using multivariate discriminant analyses has shown that different species of emergent macrophytes are quite distinct in their leaf mineral composition. His results indicated that the macrophytes possess an adaptive zone of mineral element concentrations enabling each species to survive in the environment. He further suggests that phenotypic variation in mineral element composition of a species reflects its niche size in the nutrient environment. Two important factors follow from the above. (1) Given that a species of emergent macrophyte occurs in a location, its elemental concentration may be expected to fall with definite limits for the major essential elements, regardless of their concentration in the environment. This has been suggested by Boyd and Hess (1970) and shown to occur in certain species in 2 minerally divergent riverine marshes in Wisconsin (Lindsley et al., 1976). (2) The nutrient content of emergent macrophytes depends on their phenotypic stage; the differences between the beginning and end of the growing season may vary 4 to 5 fold. Maximum concentration of nutrients found in aboveground structures of the emergent macrophytes in Theresa Marsh were N—3.25%, P—0.81%, K—3.82%, Ca—1.58%, Mg—0.33%, and total ash—13.4% and usually occurred at the beginning of the growing season. For belowground structures the maximum concentrations found were N—2.88%, P—0.75%, K—2.65%,

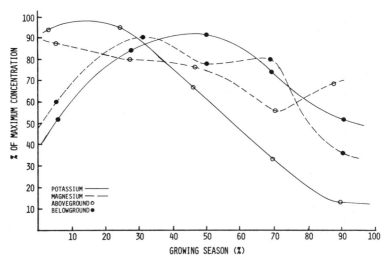

Fig. 3.   The relative change in the concentration of K and Mg in above- and belowground structures of emergent macrophytes during the growing season.

Ca—1.14%, Mg—0.59% and total ash—13.5% and occurred at the beginning or end of the growing season.

Figure 2 displays the typical nutrient concentration trends of *Typha* based on data from Theresa Marsh and other studies where both above- and belowground nutrient concentrations were available (Klopatek, 1974; Bayly and O'Neill, 1972; Mason and Bryant, 1975). At the beginning of the growing season both N and P in aboveground structures are at their maximum concentrations. Both decline rapidly early in the season and then gradually as the growing season progresses. At the end of the growing season leaf material in water may often become invaded by fungi and bacteria (Kaushik and Hynes, 1968; Mason and Bryant, 1975). At a point during decomposition processes the microorganisms find it energetically more efficient to obtain their needed N from the water. This in turn raises the N concentration of the plant (Boyd, 1970*b*; Klopatek, 1975; Mason and Bryant, 1975).

The underground structures of *Typha* undergo a decrease and then an increase in their N and P concentrations, a pattern typical for *Typha* (Prentki et al., 1978) and other emergent macrophytes including *Scirpus* sp., *Carex lacustris* and *Sparganium eurycarpum* (Klopatek, 1975; Lindsley et al., 1976; Bernard and Solsky, 1977). Although the absolute concentration of N > P in the plant tissues—≈3:1 (Westlake, 1975)—there appears to be a stronger relative retention of P in the vegetation. It can be hypothesized that the marsh ecosystem, similar to terrestrial systems (TIE, 1972), has evolved retentive mechanisms to maintain P within its internal cycle thereby slowing the flux from its boundaries.

The patterns exhibited for K, and Mg contrast with N and P for the macrophytes examined in Theresa Marsh (Fig. 3). Potassium exhibits a pattern similar to that for N and P in the aboveground structures, but the pattern for the belowground structures is exactly opposite. The maximum mid-season values for K may depict high metabolic demand during the periods of rapid growth and active, upward translocation of solutes along with the probability of luxury consumption. The depression of K values at the end of the growing season may indicate that there is no mechanism for major storage or retention of K due to its ready availability in the soil. Magnesium follows a pattern much like that of K, but with more subdued changes. The increase of Mg in aboveground structures (at the end of the growing season) probably represents a greater incorporation of Mg into structural components and its slower leaching compared to other elements (Planter, 1970b). Calcium concentrations were erratic for most macrophytes as observed by Bayly and O'Neill (1972) and Prentki et al. (1978) with peaks at the end of the growing season due to microbial activity (Mason and Bryant, 1975).

The analysis of emergent macrophyte ontogeny and associated changes in elemental concentrations yields information pertaining to internal nutrient requirements, but it does not provide a complete picture of the role vegetation plays in marsh ecosystem function. The pattern of accumulation is dependent on both the phenophase of the plant and the structure being considered. Peak accumulation of nutrients may or may not correspond to peak standing crop. In the aboveground standing crop, the greatest accumulation of N and P typically coincides with peak biomass (Bernard and Solsky, 1977; Boyd and Vickers, 1971; Klopatek, 1975; Lindsley et al., 1976; Mason and Bryant, 1975). However, the period of peak N uptake often precedes that of P. Maximum accumulations of K are generally reached prior to peak standing crop with the highest rate of accumulation occurring prior to that of N and P. Calcium and Mg levels in the aboveground standing crop appear to be more inconsistent and often reach their maximum after peak standing crop has been reached. In the belowground structures, the maximum accumulation of nutrients almost always coincides with the peak standing crop. The major exceptions are Ca and Mg whose peak precedes the peak belowground standing crop. Maximum rates of accrued nutrients in the belowground biomass occur after the maximum requirements of the aboveground structures have been met.

Table 3 lists the maximum amounts of nutrients accrued in the current year's standing crop of the below- and aboveground structures. Averaging of the dominant species in the marsh yields an uptake per hectare of 176 kg N, 37 kg P, 234 kg K, 144 kg Ca, 45 kg Mg, and 986 kg total ash constituents. Very similar values have been reported by Prentki et al. (1978) for lakeshore marshes in Wisconsin. These values show both the high nutrient status of the marsh and the high level of primary production ($\approx 1{,}200\text{--}2{,}900 \; \text{g} \cdot \text{m}^{-2} \cdot \text{yr}^{-1}$ for the species examined).

Figure 4 depicts nutrient models for a *Scirpus fluviatilis* stand. The annual nutrient budgets depicted in the models are primarily based on changes in the nutrient accumulations in the standing crops. Input into the P model was

Table 3. List of the Maximum Amount of Nutrients Accrued at any One Time in the Structures of Various Emergent Macrophytes Examined at Theresa Marsh, Wisconsin in 1972[a]

Location and total amount of nutrients

Species	N Below-ground	N Above-ground	N Total[b]	P Below-ground	P Above-ground	P Total[b]	K Below-ground	K Above-ground	K Total[b]	Ca Below-ground	Ca Above-ground	Ca Total[b]	Ash Below-ground	Ash Above-ground	Ash Total[b]	Mg Below-ground	Mg Above-ground	Mg Total[b]
Typha latifolia[b]	22.07	9.45	31.52	4.39	2.86	4.25	20.55	21.71	37.94	12.44	22.40	34.84	99.90	136.80	227.21	6.49	3.65	9.84
Typha latifolia-Sparganium eurycarpum[c]	8.05	8.30	16.36	1.44	1.65	3.09							49.70	88.30	158.00			
Scirpus fluviatilis	5.32	15.35	17.26	2.00	3.18	3.39	3.99	14.81	15.55	0.60	2.52	2.60	20.28	47.85	65.31	0.39	0.96	1.16
Carex lacustris	1.07	7.71	8.78	0.24	1.97	2.21	1.15	7.12	7.96	0.34	4.05	4.39	4.87	31.40	36.27	0.22	0.98	1.20
Phalaris arundinacea[d]	3.24	12.31	12.44	0.46	1.92	2.14							13.21	64.73	75.60			
Polygonum spp. mix[e]	0.92	8.66	9.42	0.15	1.51	1.63							10.63	60.43	71.06			

[a] Sum of below- and aboveground may not equal total accumulation, as figures indicate maximum accrued amounts which may have occurred on different dates. All values expressed as grams per square metre.

[b] Belowground productivity assumed to be 50% of aboveground productivity.

[c] Figures represent a mixed *Typha–Sparganium* stand.

[d] Belowground productivity assumed to be 25% of aboveground productivity.

[e] Figures represent a *Polygonum lapathifolium*-dominated wet soil, transient community (see Klopatek and Stearns, 1977).

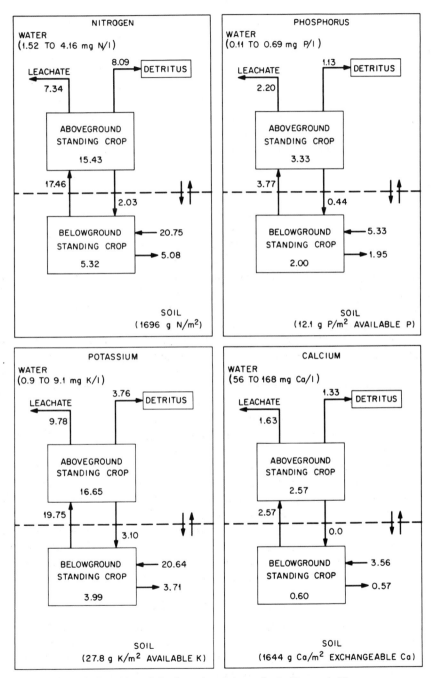

Fig. 4. Flow of N, P, K, and Ca through a *Scirpus fluviatilis* stand. Flows are grams per square metre per year and compartments are g/m² in standing crop.

obtained using radiophosphorus, ^{32}P. Information in the models was also based on biomass dynamics, the structural incorporation of the element in the plant, and from the estimates of leaching loss and translocation literature. These models show that nutrients taken from the soil are returned to the environment by 3 processes: leaching of the aboveground tissues into the surrounding water; release via decomposition of detritus falling into the water; and translocation down to the belowground structures where they are either stored for the growing season or released through death and decomposition. All the elements vary both in terms of the rates and amounts transferred. Decomposition is certainly the slowest of the nutrient transferring processes for *Scirpus fluviatilis*. For example, in the *Scirpus* stand which had a total productivity of 1,533 g·m^{-2}·yr^{-1}, >1,500 g/m^2 of undecomposed roots and rhizomes were present from previous years' growth.

Although the dominant vegetation characterizing freshwater marshes is herbaceous, some arborescent vegetation is usually present. Surprisingly, few data have been accumulated on the nutrient status of arborescent wetland vegetation although they may make up a significant proportion of the biomass present in the system (Jervis, 1969; Tilton and Bernard, 1975). Species of *Salix* were the most common woody vegetation in Theresa Marsh (J. M. Klopatek *personal observation*). Seasonal nutrient levels in *Salix interior* followed the patterns of other deciduous species (Rodin and Bazilevich, 1968) with highest levels of nutrients in leaves and new wood occurring early in the season. However Ca was the exception probably due to its incorporation into structural components and the disproportionate loss of metabolites and other compounds. Root nutrient levels remained fairly constant throughout the year. Wood produced in previous years showed the greatest variability, decreasing in N and increasing in P and Ca at the end of the growing season while K levels were highest in mid-season.

Combining the nutrient levels with the estimated annual production (1,902 g·m^{-2}·yr^{-1}) illustrates that shrubs may play an important role in the overall nutrient cycle of a marsh ecosystem. Table 4 lists the total annual accrual of nutrients by *Salix interior*. Although the values are an underestimate because litterfall was not accounted for, the magnitude of the nutrient pump effect of *Salix* is obvious. Due to the woody nature of *Salix* the turnover times for the nutrients accrued in its biomass (except in leaves and fruits) will be significantly longer than for emergent macrophytes.

PLANT–SOIL WATER INTERACTIONS

With the possibility of utilizing vegetation for pollution control, there have been several efforts to determine the relationships between the nutrient status of emergent macrophytes and their environment. Attempts to compare levels of soil and water nutrients with those in emergent macrophytes have resulted in weak or nonsignificant correlations (Stake, 1967, 1968; Boyd and Hess, 1970; Boyd, 1971*b*). The reasons for the lack of significant

Table 4. Estimated Annual Accrual of Nutrients by *Salix interior*

Component	\multicolumn{6}{c}{g/m²}
	N	P	K	Ca	Mg	Ash
Roots	2.13	0.69	2.22	3.36	0.57	18.62
Bole wood	1.07	0.17	0.44	1.99	0.18	5.80
Branches	12.01	1.50	12.38	14.21	2.82	54.97
Leaves	3.10	10.33	1.90	3.10	0.52	12.18
Total	18.31	12.69	16.94	22.69	4.09	91.57

correlations (e.g., specific nutrient regimes, seasonal changes in nutrient concentrations) have already been touched on previously.

Results of regression analyses correlating monthly averages of total N and available P in the soil and N and P accumulated in the below- and aboveground standing crops of pure stands of *Typha latifolia*, *Scirpus fluviatilis*, and *Carex lacustris* are presented in Table 5. Only the *T. latifolia* N relationships were not significant probably due to microbial enrichment.

The highly significant correlations between both *S. fluviatilis* and *C. lacustris* (and to a lesser extent *T. latifolia*) and total soil N is somewhat surprising as NH_4^+ is regarded as the sole plant-available N in continually waterlogged soils (Patrick and Mahapatra, 1968). A probable explanation is that NH_4^+ in the anaerobic marsh soil is in constant proportion to the total N present. In submerged soils, mineralization of N results in the formation of NH_4^+ (Ponnamperuma, 1972). Ellenberg (1971) stated that the rate of mineralization of N in beech forest soils is controlled by the N uptake rate of the beech tree roots. Extending this concept to the submerged marsh soils, we can postulate that the emergent macrophytes affect the mineralization rate and, thus, the amount of NH_4^+ present. Conversely N may be one of the more limiting elements to the emergent macrophytes, thus the closer relationships. The change in total soil N was insufficient to be statistically significant with the uptake by *T. latifolia*, *S. fluviatilis*, and *C. lacustris* accounting for 1.0, 0.6, and 0.3%, respectively, of the total soil N at each study location.

Regression analyses correlating the monthly accruement of nutrients were also done for K, Ca, and Mg, but no significant relationships were found. When analyses were performed correlating the monthly changes in soil nutrients (Z) with the monthly changes of accumulated nutrients in the above- (X) and belowground (Y) structures only 2 significant ($P < .05$) correlations were found, 1 for K and *S. fluviatilis*: $Z = 73.1 + 52.0X - 1.1Y$ ($r^2 = .97$); and 1 for Mg in *C. lacustris*: $Z = 2,800 - 45,717X + 8,309Y$ ($r^2 = .97$). Intuitively one would expect more significant correlations based on the fact that significant changes in the soil cations occurred during the same period of peak accumulation by the emergent macrophytes. However, the en-

Table 5. Correlation Coefficients (r) and Regression Equations for Relationships Between Soil Nutrients and Nutrients Accumulated in Above- and Belowground Standing Crops of 3 Emergent Macrophytes[a]

Species	Nutrient	Regression equation[b]	Correlation coefficient (r)
Typha latifolia	Nitrogen	$Z = 0.476 - 0.070X + 0.326Y$	.87[c]
	Phosphorus	$Z = -0.12 - 28.30X + 109.91Y$	.98[d]
Scirpus fluviatilis	Nitrogen	$Z = 1.005 - 0.026X + 0.006Y$	.99[c]
	Phosphorus	$Z = 22.984 - 60.848X + 0.334Y$	.84[d]
Carex lacustris	Nitrogen	$Z = 1.798 + 0.414X + 0.091Y$	.99[c]
	Phosphorus	$Z = 13.764 + 1354.46X - 107.80Y$	.99[c]

[a] Soil nutrients are total N (%) and available P (ppm); nutrients in standing crops are g/m².
[b] Variables are Z, soil nutrients; Y, nutrients in aboveground standing crop; and X, nutrients in belowground standing crop.
[c] Significant at the .01 level of probability.
[d] Significant at the .10 level of probability.

vironmental parameters influencing cation uptake are extremely complex (Boyd, 1971*b*). Major factors attributable to the lack of emergent macrophyte–soil correlations are luxury consumption, leaching losses to the overlying water, and a lack of an internal regulation of Ca, Mg, and K concentrations that occurs with N and P.

One might conjecture that the depletion of cations in the soil complex is a result of uptake by the macrophytes. However, the decline in Ca and Mg in the soil is 2 orders of magnitude (by weight) greater than that attributable to the macrophytes. Therefore, it is only possible to state that the changes in soil concentrations are probably a result of complex biological and physical processes indicated in part by emergent macrophyte nutrient dynamics. It should be noted, however, that the soil–vegetation relationships are more than trivial as Lindsley et al. (1976) reported similar results from a riverine marsh in northern Wisconsin which possessed significantly lower available nutrients plus a more restrictive growing season.

The significant correlations with P are expected. Patrick and Mahapatra (1968) and Chang (1965) reported that Bray P-2 available P is perhaps the best measure of P availability to lowland rice grown on submerged soils. What is surprising is the magnitude of the correlation coefficients indicating that the emergent macrophytes may significantly affect the flux of available P in the soil during the growing season. It may further be suggested that the emergent macrophytes affect available N and P by stimulating the growth of microbes in the rhizosphere. This enhancement, due to readily metabolized substrates, increases the mineralization of more resistant P-containing compounds (e.g., phytin).

MARSH DRAWDOWN

A common phenomenon in marsh systems is the periodic exposure of soils that are normally inundated. This may result from natural climatic fluctuations or from man-induced changes, such as drawdown of marsh water level to influence waterfowl habitat and food production (Linde, 1969). Early in the spring of 1971 Theresa Marsh was drained to facilitate a fish kill and removal program. Numerous mud flats virtually devoid of rooted vegetation were exposed by the drawdown. These mud flats remained unvegetated through May, but in June began to develop luxuriant growths of annual weeds dominated by *Polygonum lapathifolium*, *Acnida altissima*, *Bidens cernua*, and *Phalaris arundinacea*. (Perennial grasses developed on some of the moist soils.) This mud flat community was transient as most of the species, except for a few emergent aquatics, died after inundation.

Table 3 lists nutrient accruals for a *Polygonum lapathifolium* dominated community present during the drawdown. This community, with a net productivity of 1,117 g/m², represents a possible significant source of nutrient export out of the marsh ecosystem. Because >80% of its productivity was aboveground (most species were annuals not translocating nutrients to the roots for storage), the large majority of the nutrients were lost to the water upon reflooding.

In addition to the export of nutrients through plants, sizable amounts of nutrients can be leached from the exposed soils. Drainage of marsh soils can markedly disrupt the N balance of the marsh due to the more rapid decomposition of organic matter and resultant release of large quantities of soluble organic-N and NO_3^-. Nitrate levels in the exposed soils during June and July were >100 ppm in some locations. The release of N from drained soils and its implications to downstream loadings have been discussed by Amundson (1970), Ponnamperuma (1972), and Nicholls and MacCrimmon (1974). Total soil N in July 1971 was significantly ($P < .05$, df = 25) lower than in July 1972. Soluble organic-N of the outflow water in Theresa Marsh during drainage was 3× that experienced under flooded conditions. Furthermore, the only time the concentration of all N species in the marsh outflow was less than those of the inflow waters was during the drawdown. Amundson (1970), Bentley (1969), and Lee et al. (1975) have also noted that the drainage of marshes results in a significant loss of N from the system. Although drainage of the marsh resulted in the discharge of greater than normal loads of N from the basin, the effect of the drainage on the long-term N budget appears limited. Inundation following drainage resulted in a net input of N into the marsh with soil N increasing significantly within 1 year. Outflow from the marsh during drainage failed to show any noticeable increase in P although Amundson (1970) and Lee et al. (1975) indicated that it may be released from marsh soils upon drainage. Increases were noted for Ca, P, and Mg. Sulfur, in contrast, increased noticeably in outflow concentration during the drawdown. It is probable that short periods of drawdown when the marsh is vegetated have negligible effects on the marsh system. However, prolonged drawdown or complete drainage of the marsh would undoubtedly drastically alter the existing mineral cycles.

CONCLUSIONS

The preceding discussion has shown that the water, soil, and biotic components of a marsh ecosystem are inextricably interconnected. Many of physical–chemical processes involved in the nutrient dynamics of a riverine marsh are typical of those occurring in river or lake ecosystems, but, because of the physical morphometry of the marsh, the cycles of material uptake and release may be quite different.

It is hypothesized that the emergent macrophytes are a key to the seasonal changes in available N, P, and K in the soil. The macrophytes are visualized as nutrient pumps, taking in nutrients from the soil and immobilizing them (at least temporarily) in below- and aboveground structures. By taking in P, the plants may facilitate a greater exchange of P from the water to the underlying soil. However, this rate may be extremely slow (Pomeroy et al., 1969). Through plant uptake of cations, base exchange sites in the soil are freed and can, in turn, accept new cations moving through the interstitial water. Changes in soil-nutrient levels, water-nutrient levels, and accumulation of plant nutrients all tend to corroborate this hypothesis.

Although changes in nutrient levels in the water flowing through a marsh are undoubtedly influenced by soil–water interactions, the direct relationships are difficult to document. This is complicated by uptake of nutrients directly from the water by phytoplankton and submergent and free-floating macrophytes. Reduction of inorganic P levels in the marsh outflow during the growing season were shown to correlate ($r^2 = .90$, $P < .01$) with changes in the total P in the aboveground biomass of the emergent macrophytes (Klopatek, 1975). However, diurnal changes may be significant (Planter, 1970a; Lindsley et al., 1976; and Klopatek, 1974). This is indicative of the problems associated with analyzing time series data on a monthly or biweekly basis. Potassium, Ca, and Mg changes in the outflow water during the growing season also show a close correlation with emergent macrophyte uptake and leaching loss (J. M. Klopatek, *personal observation*). Documentation of N changes due to the vegetation is exacerbated by the volatile nature of N and the complexity of its cycle as pointed out by Mitsch (1976). However, N fluxes from the marsh system, in the form of organic-N, are evident at the end of the growing season.

The marsh is more than just a simple emergent macrophyte–soil–water system. To assess accurately the effects of the marsh on water quality, a quantified hydrology is required along with frequent sampling to establish total elemental concentrations in the soil and the relationship to their available form. Although emergent macrophytes obviously play a major role in the overall nutrient cycle, the roles of the other primary producers need to be quantified for a more complete picture of the nutrient dynamics of fresh-water riverine marshes.

ACKNOWLEDGMENTS

Appreciation is expressed to C. S. Henderson and E. A. Bondietti for their assistance in reviewing this manuscript. Barbara A. Klopatek and Forest Stearns provided invaluable assistance throughout the course of investigation. This research was funded in part by the Wisconsin Department of Natural Resources and the Department of Botany, University of Wisconsin–Milwaukee.

LITERATURE CITED

Amundson, R. W. (1970). Nutrient availability of a marsh soil. M.S. Thesis. Department of Water Chemistry, University of Wisconsin, Madison, Wisconsin. 56 p.

Bayly, I. L., and O'Neill, T. A. (1972). Seasonal ionic fluctuations in a *Typha glauca* community. *Ecology* 53, 714–719.

Bentley, E. M., III. (1969). The effect of marshes on water quality. Ph.D. Thesis. Department of Water Chemistry, University of Wisconsin, Madison, Wisconsin.

Bernard, J. M., and Solsky, B. A. (1977). Nutrient cycling in a *Carex lacustris* wetland. *Can. J. Bot.* 55, 630–638.

Blum, J. L. (1972). Plant ecology in flowing water. In "River Ecology and Man" (R. T. Oglesby, C. A. Carlson, and J. A. McCann, eds.), pp. 53–62. Academic Press, New York.

Boyd, C. E. (1969). Production, mineral nutrient absorption and biochemical assimilation by *Justicia americana* and *Alternanthera philoxeroides*. *Arch. Hydrobiol.* 66, 139–160.

Boyd, C. E. (1970a). Production, mineral accumulation, and pigment concentrations in *Typha latifolia* and *Scirpus americanus*. *Ecology* 51, 285–290.

Boyd, C. E. (1970b). Losses of nutrients during decomposition of *Typha latifolia*. *Arch. Hydrobiol.* 66, 511–517.

Boyd, C. E. (1970c). Influence of organic matter on some characteristics of aquatic soils. *Hydrobiologia* 36, 17–21.

Boyd, C. E. (1971a). Further studies on productivity, nutrient and pigment relationships in *Typha latifolia* populations. *Bull. Torrey Bot. Club* 98, 144–150.

Boyd, C. E. (1971b). The dynamics of dry matter and chemical substances in a *Juncus effusus* population. *Am. Midl. Nat.* 86, 28–45.

Boyd, C. E. (1978). Chemical composition of wetland plants. In "Freshwater Wetlands: Ecological Processes and Management Potential" (R. E. Good, D. F. Whigham and R. L. Simpson, eds.), pp. 155–167. Academic Press, New York.

Boyd, C. E., and Hess, L. W. (1970). Factors influencing shoot production and mineral nutrient levels in *Typha latifolia*. *Ecology* 51, 296–300.

Boyd, C. E., and Vickers, D. H. (1971). Relationships between production, nutrient accumulation, and chlorophyll synthesis in an *Eleocharis quadrangulata* population. *Can. J. Bot.* 49, 883–888.

Bozniak, E. G., and Kennedy, L. L. (1968). Periodicity and ecology of the phytoplankton in an oligotrophic and eutrophic lake. *Can. J. Bot.* 46, 1259–1271.

Brezonik, P. L. (1968). The dynamics of the nitrogen cycle in natural waters. Ph.D. Thesis. Department of Water Chemistry, University of Wisconsin, Madison, Wisconsin.

Buckman, H. O., and Brady, N. C. (1969). The nature and properties of soils (7th Edition). The MacMillan Company, New York.

Chang, S. C. (1965). Phosphorus and potassium tests in rice soils. In "The Mineral Nutrition of the Rice Plant," pp. 271–293. International Rice Research Institute, Johns Hopkins Press, Baltimore, Maryland.

214   JEFFREY M. KLOPATEK

Cook, A. H., and Powers, C. F. (1958). Early biochemical changes in the soils and water of artificially created marshes in New York. *N.Y. Fish and Game J.* 5, 9–65.

Ellenburg, H. (1971). Nitrogen content, minerals and cycling. *In* "Productivity of Forest Ecosystems" (P. Duvigneaud, ed.), pp. 509–513. Proceedings of the Brussels Symposium, UNESCO and IBP, 1969.

Garten, C. T., Jr. (1977). Multivariate perspectives on the ecology of plant mineral element composition. *Am. Natur. (In press).*

Golterman, H. L. (1975). Chemistry. *In* "River Ecology" (B. A. Whiton, ed.). University of California Press, Berkeley, California.

Granert, W. G., Jr. (1973). The algal flora of Theresa Marsh, Washington and Dodge counties, Wisconsin. M.S. Thesis. Department of Botany, University of Wisconsin-Milwaukee, Milwaukee, Wisconsin.

Harter, R. D. (1966). The effect of water levels on soil chemistry and plant growth of the Magee Marsh wildlife area. Ohio Game Monographs. July 1966, No. 2.

Isirimah, N. O. (1972). Nitrogen cycling in Lake Wingra. Ph.D. Thesis. Dept. Soil Science, Univ. Wisconsin-Madison, Wisconsin. 224 p.

Jervis, R. A. (1969). Primary production in the freshwater marsh ecosystem of Troy Meadow, New Jersey. *Bull. Torrey Bot. Club* 96, 209–231.

Jorgensen, S. E., Kamp-Nielsen, L., and Jacobsen, O. S. (1975). A submodel for anaerobic mud-water exchange of phosphate. *Ecol. Modelling* 1, 133–146.

Kaushik, N. K., and Hynes, H. B. N. (1968). Experimental study on the role of autumn-shed leaves in aquatic environments. *J. Ecol.* 56, 229–244.

Keeney, D. R. (1973). The nitrogen cycle in sediment-water systems. *J. Environ. Quality* 3, 151–162.

Kiefer, W. (1968). Biological waste water treatment with plants. *Umschau No. 7*, 210. Trans. from German for the Great Lakes Basin Comm. and N. E. Wisconsin Regional Plan. Comm.

Kitchens, W. M., Jr., Dean, J. M., Stevenson, L. H., and Cooper, J. H. (1975). The Santee Swamp as a nutrient sink. *In* "Mineral Cycling in Southeastern Ecosystems" (F. G. Howell, J. B. Gentry, and M. H. Smith, eds.), pp. 349–366. ERDA Symposium Series (CONF-740513).

Klopatek, J. M. (1974). Production of emergent macrophytes and their role in mineral cycling within a freshwater marsh. M.S. Thesis. Department of Botany, University of Wisconsin-Milwaukee, Milwaukee, Wisconsin.

Klopatek, J. M. (1975). The role of emergent macrophytes in mineral cycling in a freshwater marsh. *In* "Mineral Cycling in Southeastern Ecosystems" (F. G. Howell, J. B. Gentry, and M. H. Smith, eds.), pp. 357–393. ERDA Symposium Series (CONF-740513).

Klopatek, J. M., and Stearns, F. W. (1977). Primary productivity of emergent macrophytes in a freshwater marsh ecosystem. *Am. Midl. Nat. (In press).*

Konenko, A. D., Garasevich, I. G., and Yenaki, I. G. (1974). Nitrogen, phosphorus and potassium in the water of the small tributaries of the Pripyat' in the Ukranian Poles'ye. *Hydrobiological Journal* 10, 8–12.

Korelyakova, I. L. (1970). Chemical composition of the higher aquatic vegetation of Kiev Reservoir. *Hydrobiological Journal (Giorobiologicheskiy Zhvrnnaz)* 6, 15–21.

Kramer, J. R., Herbes, S. E., and Allen, H. E. (1972). Phosphorus: Analysis of water, biomass, and sediment. *In* "Nutrients in Natural Waters" (H. E. Allen and J. R. Kramer, eds.), pp. 51–100. Wiley-Interscience, New York.

Lee, G. F., Bentley, E., and Amundson, R. (1975). The effect of marshes on water quality. *In* "Coupling of Land and Water Systems" (A. D. Hasler, ed.), pp. 105–127. Ecological Studies, Vol. 10. Springer-Verlag, New York.

Lieth, H. (1975). Primary production of the major units of the world. *In* "Primary Productivity of the Biosphere" (H. Lieth and R. H. Whittaker, eds.), pp. 203–215. Ecological Studies, Vol. 14. Springer-Verlag, New York.

Likens, G. E. (1975). Nutrient flux and cycling in freshwater ecosystems. *In* "Mineral Cycling in Southeastern Ecosystems" (F. G. Howell, J. B. Gentry, and M. H. Smith, eds.), pp. 314–348. ERDA Symposium Series (CONF-740513).

Linde, A. F. (1969). Techniques for wetland management. Wisconsin Department of Natural Resources Research Report #45, Madison, Wisconsin.

Lindsley, D. L., Shuck, T., and Stearns, F. W. (1976). Productivity and nutrient content of emergent macrophytes in two Wisconsin marshes. *In* "Freshwater Wetlands and Effluent Disposal" (D. L. Tilton, R. H. Kadlec and C. J. Richardson, eds.), pp. 51–75. The University of Michigan, Ann Arbor, Michigan.

Mason, C. F., and Bryant, R. J. (1975). Production, nutrient content and decomposition of *Phragmites communis* Trin. and *Typha angustifolia* L. *J. Ecol.* **63**, 71–95.

Maystrenko, Yu. G., Denisova, A. I., Bagnyuk, V. M., and Arymaova, Zh. M. (1969). The role of higher aquatic plants in the accumulation of organic and biogenic substances in water bodies. *Hydrobiological Journal* **5**, 20–31.

McNelly, J., and Klopatek, J. M. (1973). Submergent macrophytes in Theresa Marsh. *Univ. Wis.-Milwaukee Field Stations Bull.* **6**, 9–14.

Mitsch, W. J. (1976). Ecosystem modeling of water hyacinth management in Lake Alice, Florida. *Ecol. Modelling* **2**, 69–89.

Nicholls, K. H., and MacCrimmon, H. R. (1974). Nutrients in subsurface and runoff waters of the Holland Marsh, Ontario. *J. Environ. Quality* **3**, 31–35.

Niering, W. A. (1968). The ecology of wetlands in urban areas. *Garden Journal* **18**, 177–183.

Patrick, W. H., Jr., and Mahapatra, I. C. (1968). Transformation and availability to rice of nitrogen and phosphorus in water logged soils. *Advan. Agron.* **20**, 323–359.

Patrick, W. H., Jr., and Khalid, R. A. (1974). Phosphate release and sorption by soils and sediments: effect of aerobic and anaerobic conditions. *Science* **186**, 53–55.

Patrick, W. H., Jr., and Tusneem, M. E. (1972). Nitrogen loss from a flooded soil. *Ecology* **53**, 735–737.

Patrick, W. H., Jr., and Reddy, K. R. (1976). Nitrification-denitrification reactions in flooded soils and water bottoms: dependence on oxygen supply and ammonium diffusion. *J. Environ. Quality* **5**, 469–472.

Planter, M. (1970a). Physico-chemical properties of the water of reed belts in Mikolajskie, Taltuwisko, and Sniardwy Lakes. *Pol. Arch. Hydrobiol.* **17**, 337–356.

Planter, M. (1970b). Elution of mineral components out of dead reed *Phragmites communis* Trin. *Pol. Arch. Hydrobiol.* **17**, 357–362.

Pomeroy, L. R., Johannes, R. E., Odum, E. P., and Roffman, B. (1969). The phosphorus and zinc cycles and productivity of a salt marsh. *In* "Proc. 2nd Symp. on Radioecology" (D. J. Nelson and F. C. Evans, eds.). Clearing House Fed. Sci. Tech. Info., TID, Springfield, Virginia.

Ponnamperuma, F. N. (1965). Dynamic aspects of flooded soils and the nutrition of the rice plant. *In* "The Mineral Nutrition of the Rice Plant" pp. 295–328. International Rice Research Institute, Johns Hopkins Press, Baltimore, Maryland.

Ponnamperuma, F. N. (1972). The chemistry of submerged soils. *Advan. Agron.* **22**, 29–96.

Prentki, R. F., T. D. Gustafson, and H. S. Adams. (1978). Nutrient movements in lakeshore marshes. *In* "Freshwater Wetlands: Ecological Processes and Management Potential" (R. E. Good, D. F. Whigham, and R. L. Simpson, eds.), pp. 169–194. Academic Press, New York.

Pringle, W. L., and van Ryswyk, A. L. (1965). Response of water sedge in the growth room to fertilizer and temperature treatments. *Can. J. Plant Sci.* **45**, 60–66.

Reitemeier, R. F. (1957). Soil potassium and fertility. *In* Soil, The Yearbook of Agriculture" (A. Stefferud, ed.), pp. 101–106. U.S. Government Printing Office, Washington, D. C.

Rodin, L. E., and Bazilevich, M. I. (1968). "Production and Mineral Cycling in Terrestrial Vegetation." Oliver and Boyd, London.

Ruttner, F. (1963). "Fundamentals of Limnology." (Third Edition). University of Toronto Press, Toronto, Canada.

Sculthorpe, C. D. (1967). "The Biology of Aquatic Vascular Plants." St. Martin's Press, New York.

Spangler, F. L., Sloey, W. E., and Fetter, C. W., Jr. (1976). Waste-water treatment by natural and artificial marshes. EPA-600/2-76-207. September 1976.

Stake, E. (1967). Higher vegetation and nitrogen in a rivulet in Central Sweden. *Schweiz. Z. Hydrol.* **29**, 107–125.

Stake, E. (1968). Higher vegetation and phosphorus in a small stream in Central Sweden. *Schweiz. Z. Hydrol.* **30**, 353–373.

Stumm, W., and Leckie, J. O. (1971). Phosphate exchange with sediments; its role in the productivity of surface waters. *Proc. 5th Int'l Water Pollution Res. Conf.* III-26/1-26/16. San Francisco, July–August, 1970.

Syers, J. K., Harris, R. F., and Armstrong, D. E. (1973). Phosphate chemistry in lake sediments. *J. Environ. Quality* **2**, 1–14.

Takahashi, J. (1965). Natural supply of nutrients in relation to plant requirements. *In* "The Mineral Nutrition of the Rice Plant." pp. 271–294. International Rice Research Institute, Johns Hopkins Press, Baltimore.

The Institute of Ecology (TIE). (1972). "Man in the Living Environment." University of Wisconsin Press, Madison, Wisconsin.

Tilton, D. L., and Bernard, J. M. (1975). Primary productivity and biomass distribution in an older shrub ecosystem. *Am. Midl. Nat.* **94**, 251–256.

van Dyke, G. D. (1972). Aspects relating to emergent vegetation dynamics in a deep marsh, northcentral Iowa. Ph.D. Thesis. Department of Botany and Plant Pathology, Iowa State University, Ames, Iowa.

van Schreven, D. A. (1970). Leaching losses of nitrogen and potassium in Polders reclaimed from Lake Ijssel. *Plant and Soil* **33**, 629–643.

Westlake, D. F. (1963). Comparisons of plant productivity. *Bio. Rev.* **38**, 385–425.

Westlake, D. F. (1975). Macrophytes. *In* "River Ecology" (B. A. Whitton, ed.). University of California Press, Berkeley, California.

Whitton, B. A. (1975). Algae. *In* "River Ecology" (B. A. Whitton, ed.). University of California Press, Berkeley, California.

NUTRIENT DYNAMICS OF NORTHERN WETLAND ECOSYSTEMS

Curtis J. Richardson,[1] Donald L. Tilton,[2] John A. Kadlec,[3] Jim P. M. Chamie,[4] and W. Alan Wentz[5]

Abstract   Nutrient information was reviewed by compartment (i.e., soils, plants, water) and at the ecosystem level among 4 northern wetland types: fens, bogs, swamps, and marshes. Total soil N, P and Ca were lowest in bog peats but the Mg content did not vary appreciably among wetland types. Cation exchange capacity was >100 meq/100 g for all wetlands. Concentrations of N, P, Ca and Mg were significantly lower in bog plants than fen plants. Seasonal patterns of nutrient concentrations (N and P) in leaves and stems decreased through the growing season in fen plants. Translocation of nutrients from plant parts prior to abscission did not occur in the plants studied. Minerotrophic fen water chemistry differed from that of ombrotrophic bogs in that fen waters were dominated by Ca^{++} and HCO_3^- ions and bog waters were dominated by H^+ and SO_4^{++} ions (Moore and Bellamy, 1974). Seasonal variations in NH_4-N, NO_3-N and PO_4-P were closely related to peatland hydrology, organism uptake and peat exchange characteristics. A cycling study for a central Michigan fen revealed that >97% of the N, P and Ca was in the peat compartment. The turnover time for N in aboveground biomass was ≈2 years. An examination of plant uptake rates of N and P in the Michigan fen revealed that low plant productivity may be related to low N and P availability. Nutrient outputs for wetland ecosystems, when compared to yields from terrestrial forest systems, indicate that natural outputs for some wetland types are well within the range, or in some cases exceed, outputs from upland terrestrial ecosystems. The capacity of acid peatlands to store or assimilate

[1] School of Natural Resources, The University of Michigan, Ann Arbor, Michigan 48109 (Present address: School of Forestry and Environmental Studies, Duke University, Durham, North Carolina 27706).

[2] Rockefeller Fellow in Environmental Affairs, The University of Michigan, Ann Arbor, Michigan 48109.

[3] Department of Wildlife Sciences, Utah State University, Logan, Utah 84321.

[4] Botany-Biology Department, South Dakota State University, Brookings, South Dakota 57007.

[5] Department of Wildlife Sciences, South Dakota State University, Brookings, South Dakota 57007.

additional P or K on a long-term basis appears limited. Nutrient dynamics in wetlands at the organism, community and ecosystem level are all poorly understood and are deserving of further study.

Key words   *Available nutrients, bog, cation exchange capacity, decomposition, ecosystem yield, fen, foliar concentrations, litter fall, marsh, Michigan, N and P flux, nutrient budget, nutrient cycling, peat soil, swamp, Wisconsin.*

INTRODUCTION

Wetlands are ecosystems with complex hydrologic and biogeochemical cycles which can transform various elements into compounds that may or may not, depending on the chemical variable, improve water quality for downstream ecosystems. In this paper we have attempted to compile the most complete data sets available for a comparison of nutrient dynamics in northern wetlands. For convenience we have grouped these wetlands into 4 general formations; fens, bogs, swamps and marshes (Jeglum et al., 1974). For a complete explanation of the specific differences or similarities between wetland terms [e.g., ombrogeneous = ombrotrophic = ombrophilous = wetlands dependent on rainfall for water and minerals; soligeneous = minerotrophic = rheophilous = wetlands which receive terrestrial nutrients and water in addition to rainfall] used by various researchers cited herein, and classification types [e.g., raised bog = highmoor = ombrogeneous swamp = forested fen; blanket bog = bog covering undulating semi-uplands] the reader is referred to Heinselman (1963), Moore and Bellamy (1974) and Zoltai et al. (1975).

Our approach to analyzing nutrient dynamics in northern wetlands is (1) to compare nutrient information by compartments (soils, plants and water) for various wetland types, (2) give an example of internal cycling processes within one of these wetlands, (3) compare nutrient yields (output) of wetlands to other ecosystem types, and (4) compare ecosystem nutrient budgets from a natural and perturbed wetland.

We have used data from our research site, the Houghton Lake fen, a 716-ha peatland located in central Michigan, as the main source of data for plant nutrients, seasonal water chemistry variations, and the internal cycling dynamics discussion. Because of this we have included a methods section to aid the reader when comparing our data to other site chemistry. The Houghton Lake vegetation is representative of fens in higher latitudes of North America and is composed of 2 main cover types, leatherleaf–bog birch (*Chamaedaphne calyculata* and *Betula pumila*; 19% of the area) and sedge–willow (*Carex* spp. and *Salix* spp.; 68% of the area). Stands of *Typha latifolia* (1.8%), *Populus tremuloides* (2.5%), *Alnus rugosa* (3.4%), plus open water (5.0%) make up the remaining area of the wetland. For a more complete analysis of this ecosystem see Chamie (1976), Richardson et al. (1976), and Wentz (1976).

SAMPLING PROCEDURES AND METHODS

The data sets chosen for comparisons were selected on the basis of similarity in methods as well as completeness of analyses. Because many of the units employed by other authors were transformed for comparisons, we are responsible for any conversion errors. The following is a brief description of the procedures and methods utilized by researchers at the Houghton Lake fen site. For a complete description of water and plant sampling procedures and chemical analyses techniques see Chamie (1976), Richardson et al. (1976), and Wentz (1976).

Water samples were collected monthly from 45 locations following standard limnological techniques (J. Kadlec, 1976). All H_2O samples were filtered through 0.45μm pore membrane filters. Chemical analysis of rainfall included dry fall contributions. An auto-analyzer (Technicon®) was used to analyze for NO_3^-, NH_4^+, and PO_4^{---}. Nitrate ($NO_3^- + NO_2^-$) was analyzed with the cadmium reduction reaction and is reported as NO_3-N and the Berthelot reaction was used to analyze for NH_4-N (E.P.A., 1974). After persulfate digestion P was analyzed using the molybdate reaction (Technicon, 1975). Cation analyses (H_2O, plant and soil) were completed following standard methods (Perkin-Elmer, 1973).

Plant samples were harvested monthly from May through September (1973) for biomass and nutrient analysis by species (Richardson et al., 1976). Wet ashing ($HClO_4 + HNO_3$) was carried out according to the method of Behan and Kinraide (1973). Total N was determined using semimicro Kjeldahl digestion (Black, 1965). Phosphorus was determined colorimetrically utilizing the molybdate–vanadate technique following wet acid digestion (Wolfe, 1962).

Soil cores were taken annually (1973–1976) at 16 sites at 20-cm intervals up to depths of 120 cm below the peat surface. Total and exchangeable values reported here are for the top 20 cm. This depth included most of the available plant nutrients as Chamie (1976) reported 83% of the roots are found in the upper 15 cm and R. Kadlec (1976) determined a lack of vertical water movement in peat at our site. Wet samples were transferred immediately to polyethylene bags to avoid excessive air contact. Determinations of bulk density followed (Blake, 1965). Percent H_2O content, ash and organic content were determined by methods described in Thorpe (1967, 1968). Peat samples for total nutrient analysis were dried and sieved (2-mm mesh). Following Caro's digestion (sulfuric acid–hydrogen peroxide, H. S. Lowendorf and A. S. Dominski, *personal communication*) the analyses of total N and P were determined via the ammonia–salicylate complex and phosphomolybdenum complex, respectively (Technicon, 1975).

Determinations of cation exchange capacity, exchangeable cations, NH_4-N, NO_3-N, and available P were made using wet peat soils. This procedure provides more realistic exchange values for natural waterlogged peat soils because the drying of peat has been shown to (1) increase cation exchange capacity (Puustjärvi, 1956), (2) affect ammonia fixation (Burge and Broadbent, 1961), and (3) influence chemical mobility (Ponnamperuma, 1972). Approximately 6 g fresh weight peat was needed to produce 1 g of dry peat for extraction analysis. The extraction filtrate from the wet peat was used for chemical analysis and the actual dry weight determined as follows: dry weight = filter paper weight + dry peat weight − filter paper weight − chemical additions. Cation exchange capacity procedures (0.5 N HCl at pH 7.0) followed Puustjärvi (1956) and Thorpe (1973). Exchangeable cation procedures (1 M NH_4OAC at pH 4.8) were adopted from Andersson (1975). Ammonium-nitrogen (2 N KCl) and NO_3-N (2 N KCl) were determined by procedures outlined by Bremner (1965). Available P techniques (0.10 N HCl and 0.03 N NH_4F) were modified from Bray and Kurtz (1945) and Olsen and Dean (1965).

RESULTS AND DISCUSSION
Soil Chemistry in Wetlands

Although there is considerable information on the acidity, water chemistry, and nutrient content for a few plants in northern wetland types, specific information on chemical composition, availability, and sorption of nutrients by waterlogged organic soils (histosols) in the USA is almost completely lacking. Most chemical data on organic soils have been limited to those of agricultural or forestry studies initiated after drainage and/or fertilization had occurred (Aandahl, 1974; Stanek, 1975). No attempt is made here to review all the chemistry of organic and waterlogged soils. The reader is referred to articles by Dawson (1956), Lucas and Davis (1961), Ponnamperuma (1972), Tusneem and Patrick (1972) and Patrick and Khalid (1974).

Peat soils from wetlands have many characteristics which distinguish them from mineral soils. Organic soils, as compared to mineral soils, have very low bulk density (Boelter, 1974), high water holding capacity and content by percent volume (Thorpe, 1968), low hydraulic conductivity (Boelter, 1965), low percent ash and high organic matter content (Thorpe, 1967), high organic nutrient content (Pollett, 1972) and extremely high cation exchange (Gore and Allen, 1956; Puustjärvi, 1956).

The nutrient content of peat is often an indicator (especially if the peat is drained) of its nutritive value for plant growth (Stanek, 1975). Malmstrom (1956) gave minimum nutrient values (percent dry weight at depth 20 cm below the surface) required for sustaining forest production on peat soils with a pH > 3.5 as follows: 1.0 (N), 0.04 to 0.09 (P), 0.08 (K), and 0.14 to 0.29 (Ca). With Malmstrom's values as a rough guideline, it is possible to compare different wetland peats in terms of both nutrient content and potential

Wetland type[b] and location

Characteristics	Fen (rich)[c] Houghton Lake, Michigan, USA	Marsh[d] Theresa, Wisconsin, USA	Fen (marginal)[e] Aneboda, Sweden	Bog (raised)[f] Kilmacshane, Galway, Ireland	Bog (blanket)[g] Northern Pennines, England	Bog (ombrotrophic)[h] Aneboda, Sweden	Bog (black spruce)[i] Fairbanks, Alaska, USA
Vegetation type	Chamaedaphne—Betula	Typha—Scirpus	Carex—Sphagnum	Trichophorum—Carex	Sphagnum—Calluna	Rhynchospora—Sphagnum	Picea—Sphagnum
Chemical composition							
pH (range)	5.1 ± 5.9[d]	6.4	5.2	4.7	...	3.2	3.6-4.0
ash (%)	28.6 ± 12.0	...	11.6	3.7	3.2	3.0	...
organic matter (%)	71.4 ± 12.0[k]	40.4 ± 15	...	...	96.8	...	...
Total nutrients (% dry wt)							
N	$2.54 \pm .03$	1.75 ± 0.6	2.5	1.8	...	1.1	0.43-0.59
P	$0.09 \pm .02$	...	0.07	0.02	...	0.03	...
K	$0.08 \pm .03$	...	...	0.40	0.04	...	...
Mg	$0.14 \pm .01$	...	...	0.10	0.08	...	...
Ca	$1.33 \pm .33$	...	...	0.07	0.12	...	...
Capacity (meq/100 g)	124.00 ± 16	...	114	...	155	104	100-107
Exchangeable cations (meq/100 g)							
K	0.45 ± 0.40	0.34-0.59	0.49	0.64	$1.02 \pm .33$	1.23	3.8-4.2
Ca	52.40 ± 9.40	28.6-63.6	19.5	2.0-4.0	$4.35 \pm .73$	5.0	15-16
Mg	7.70 ± 1.40	10.1-19.3	...	...	4.56 ± 1.0	1.83	5.3-6.4
Na	1.20 ± 0.39	...	0.43	...	0.63 ± 0.07	0.87	...
H (by difference)	(61.3)	...	55	...	144.10 ± 16	89	...
Available P (μg/g)	16.9 ± 2.9	50-203	12	14	...	...	30-103

[a] 0–20 cm depth.
[b] Names from authors.
[c] Richardson et al. (1976).
[d] Klopatek (1975).
[e] Malmer and Sjörs (1955).
[f] Walsh and Berry (1958).
[g] Gore and Allen (1956).
[h] Malmer and Sjörs (1955).
[i] Heilman (1968).
[j] Surface H_2O pH (5.8-6.9).
[k] Bulk density in grams per cubic centimetre ($0.13 \pm .02$) and H_2O content by volume percent (77.8 ± 9.1).

productivity. For data on production values for different wetland types see
Reader (1978).

Total N is lowest in the ombrotrophic peat bogs that are *Sphagnum* dominated (Table 1) and is in a nonavailable form (Weetman, 1962, 1968; Tamm
and Holmen, 1967). The peatlands that are highly acid show a low content of
Ca and are also lower in P than are the fens (Table 1). This trend was also
noted by Brune (1948) and supported by Lucas and Davis (1961) who reported a decrease of available and total P when pH dropped below 5.5. They
noted that very acid peats often contained as little as 0.01% P and 0.01% K.
The total Mg content does not vary as appreciably as Ca between wetland
types but bogs have a slightly lower percentage (Table 1). The nutrient data
from Table 1, when coupled with Malmstrom's guidelines, suggest that bogs
in terms of plant growth needs are deficient in P and Ca, and in some bog
systems, K and N may also be limiting.

The cation exchange capacity (CEC) is >100 meq/100 g for all the wetlands types. Exchangeable Ca dominates the fens and marsh systems while
H^+ ions comprise the largest proportion of the bog cations. Available P information is difficult to interpret because different extraction procedures were
employed by the authors and the data do not take into account variations in
bulk density. The variations in peat nutrients from one wetland type to
another as noted in Table 1 are controlled, depending on the wetland type,
by inputs from parent material, runoff, rainfall, geographic location, etc. For
a review of these factors see Moore and Bellamy (1974).

Toth (1968) has reported a Cu deficiency on some peatlands but in general
the study of micronutrients has not had much attention. For a review on the
available literature concerning nutrients in peatlands see Stanek (1975).

The measured relationship between exchangeable and total nutrients is
dependent on the extraction procedure used, the time of year (especially
during the growing season) when the extractions are conducted (Klopatek,
1975), and the amount of mineral content present (Gore and Allen, 1956).
The ratio between exchangeable and total nutrients for peats low in mineral
content is shown to be high for most cations (Table 2). The percent of
exchangeable K to total K ranges from 100% on a blanket bog to 22% in the
Houghton Lake fen. More than half the Ca, and >60% of the total Mg are
exchangeable in most of the peat soils examined (Table 2). Sodium also
shows a high percent of exchangeability. The percentages of the total N, P
and Fe which are available are extremely low in comparison to other elements, but the percentages of available P and Fe in the organic soils are
several orders of magnitude greater than those measured in mineral soils.
Gore and Allen (1956) found that exchangeable iron in peats is largely in the
ferrous state. The percentages of total N and P are lower in bogs than fens
(Table 1) but it appears that a higher percentage of the total concentrations of
these nutrients is available in bog ecosystems (Table 2). An actual comparison between peatland systems of the total N and P available is not possible

Table 2. A Comparison of Extractable and Total Nutrients on Peat Soils vs. Mineral Soils[a]

Element	Peat soil			Mineral soil	
	Fen[b]	Blanket bog[c]	Sphagnum bog[d]	Spruce forest[e]	Deciduous forest[f]
N (as NO_3-N + NH_4-N)	0.1	...	0.4–0.5	...	...
P	1.9	...	4.6–8.8	2.5	.01–.05
K	21.9	100.0	70.2–74.1	12.2–13.6	...
Ca	80.3	71.5	41.3–56.3	24.7–40.9	5.8–50
Mg	65.9	69.9	61.0–71.5	2.2–5.5	.02–.06
Na	100.0	88.7	...	...	...
Fe	...	7.9	...	...	.0001–.0008

[a] Values in table are percentages of availability and reflect the ratio between exchangeable (or extractable) nutrient and total nutrient present.
[b] Richardson et al. (1976).
[c] Gore and Allen (1956).
[d] Pollett (1972); ranges from 1 raised and 2 blanket bogs.
[e] Heilman (1968).
[f] Enfield and Bledsoe (1975); mineral soil with pH of 5.3 and organic matter 1.8%—Entisol, Roscommon, Michigan, $n = 3$.

because bulk density values (which are needed to calculate total and available nutrients mass) are not available.

The chemistry of waterlogged peat soils is not well understood because analyses are few and most of the chemical analysis was done on samples exposed to air. Data are lacking on natural systems, especially in relation to seasonal availability, sorption potentials, microbial activity, interactions of Al and Fe on phosphorus availability, etc. Soil studies in the Houghton Lake peatland as well as those of Gore and Allen (1956) and Heilman (1968) suggest that future work on nutrient content and availability should be expressed on a volume basis (due to bulk density problems) in order to appropriately express relationships of nutrients to plant growth. The relationship between anaerobic peat soils and native wetlands species also deserves considerable study.

Vegetation and Nutrients

Less is known about nutrient dynamics of the vegetation of northern temperate wetlands than any other major ecosystem. This is especially true in the United States and Canada, but Britain and northern Europe have only slightly better information on nutrient dynamics in wetland ecosystems.

Wetland sites vary in the degree to which they are influenced by mineral soil waters. Wetlands under such an influence (i.e., fens) tend to have different floristic and water chemistry characteristics compared to wetlands not affected by groundwaters and surface runoff (i.e., bogs) (Gorham, 1956;

Heinselman, 1970). Concentrations of certain elements in foliage of the same species on a single sampling date vary among wetland types. Elemental concentrations (N, P, Ca and Mg as percent dry weight) in foliage of *Larix laricina* collected in August were higher in a fen (1.95, 0.26, 0.52 and 0.19) than in a conifer swamp (1.06, 0.25, 0.42 and 0.14) or bog (1.24, 0.17, 0.37, 0.09) (Tilton, 1977). Elemental concentrations in "shoots" of herbaceous plant species such as *Erica tetralix, Calluna vulgaris* and *Molinia caerulea* were lowest in a valley bog compared to 2 other peatland sites (Loach, 1968) and concentrations of N, P, K, Na, Mg and Ca in leaves of *Narthecium ossifragum* were lowest in a bog site compared to several fen sites in Sweden (Malmer, 1962). Similarly, Small (1972) found lower concentrations of N and P in foliage of 17 woody and herbaceous bog plant species than in the foliage of different species in a forest and a marsh. Boyd (1978) also confirmed that elemental composition of individual wetland species may vary greatly between sites.

Low supplies of elements in peats of ombrotrophic wetlands have been considered the cause of low concentrations of N, P, Ca, Mg and Na in plant species. However, an additional factor which aggravates ion uptake by wetland plant species is that soils are usually waterlogged in these sites (Loach, 1968; Tilton, 1975). Absorption of many ions is inhibited under anaerobic conditions (Epstein, 1972) and organic matter mineralization is restricted in poorly aerated peats (Latter et al., 1967; Avnimelech, 1971; Chamie, 1976). Poor soil aeration may also reduce root absorptive capacity by retarding root growth (Gore and Urquhart, 1966).

Elemental concentrations in vascular plant species vary among wetlands because of variable supplies of soil nutrients and anaerobic soil conditions, but element concentrations in nonvascular plant species (mosses, lichens, etc.) also vary considerably due to differences in the quantity and quality of atmospheric inputs of elements. *Sphagnum fuscum* typically occupies the uppermost portion of *Sphagnum* moss hummocks and in this position is dependent on dry fallout and precipitation for nutrient supply. Elemental concentrations in live samples of this and other *Sphagnum* species are proportional to bulk precipitation inputs (Ruhling and Tyler, 1969, 1970, 1971; Gorham and Tilton, 1972). Concentrations of some elements, especially Ca, Mg, Al and Fe were lower in *Sphagnum fuscum* from forested regions than from cultivated or urban areas (Gorham and Tilton, 1972).

Seasonal patterns of nutrient concentrations in some common wetland plant species are similar to patterns reported in the literature for other plant species (Guha and Mitchell, 1966; Woodwell, 1974; Chapin et al., 1975; Klopatek, 1975; Boyd, 1978). Nitrogen and P concentrations in sedge, and leaves, stems and twigs of woody plants decrease through the growing season (Table 3). Current year leaves and stems have higher concentrations of N and P than similar older structures and leaves have higher nutrient concentrations than stems.

Table 3. Seasonal Patterns of Nitrogen and Phosphorus Concentration in Various Wetland Plant Species[a]

Component	Element	Percent dry wt of N and P ($\bar{x} \pm$ SD; $n = 16$)			
		May	June	July	August
		Sedge (*Carex* spp.[b])			
	N	1.97 ± .12	1.71 ± .14	1.36 ± .35	1.22 ± .15
	P	.16 ± .02	.12 ± .02	.10 ± .02	.09 ± .02
		Willow (*Salix* spp.[c])			
Leaves	N	3.37 ± .37	2.36 ± .26	2.21 ± .24	2.00 ± .14
	P	.33 ± .07	.22 ± .05	.14 ± .02	.12 ± .01
Stems (new)	N	...	...	.87 ± .12	.86 ± .13
	P	...	...	.11 ± .02	.09 ± .02
Stems (old)	N	.68 ± .12	.62 ± .06	.55 ± .08	.55 ± .08
	P	.06 ± .01	.06 ± .01	.05 ± .08	.05 ± .01
		Bog birch (*Betula pumila*)			
Leaves	N	3.49 ± .24	2.41 ± .24	2.06 ± .26	1.87 ± .24
	P	.30 ± .05	.15 ± .02	.11 ± .02	.09 ± .02
Stems (new)	N	...	2.17	1.24 ± .14	.90 ± .12
	P	...	...	.10 ± .01	.07 ± .01
Stems (old)	N	.65 ± .08	.69 ± .07	.57 ± .05	.57 ± .07
	P	.04 ± .01	.05 ± .01	.05 ± .02	.04 ± .01
		Leatherleaf (*Chamaedaphne calyculata*)			
Leaves (new)	N	...	2.59 ± .20	1.88 ± .11	1.71 ± .18
	P	...	.22 ± .03	.10 ± .01	.09 ± .01
Leaves (old)	N	1.83 ± .14	1.76 ± .16	1.50 ± .10	1.33 ± .41
	P	.12 ± .01	.10 ± .04	.07 ± .01	.07 ± .02
Stems (new)	N	...	1.92	.99 ± .06	.71 ± .07
	P	...	...	.10 ± .01	.07 ± .01
Stems (old)	N	.56 ± .06	.55 ± .10	.44 ± .05	.44 ± .05
	P	.04 ± .01	.04 ± .01	.04 ± .01	.03 ± .01

[a] Houghton Lake, Michigan, USA, 1973.
[b] Includes *Carex lasiocarpa*, *Carex aquatilis*, *Carex oligosperma*, *Carex rostrata*, *Carex comosa*, and *Carex lacustris*.
[c] Includes *Salix subserica*, *Salix lucida*, *Salix pedicellaris*, and *Salix discolor*.

Translocation of large amounts of nutrients from plant parts of wetland plant species prior to abscission or senescence does not seem to be occurring as much as might be expected (Small, 1972). Phosphorus concentrations in some sedges (*Carex* spp.) and leaves of certain shrubs (*Salix* spp. and *Betula pumila*) did not decrease markedly in late summer and their stems did not increase in P concentrations to suggest translocation of P to these perennial structures (Table 3). Furthermore, P concentrations in needles of *Larix laricina* (tamarack), an important wetland tree species, did not decrease significantly prior to needle fall (Tilton, 1977). These studies suggest that translocation of P does not occur from foliage prior to abscission in these plant species.

An exception to the absence of retranslocation of foliar nutrients may be the evergreen ericad *Chamaedaphne calyculata*. Leaves of this species remain attached to the plant during winter and are shed in the next growing season. Phosphorus concentrations in old leaves decrease prior to abscission, but it is not clear whether the decrease is a result of translocation or leaching of soluble P compounds.

In contrast to the decrease in concentrations of N and P in aboveground tissues of sedges during 1973, concentrations of these nutrients remained fairly constant from late May to late August in belowground structures (Wentz, 1976). Due to the role of belowground plant organs (roots and rhizomes) in the growth of sedges, a spring depletion and a fall recharge of nutrients in these structures might be expected, and has been shown in other studies (Klopatek, 1975, 1978). That such a pattern was not found by Wentz may have been a result of late sampling in the spring, insufficient sampling in the fall, or the expression of nutrient concentrations on a percent dry weight basis. Standing crops of green material in a sedge meadow in Minnesota were still changing in November (Bernard, 1974) and P absorption in tundra graminoids continued well into September (Chapin and Bloom, 1976). Studies of nutrient dynamics in wetland plants should begin in early spring and continue into fall and early winter.

The mass of N and P in aboveground live plant material in the Houghton Lake fen varies considerably during the growing season. Maximum standing crops of 78.8 ± 46.8 kg/ha of N and 5.2 ± 3.3 kg/ha of P occurred in July in the leatherleaf and bog birch cover type compared to the 47.9 ± 18.0 kg/ha of N and 3.7 ± 2.1 kg/ha of P in August in the sedge and willow cover type. The lowest amounts of N and P were generally found early in the growing season. In May, 57.8 ± 19.2 kg/ha of N and 4.3 ± 1.7 kg/ha of P were distributed among live aboveground biomass in the leatherleaf and bog birch cover type compared to 16.5 ± 9.8 kg/ha of N and 1.4 ± 1.0 kg/ha of P in the sedge and willow area.

These nutrient data are difficult to compare to other wetlands without knowing standing crops of plant biomass and species composition. For example, a tundra wet meadow composed of sedges, grasses and shrubs had 17.9 ± 2.0 kg/ha of N and 1.2 ± 0.1 kg/ha of P in foliage at the time of peak standing crop (Chapin et al., 1975). Standing crop in August at the tundra site was 1,015 kg/ha (Tieszen, 1972), compared to 4,416 kg/ha in the sedge and willow cover type and 8,572 kg/ha in the leatherleaf and bog birch cover type in Michigan. On a dry weight comparison, the tundra meadow has a higher percentage of N and P than both the Houghton Lake fen cover types.

At the time of peak standing crop, 50% of the total aboveground mass of N in live plant material in the leatherleaf and bog birch cover type was in stemwood and bark produced in previous years. A lower value (13%) in the sedge and willow cover type was due to the high proportion of sedge biomass (60%). Of the total aboveground P in the leatherleaf and bog birch cover type, 60% was in previous stemwood and bark compared to 17% in the same component in the sedge and willow cover type. In contrast to the high proportion of N and P in stemwood and bark, leaves produced in the previous years represented only 1% of the total aboveground mass of these elements on the leatherleaf and bog birch system. If evergreeness provides an advantage through nutrient storage in past year's leaves (Small, 1972), then the proportion of nutrients in this component of leatherleaf might be expected to represent a larger proportion of the aboveground mass of N and P.

Water Chemistry in Wetlands

The chemical composition of waters entering and flowing across a peatland have profound effects on peatland development, floristics, and wetland types. Excellent discussions of these effects and examples of ion concentrations in different wetland types can be found in Gorham (1956, 1967), Sjörs (1959, 1961), Malmer (1962), Heinselman (1970, 1975) and Moore and Bellamy (1974).

Absolute comparisons among various wetland types are difficult because of differences in vegetation, substrates and methods of study. However, differences in water chemistry between such diverse ecosystems as a minerotrophic fen and an ombrotrophic bog are easily distinguishable in that fen waters are dominated by Ca^{++} and HCO_3^- ions and bog waters are dominated by H^+ and SO_4^{--} ions (Moore and Bellamy, 1974). Nutrient dynamics in peatland waters, unfortunately, have not received much attention and are not well known.

Seasonal variation in the chemical composition of peatland waters is closely related to peatland hydrology and, as a result, varies from year to year depending on climatological conditions (J. Kadlec, 1976). The presence of standing water generally induces anaerobic conditions, but lower water levels in late summer may allow for improved aeration within surface litter or peat. Reducing conditions should increase the solubility of Fe and P while increasing aeration should produce the opposite effect.

Concentrations of the inorganic forms of N are also related to the moisture–aeration regime. High NH_4 and NO_3 concentrations in aerated zones created by water level fluctuations result from higher rates of organic N mineralization and oxidation (Avnimelech, 1971; Klopatek, 1975). In addition, NO_3 concentrations may increase during periods of low water because denitrification is inhibited in aerobic conditions (van Cleemput et al., 1975).

Seasonal changes in dissolved nutrients of the surface water in the Houghton Lake fen (Fig. 1) reflect the influence of hydrology. As the water level decreased (in late summer) to below the litter zone (Richardson et al.,

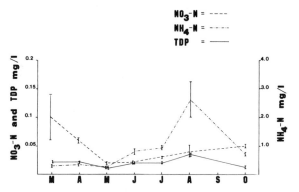

Fig. 1.   Seasonal background NO$_3$-N, total dissolved phosphorus (TDP) and NH$_4$-N concentrations of surface water from the Houghton Lake fen, 1973. Error bars denote 1 SE of the $\bar{x}$.

1976), NH$_4$-N concentrations increased as a result of increased organic matter mineralization. Nitrate-nitrogen was high in early spring (perhaps as a result of the inhibitory effect of low water temperatures on denitrification), decreased in late spring, and increased in late summer. Decreases in NO$_3$ concentrations during late spring and early summer may be due to vascular plant and algal uptake in addition to changes in denitrification rates in the peat (Klopatek, 1975). Total dissolved phosphorus (TDP) remained fairly stable until late summer when concentrations increased slightly (Fig. 1).

In contrast to forms of inorganic N, dissolved cation concentrations changed very little at the Houghton Lake fen through the ice-free season (J. Kadlec, 1976). Seasonal fluctuation may occur, however, if a peatland is ombrotrophic and, therefore, influenced by variation in local precipitation inputs, especially oceanic sources (Gorham, 1967).

Water chemistry also varies with the depth below the peat surface. As with seasonal variation in the chemical composition of peatland water, much of the variation with depth can be attributed to moisture–aeration conditions and plant uptake. As a result, variation in water chemistry between years at a given depth does occur (J. Kadlec, 1976). Concentrations of NH$_4$-N, NO$_3$-N, and TDP were lower in surface waters than in interstitial water (Table 4). The lower NO$_3$-N ion concentrations in surface waters were in part due to algal and microbial uptake. Nitrate-N, Ca, Mg, Fe, Cl and K concentrations in interstitial waters at 15-cm depths were not different from those found at 45-cm depths. Total dissolved phosphorus concentrations were significantly different at all depths with higher concentrations below the surface due to reduced soil conditions. Ammonium ion concentrations tend to be lower in the surface water because as decomposition of organic N occurs, nitrification of the product proceeds rapidly in this relatively well-aerated zone (Tusneem and Patrick, 1972). In the anaerobic zones below the

Table 4. Comparison of Dissolved Nutrient Status Changes with Depth[a, b]

Nutrient	Nutrient concentration		
		Depth below peat surface (cm)	
	Surface H_2O	15	45
NH_4-N (μg/l)	**728** ± 818 (132)	**2,099** ± 1,572 (150)	**1,889** ± 1,532 (157)
NO_3-N (μg/l)	**39** ± 24 (132)	59 ± 28 (151)	57 ± 33 (157)
Ca (mg/l)	**19.3** ± 10.7 (132)	30.2 ± 16.5 (146)	32.3 ± 17.8 (148)
Mg (mg/l)	3.9 ± 1.8 (132)	5.6 ± 2.8 (148)	6.0 ± 3.4 (152)
Cl (mg/l)	27.9 ± 24.6 (127)	28.5 ± 21.2 (149)	23.3 ± 17.5 (155)
TDP (μg/l)	**19.6** ± 9.5 (132)	**40.8** ± 35.5 (147)	**29.3** ± 23.5 (154)
Fe (mg/l)	**0.5** ± 1.6 (132)	1.8 ± 1.5 (149)	1.8 ± 1.4 (154)
K (mg/l)	0.7 ± 0.6 (132)	0.8 ± 0.6 (147)	0.5 ± 0.4 (153)

[a] Houghton Lake peatland, 1973.

[b] Values are $\bar{x}$ ± SD. Sample sizes are in parentheses and were summed over times and cover types. **Within a row, values in boldface are significantly different.**

surface, ammonium ions accumulated as nitrification was inhibited (Table 4).

Peatlands have a profound influence on the chemical composition of waters which enter as streamflow, runoff or fall on the surface as rainfall. The nutrient status of these waters is modified by several processes: microbial activity, sorption and exchange by peat soil, and plant uptake.

The predominant microbial process which modifies the chemical composition of N in incoming waters is denitrification. The rate of NO_3 removal from surface waters is dependent on the amount of water-soluble carbon (Burford and Bremner, 1975), soil pH, redox potential, moisture status, and temperature (Avnimelech, 1971; van Cleemput et al., 1975). Nitrate removal rates vary greatly, but laboratory studies at constant temperature (30°C) of intact flooded soil cores from a freshwater swamp and a salt marsh in Louisiana showed NO_3 removal rates of 2.50 and 7.64 mg $N \cdot l^{-1} \cdot day^{-1}$, respectively (Engler and Patrick, 1974). In the Houghton Lake fen in Michigan, NO_3-N (2.8 mg/l) was added weekly to study plots. Nutrient additions began in May and after 10 weeks of such additions NO_3-N concentrations in surface waters

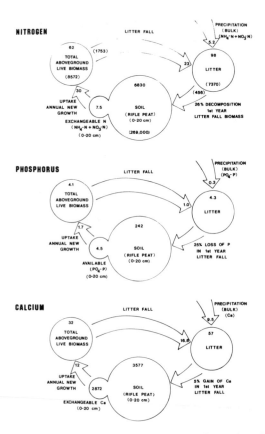

Fig. 2.   A preliminary model depicting annual N, P and Ca flux and pools for the leatherleaf
and bog birch vegetation and soils components of a central Michigan wetland (1973). All values
are in kg/ha. Total biomass for each compartment is in brackets.

of the plots averaged 0.10 mg/l (J. Kadlec, 1976). The greater part of this
decrease in NO_3 concentration was probably due to denitrification.

In addition to plant uptake and microbial processes, sorption and precipitation of ions in the peat are also involved in the dynamics of wetland water
chemistry. Dissolved P is adsorbed onto aerated peat particle surfaces quite
well (Farnham, 1974; Stanlick, 1976). Fulvic and humic acids, which are
common in organic soils and surface waters, are known to form stable or-
ganometallic phosphates (Sinha, 1971). Nutrient addition experiments in the
Houghton Lake fen in which phosphorus was added to vegetation plots at a
concentration of 5.0 mg/l, showed that after 10 weeks of nutrient additions,
surface water concentrations of PO_4-P had remained within background
concentrations of 0.02 mg/l (J. Kadlec, 1976). Klopatek (1975) reported plant
uptake of P during the growing season in a Wisconsin marsh.

Cycling of N, P and Ca in a Peatland

An understanding of biogeochemical cycling in any ecosystem is predicated on information concerning nutrient sources and sinks, transfer rates between compartments and controlling factors for the system. An annual estimate of flux, key transfers and reservoirs for N, P and Ca for the vegetation in the bog birch and leatherleaf cover type in the Houghton Lake fen for 1973 is shown in Fig. 2.

The compartmental pools and transfer rates were calculated on an annual basis for a steady state system. The data base was as follows: (1) litter and litter fall values were based on monthly and biweekly measurements, respectively, and were totaled for the year (Chamie, 1976); (2) soils information was obtained from 3 sampling periods but estimates of exchangeable elements were made only in June; (3) net uptake was calculated from all aboveground new growth through August; (4) total aboveground live biomass was based on mean values in August (N = sixteen 1-m^2 plots) which were considered representative of peak biomass; (5) litter fall biomass, nutrient composition and decomposition values were weighted according to percent biomass contribution (leaves, stems, etc.) and percent nutrient content by species.

The upper 20 cm of the fen peat averaged 6,830, 242 and 3,577 kg/ha of total N, P and Ca, respectively (Fig. 2). This compartment thus represents >97% of the total N, P and Ca. As expected, it is the major pool of nutrients. The N and P content for our site places it in the rich fen classification (Pollett, 1972). Corresponding ranges of N, P and Ca (kilograms per hectare) in the upper 20 cm for raised and blanket bogs in Canada are 370–2,240, 28–102, and 126–799, respectively (Pollett, 1972). The Houghton Lake fen has >3× the total N, 2× the total P, and >4× the Ca values of Pollet's bogs. The differences in nutrient mass between rich fens and bogs are considerable but care must be taken when generalizing between systems because some wetland systems (e.g., minerotrophic fens and blanket bogs) can show considerable overlap in N, P, K, Mg and even Ca content.

An estimate of yearly N nutrient flux reveals that 30 kg/ha and 23 kg/ha are taken up and returned through new growth and litter fall, respectively (Fig. 2). Annual aboveground uptake of N equalled 27% of the total N in aboveground biomass (live and dead). The 7 $kg \cdot ha^{-1} \cdot yr^{-1}$ increase of N in the aboveground biomass compartment is an overestimate since the following losses are not taken into account: (1) belowground transfers, (2) a large dead woody compartment ($\approx$43% of total live, Richardson et al., 1976) which irregularly transfers material to the litter compartment, (3) herbivory effects, and (4) leaching losses. A crude estimate of turnover time for N in the aboveground compartment is $\approx$2 years (i.e., 62/30).

Belowground estimates of biomass for the bog birch and leatherleaf cover types are $\approx$30% of aboveground standing crop (Chamie, 1976; Wentz, 1976). Using this estimate and an average root concentration of 1.5% N we calculate belowground N in live tissue to be $\approx$39 kg/ha. Total peak (aboveground + belowground) N is 101 kg/ha for the bog birch and leatherleaf cover type. Living biomass in this wetland represented a smaller percentage ($\approx$1%) of the total ecosystem N pool than the 6% reported by Odum (1971) for most northern terrestrial ecosystems. The transfer of N from the biomass compartment to the litter compartment via litter fall represents $\approx$23% of the recipient pool (i.e., 23/98). Even if all the annual rainfall N was absorbed by vascular plants, the contributions would be only 17% of the annual uptake of N and far short of yearly growth needs.

Decomposition of the annual litter fall added to the litter compartment averaged 26% or 456 kg/ha (Fig. 2). The percent of dry weight loss did not exceed 42% for any plant parts (even after 20 months) and thus 58% of the biomass of litter fall is added to the peat compartment. The N content of new litter fall decreased steadily for 7 months and then rose to nearly 10% above original values because of increased microorganism activity (Chamie, 1976). For a more detailed analysis of decomposition at this site see Chamie (1976) and Chamie and Richardson (1978).

The amount of P in the bog birch and leatherleaf cover type is an order of magnitude below that of N (Fig. 2). The annual uptake and litter fall return of P is 1.7 and 1.0 kg/ha, respectively. This indicates that aboveground P is accumulated at an annual rate of 0.7 kg/ha, minus, of course, leaching, herbivory, and other losses. Belowground live woody storage of P is estimated to be 2 $kg \cdot ha^{-1} \cdot yr^{-1}$.

The percent of P lost from new litter fall due to decomposition is 25% or 0.25 $kg \cdot ha^{-1} \cdot yr^{-1}$. This may be directly sorbed by the peat, taken up by plants or removed through water flow during high levels.

An examination of plant uptake rates of N and P, the total mass of these nutrients in the soils reservoir and the exchange pool reveals several interesting relationships between low plant productivity at this site (Wentz, 1976) and nutrient availability. It is possible that one of the factors controlling plant growth at our site may be the low level of usable nutrients as indicated by the small amount of exchangeable N and P (7.5 and 4.5 kg/ha to 20 cm depth) (Fig. 2). Nutrient studies from Wentz (1976) do support this hypothesis for the sedge plants because significant uptake of N and P occurred and plant productivity increased significantly on fertilized sedge plots in the wetland.

The cycle for Ca (Fig. 2) and the percent exchangeable in the soil (Table 2) indicate that this cation is readily available and is not in limited supply. The percentage of exchangeable Ca in the system is extremely high when compared to the percentage available in an adjacent mineral soil (Table 2). The slight increase in Ca content in decomposing litter fall after 1 year (Fig. 2) is probably due to lack of removal from structural tissue, the sorption of Ca from calcium-rich surface waters (i.e., 9.5 kg/ha of Ca input from rainfall,

Richardson and Merva, 1976), and not expressing nutrient changes on an ash-free weight basis.

The net flux of Ca ($\approx$5 kg/ha greater than measured uptake) from the biomass compartment via litter fall may in part be indicative of the high percentage (>91%) of leaves in the litter fall and their higher calcium content (0.68%). It may also be the result of an underestimation of uptake as our data are only for the May–August sampling period. An estimate of total live belowground root Ca is 15 kg·ha^{-1}·yr^{-1}. Assuming that belowground new growth increases are proportional to aboveground new growth (annual aboveground total new growth is 25% of total live) allows us to estimate belowground new growth uptake of calcium to be 4 kg·ha^{-1}·yr^{-1}.

Potassium, although in our soils not as exchangeable as Ca (Table 2), is readily leached (95%) from the new litter fall during the 1st year and is the most mobile element in the fen. This is expected because K is soluble and is not a structural part of plant tissue (Epstein, 1972).

Nutrient Budgets at the Ecosystem Level

It has been suggested that wetland ecosystems may "function" as biotic nutrient filters (Surakka and Kamppi, 1971; Richardson et al., 1976; Tourbier and Pierson, 1976) and as sediment traps (Lee et al., 1975). However, complete hydrologic and nutrient budget data are needed to obtain a conclusive answer to the question of whether or not wetlands act as a nutrient trap (natural biological filter) for runoff water from adjacent ecosystems.

The most complete budget for any wetland is Crisp's (1966) study on an eroding blanket bog (Table 5). This study showed that losses due to the fauna were minimal. The highest percentage of total N (83%) and P (52%) losses from the system were due to peat (particulates) in the stream output. Dissolved matter in stream output accounted for the vast majority of K (>81%) and Ca (>91%) removal. However, unknown inputs of cations from springs and bedrock under the stream render Crisp's budget data for those ions suspect. Stream particulate losses were >65% of the annual rainfall input of P and K, while N output considerably exceeded input by precipitation (Table 5). The output of particulate N and P in Crisp's bog was far greater than particulate losses in a New England woodland ecosystem but particulate P yield in the woodland, like the bog, comprised the majority (63%) of the P loss (Bormann et al., 1969; Likens et al., 1977). However, particulate output from this blanket bog was probably higher than normal because of erosion in 11–20% of the catchment area (Crisp, 1966).

Budget data from a noneroding blanket bog in Ireland (Burke, 1975) showed a higher output of dissolved N and approximately half of the P yield of the English bog (Table 5). This output measured as annual surface flow represented 75%, 37%, 21%, 26% and 95% of the precipitation input of NO$_3$-N, NH$_4$-N, P, K and Ca, respectively (Burke, 1975). Dissolved elemental yields

Table 5. A Comparison of Nutrient Budgets for Blanket Bogs, Perched Bogs and Forested Watersheds

Parameters	Nutrients (kg·ha^{-1}·yr^{-1})			
	N	P	K	Ca
Pennine, England, blanket bog[a] (blanket bog vegetation)				
Input				
Precipitation	8.20	0.69	3.07	8.98
Output				
Stream (dissolved)	2.94	0.39	8.97	53.81
Peat (peat in stream)	14.63	0.45	2.06	4.83
Fauna (in stream)	0.06	0.01	0.01	0.00
Sheep (harvesting and sale)	0.05	0.01	0.01	0.02
Yield or total output	17.68	0.86	11.05	58.66
Net loss	9.48	0.15	7.97	49.68
Glenamoy, Ireland, blanket bog[b] (blanket bog vegetation)				
Yield	11.2[c]	0.15	4.7	22.0
Minnesota, USA, perched bogs[d] (watershed aspen—*Picea* forest)				
Yield				
Bog ws-2	1.91	0.08	1.26	3.46
Bog ws-4	1.97	0.08	1.41	3.55
New Hampshire, USA, hardwood forest[e] (*Acer, Betula, Fagus*)				
Yield	2.3[c]	0.01	1.7	7.5[f]
Ontario, Canada, hardwood forest[g] (*Acer, Fagus, Quercus*)				
Yield	2.3[c]	0.16	1.5	12.30
Michigan, USA, aspen forest[h] (*Populus* on good site)				
Yield (soil leachate data at 1 metre)	0.4[c]	0.3	7.6	38.8

[a] Crisp (1966).
[b] Burke (1975).
[c] NO_3-N + NH_4-N.
[d] Verry (1975).
[e] Likens and Bormann (1974).
[f] Likens et al. (1967).
[g] Schindler and Nighswander (1970).
[h] Richardson and Lund (1975).

from stream output flowing from watersheds with 2 perched bogs within them (Verry, 1975), although not as high as Crisp's and Burke's values, were well within the range of dissolved losses from terrestrial ecosystems (Table 5). Verry (1975) noted that output values for the 2 ombrotrophic peatlands watersheds were generally low and were quite similar to values for forested areas which do not contain peatlands. It is important to stress that, unlike the European bog studies, peatlands comprised only 21% of the Minnesota forest watershed and thus Verry's output values are not truly reflective of peat soils.

Table 6. Total Inputs and Losses of Nutrients in a Blanket Bog Following Fertilization[a]

Nutrient	Input			Losses		Total input loss		1973 input[b] loss	
	Rainfall	Applied	Total	Undrained	Drained	Undrained	Drained	Undrained	Drained
NH_4-N (kg/ha)	54			45	35				
NO_3-N (kg/ha)	17			19	39				
Total inorganic N (kg/ha)	71	260	331	64	74	19%	22%	41%	47%
PO_4-P (kg/ha)	2.2	260	262.2	85	84	32%	32%	58%	84%
Ca (kg/ha)	62	2,200	2,262	97	171	4%	8%	...	...
K (kg/ha)	54	520	574	228	139	40%	24%	74%	59%
H_2O (mm)	3,353	...	3,353	1,330	2,231	40%	67%	...	...

[a] January 1971 to October 1973, inclusive; after Burke (1975).
[b] Only an estimate because part of output in 1973 is from previous fertilization.

The outputs of dissolved N, P, K and Ca from the 3 bog systems are equal to or (in some cases) greatly exceed yields for northern forest ecosystems, except for soil K and Ca leachate yields measured on a very sandy aspen site in Michigan (Table 5). Total P and N outputs for the eroding bog were much higher than any of the forested systems yields.

Total inputs (runoff, parent material weathering data, etc.) were not included in the budgets for all these systems and thus a comparison of net losses was not possible. Yield data, from such a wide variety of ecosystems with different geological substrates and geographical regions must be used cautiously, but can indicate relative trends in nutrient flux. The data from Table 5 suggest that outputs from bog ecosystems can be as high as (or even exceed) yields from forested terrestrial systems. Budget data for fens, swamps and marsh systems are not available.

Some insight into the capabilities of one type of wetland ecosystem to assimilate and/or sorb nutrient inputs can be gained from a fertilization study on a drained and undrained blanket bog in Ireland (Burke, 1975). As mentioned previously, natural outputs from this wetland exceeded most of the yields for terrestrial systems. During the 1st year of fertilization (1971), only a slight increase of nutrient content over background levels was noted by Burke. However, by the 2nd and 3rd year of fertilization, outflow of P and K had increased significantly (Table 6). Losses, as percentage of total input for N and P from the undrained bog during the 2-year period, were 19% and 32%, respectively. Surprisingly the drained and undrained bog retained nearly the same percentage of the N and P input (Table 6). Burke (1975) indicated that the capacity of acid peatlands to store either K or P on a long term basis was limited. His conclusions were further supported by the increase in percent loss in 1973, when losses of inputs for P and K on undrained plots were 58% and 74%, respectively (Table 6). Two years following the budget phase of the study, Burke (1975) reported P and K concentrations of 0.7–1.7 mg/l and 4.6–6.7 mg/l, respectively, which were lower than 1973 levels but were still considerably higher than prior to fertilization.

Bogs, fens, swamps and marshes are complex ecosystems that differ among themselves and from terrestrial systems in internal processes, structure and forms of chemical output. The natural yields of bogs are well within the range of (or in some cases exceed) outputs from other systems depending on both the element and the system. It is important to be aware that on a percentage basis wetlands may be very efficient filtering systems, but because of immense nutrient reservoirs or loadings, total losses still can be high. The processing of nutrients by organisms and the cycling of elements at the community and ecosystem level is at this point poorly understood and is deserving of further research.

ACKNOWLEDGMENTS

The authors gratefully acknowledge the funding of the National Science Foundation, Research Applied to National Needs Grant GI-34812X, NIH

Biomedical Grant RR07050 for laboratory research equipment, and the Rockefeller Foundation, Fellowship Program in Environmental Affairs. We are grateful for the cooperation and facilities supplied by the Michigan Department of Natural Resources and The University of Michigan Botanical Gardens. We express special thanks to E. Kasischke, B. Leedy, M. Bergland and B. Schwegler and other graduate students for their help in the field collections, and Y. Wang, M. Quade and M. Dykstra in the School of Natural Resources Terrestrial Ecosystem Laboratory for chemical analyses. We also thank the researchers from the other wetland studies (without whose data this brief review would not have been possible) and the reviewers for their helpful comments.

LITERATURE CITED

Aandahl, H. R., ed. (1974). "Histosols: Their Characteristics, Classification and Use." SSS A: Pub. No. 6. Madison, Wisconsin.

Andersson, A. (1975). Relative efficiency of nine different soil extractants. *Swedish J. Agric. Res.* 5, 125–135.

Avnimelech, Y. (1971). Nitrate transformation in peat. *Soil Sci.* 111, 113–118.

Behan, M. J., and Kinraide, T. (1973). Rapid wet ash digestion of coniferous foliage for analysis of K, P, Ca and Mg. Dept. of Botany, University of Montana, Missoula, Montana. 9 pp. (mimeo).

Bernard, J. M. (1974). Seasonal changes in standing crop and primary production in a sedge wetland and an adjacent dry old-field in central Minnesota. *Ecology* 55, 350–359.

Black, C. A. (1965). "Methods of soils analysis. Parts I and II." Am. Soc. of Agronomy, Inc., Madison, Wisconsin.

Blake, G. R. (1965). Bulk density. *In* "Methods of soil analysis. Part I" (C. A. Black, ed.), pp. 374–390. Am. Soc. of Agronomy, Inc., Madison, Wisconsin.

Boelter, D. H. (1965). Hydraulic conductivity of peats. *Soil Sci.* 100, 227–231.

Boelter, D. H. (1974). The hydrologic characteristics of undrained organic soils in the lake states. *In* "Histosols: Their Characteristics, Classification and Use" (H. R. Aandahl, ed.), pp. 35–45. Pub. No. 6. SSS A. Madison, Wisconsin.

Bormann, F. H., Likens, G. E., and Eaton, J. E. (1969). Biotic regulation of particulate and solution losses from a forest ecosystem. *Bioscience* 19, 600–610.

Boyd, C. E. (1978). Chemical composition of wetland plants. *In* "Freshwater Wetlands: Ecological Processes and Management Potential" (R. E. Good, D. F. Whigham and R. L. Simpson, eds.), pp. 155–167. Academic Press, New York.

Bray, R. H., and Kurtz, L. T. (1945). Determination of total, organic and available forms of phosphorus in soils. *Soil Sci.* 59, 39–45.

Bremner, J. M. (1965). Inorganic forms of nitrogen. *In* "Methods of Soil Analysis. Part II" (C. A. Black, ed.), pp. 1179–1232. Am. Soc. of Agronomy, Inc., Madison, Wisconsin.

Brune, F. (1948). "Die Praxis der Moor-und Heidekultur." Verlag P. Parey, Berlin.

Burford, J. R., and Bremner, J. M. (1975). Relationships between the denitrification capacities of soils and total, water-soluble and readily decomposable soil organic matter. *Soil Biol. Biochem.* 7, 389–394.

Burge, W. D., and Broadbent, F. E. (1961). Fixation of ammonia by organic soils. *Soil Sci. Soc. Am. Proc.* 25, 199–204.

Burke, W. (1975). Fertilizer and other chemical losses in drainage water from blanket bog. *Ir. J. Agric. Res.* 14, 163–178.

Chamie, J. P. M. (1976). "The effects of simulated sewage effluent upon decomposition, nutrient status and litter fall in a central Michigan peatland" Ph.D. Dissertation, The University of Michigan, Ann Arbor, Michigan.

Chamie, J. P. M., and Richardson, C. J. (1978). Decomposition in northern wetlands. *In* "Freshwater Wetlands: Ecological Processes and Management Potential" (R. E. Good, D. F. Whigham and R. L. Simpson, eds.), pp. 115–130. Academic Press, New York.

Chapin, F. S., and Bloom, A. (1976). Phosphate absorption: adaptation of tundra graminoids to a low temperature, low phosphorus environment. *Oikos* **27**, 111–121.

Chapin, F. S., van Cleve, K., and Tieszen, L. L. (1975). Seasonal nutrient dynamics of tundra vegetation at Barrow, Alaska. *Arctic and Alpine Res.* **7**, 209–226.

Crisp, D. T. (1966). Input and output of minerals for an area of Pennine moorland: the importance of precipitation, drainage, peat erosion and animals. *J. Appl. Ecol.* **3**, 327–348.

Dawson, J. E. (1956). Organic soils. *Adv. Agron.* **8**, 377–401.

Enfield, C. G., and Bledsoe, B. E. (1975). "Kinetic model for orthophosphate reactions in mineral soils." Environmental Protection Technology Series. EPA—66012-75-022.

Engler, R. M., and Patrick, W. H., Jr. (1974). Nitrate removal from floodwater overlying flooded soils and sediments. *J. Environ. Qual.* **3**, 409–413.

Environmental Protection Agency. (1974). "Methods for Chemical Analysis of Water and Wastes" Nat. Env. Res. Center Pub. 16020: 625/6-74-003.

Epstein, E. (1972). "Mineral Nutrition of Plants: Principles and Perspectives." Wiley and Sons, New York.

Farnham, R. S. (1974). Use of organic soils for wastewater filtration. *In* "Histosols: Their Characteristics, Classification and Use" (A. R. Aandahl, ed.), pp. 111–118. Pub. No. 6. SSS A: Madison, Wisconsin.

Gore, A. J. P., and Allen, S. E. (1956). Measurement of exchangeable and total cation content for H^+, Na^+, K^+, Mg^{++}, Ca^{++}, and iron in high level blanket peat. *Oikos* **7**, 48–55.

Gore, A. J. P., and Urquhart, C. (1966). The effects of waterlogging on the growth of *Molinia caerulea* and *Eriophorum vaginatum*. *J. Ecol.* **54**, 617–633.

Gorham, E. (1956). The ionic composition of some bog and fen waters in the English lake district. *J. Ecol.* **44**, 142–152.

Gorham, E. (1967). Some chemical aspects of wetland ecology. *In* "12th Ann. Muskeg. Res. Conf. Proc. Techn" pp. 20–38. Mem. 90, N.R.C. Assoc. Comm. on Geotech. Res. Canada.

Gorham, E., and Tilton, D. L. (1972). Major and minor elements in *Sphagnum fuscum* from Minnesota, Wisconsin and northeastern Saskatchewan. *Bull. Ecol. Soc. Amer.* **53**, 33.

Guha, M. M., and Mitchell, R. L. (1966). The trace and major element composition of the leaves of some deciduous trees. II. Seasonal changes. *Pl. Soil* **24**, 90–112.

Heilman, P. E. (1968). Relationship of availability of phosphorus and cations to forest succession and bog formation in interior Alaska. *Ecology* **49**, 331–336.

Heinselman, M. L. (1963). Forest sites, bog processes, and peatland types in the Glacial Lake Agassiz Region, Minnesota. *Ecol. Monogr.* **33**, 327–374.

Heinselman, M. L. (1970). Landscape; evolution, peatland types, and the environment in the Lake Agassiz Peatlands Natural Area, Minnesota. *Ecol. Monogr.* **40**, 235–261.

Heinselman, M. L. (1975). Boreal peatlands in relation to environment. *In* "Coupling of Land and Water Systems" (A. D. Hasler, ed.), pp. 93–103. Springer-Verlag, New York.

Jeglum, J. K., Boissonneau, A. N., and Haavisto, V. F. (1974). "Toward a wetland classification for Ontario" Can. For. Serv., Sault Ste. Marie, Ont. Inf. Rep. O-X-215.

Kadlec, J. A. (1976). Dissolved nutrients in a peatland near Houghton Lake, Michigan. *In* "Freshwater Wetlands and Sewage Effluent Disposal" (D. L. Tilton, R. H. Kadlec and C. J. Richardson, eds.), pp. 25–50. The University of Michigan, Ann Arbor, Michigan.

Kadlec, R. H. (1976). Surface hydrology of peatlands. *In* "Freshwater Wetlands and Sewage Effluent Disposal" (D. L. Tilton, R. H. Kadlec and C. J. Richardson, eds.), pp. 3–24. The University of Michigan, Ann Arbor, Michigan.

Klopatek, J. M. (1975). The role of emergent macrophytes in mineral cycling in a freshwater marsh. *In* "Mineral Cycling in Southeastern Ecosystems" (F. G. Howell, J. B. Gentry and M. H. Smith, eds.), pp. 367–393. ERDA Symposium Series, Conf. 740513.

Klopatek, J. M. (1978). Nutrient dynamics of freshwater riverine marshes and the role of emergent macrophytes. *In* "Freshwater Wetlands: Ecological Processes and Management

Potential" (R. E. Good, D. F. Whigham and R. L. Simpson, eds.), pp. 195–216. Academic Press, New York.

Latter, P. M., Cragg, J. B., and Heal, O. W. (1967). Comparative studies on the microbiology of four moorland soils in the northern Pennines. *J. Ecol.* **55**, 445–464.

Lee, G. F., Bentley, E., and Amundson, R. (1975). Effects of marshes on water quality. In "Coupling of Land and Water Systems" (A. D. Hasler, ed.), pp. 105–126. Springer-Verlag, New York.

Likens, G. E., and Bormann, F. H. (1974). Linkages between terrestrial and aquatic ecosystems. *Bioscience* **24**, 447–456.

Likens, G. E., Bormann, F. H., Johnson, N. M., and Pierce, R. S. (1967). The calcium, magnesium, potassium and sodium budgets for a small forested ecosystem. *Ecology* **48**, 277–285.

Likens, G. E., Bormann, F. H., Pierce, R. S., Eaton, J. S., and Johnson, N. M. (1977). "Biogeochemistry of a Forested Ecosystem." Springer-Verlag, New York.

Loach, K. (1968). Relations between soil nutrients and vegetation in wet heaths. II. Nutrient uptake by the major species in the field and in controlled conditions. *J. Ecol.* **56**, 117–127.

Lucas, R. E., and Davis, J. F. (1961). Relationships between pH values of organic soils and availabilities of 12 plant nutrients. *Soil Sci.* **92**, 177–182.

Malmer, N. (1962). Studies on mire vegetation in the Archaean area of southwestern Gotaland. *Op. Bot.* **77**, 1–67.

Malmer, N., and Sjörs, H. (1955). Some determinations of elementary constituents in mire plants and peat. *Botaniska Notiser* **108**, 46–80.

Malmstrom, C. (1956). OM Skogsproduktionens Naringsekologiska forutsattningar och Mojligheterna att paverka dem. sven. *Skogsvard foren. Tidskr.* **47**, 123–140.

Moore, P. D., and Bellamy, D. J. (1974). "Peatlands." Springer-Verlag, New York.

Odum, E. P. (1971). "Fundamentals of Ecology." 3rd edition. W. B. Saunders, Philadelphia, Pa.

Olsen, S. R., and Dean, L. A. (1965). Phosphorus. In "Methods of Soil Analysis, Part II" (C. A. Black, ed.) pp. 1035–1048. Am. Soc. of Agronomy, Inc., Madison, Wisconsin.

Patrick, W. H., Jr., and Khalid, R. A. (1974). Phosphate release and sorption by soils and sediments: effect of aerobic and anaerobic conditions. *Science* **186**, 53–55.

Perkin-Elmer. (1973). "Analytical method for atomic absorption spectrophotometry." Perkin-Elmer, Conn.

Pollett, F. C. (1972). Nutrient contents of peat soils in Newfoundland. *Fourth Int. Peat Cong. Proc.*, Otaniemi, Finl. **3**, 461–468.

Ponnamperuma, F. N. (1972). The chemistry of submerged soils. *Advan. Agron.* **22**, 29–96.

Puustjärvi, V. (1956). On the cation exchange capacity of peats and on the factors of influence upon its formation. *Acta Agr. Scand.* **6**, 410–449.

Reader, R. (1978). Primary production in northern bog marshes. In "Freshwater Wetlands: Ecological Processes and Management Potential" (R. E. Good, D. F. Whigham and R. L. Simpson, eds.), pp. 53–62. Academic Press, New York.

Richardson, C. J., Kadlec, J. A., Wentz, A. W., Chamie, J. M., and Kadlec, R. H. (1976). Background ecology and the effects of nutrient additions on a central Michigan wetland. In "Proceedings: Third Wetlands Conference. Institute of Water Resources" (M. W. Lefor, W. C. Kennard and T. B. Helfgott, eds.) pp. 34–72. The University of Connecticut Report No. 26.

Richardson, C. J., and Lund, J. A. (1975). The effects of clearcutting on nutrient losses in the aspen forest on three soil types in Michigan. In "Mineral Cycling in Southeastern Ecosystems" (F. G. Howell, J. B. Gentry and M. Y. Smith, eds.) pp. 673–686. ERDA Symposium Series, Conf. 740513.

Richardson, C. J., and Merva, G. E. (1976). The chemical composition of atmospheric precipitation from selected stations in Michigan. *Water, Air and Soil Pollution* **6**, 385–393.

Ruhling, A., and Tyler, G. (1969). Ecology of heavy metals—a regional and historical study. *Bot. Notiser* **122**, 248–259.

Ruhling, A., and Tyler, G. (1970). Sorption and retention of heavy metals in the woodland moss *Hylocomium spendens*. *Oikos* 21, 92–97.

Ruhling, A., and Tyler, G. (1971). Regional differences in deposition of heavy metals over Scandinavia. *J. Appl. Ecol.* 8, 497–507.

Schindler, D. W., and Nighswander, J. E. (1970). Nutrient supply and primary production in Clear Lake, Eastern Ontario. *J. Fish. Res. Can.* 27, 2009–2036.

Sinha, M. K. (1971). Organic-metallic phosphates. I. Interactions of phosphorus compounds with humic substances. *Pl. Soil* 35, 471–484.

Sjörs, H. (1959). Bogs and fens in the Hudson Bay Lowlands. *Arctic* 12, 2–19.

Sjörs, H. (1961). Surface patterns in boreal peatland. *Endeavour* 20, 217–224.

Small, E. (1972). Ecological significance of four critical elements in plants of raised *Sphagnum* peat bogs. *Ecology* 53, 498–503.

Stanek, W. (1975). "Annotated bibliography of peatland forestry." Environment Canada Libraries. Bibliography Series, 76, 1.

Stanlick, H. T. (1976). Treatment of secondary effluent using a peat bed. *In* "Freshwater Wetlands and Sewage Effluent Disposal" (D. L. Tilton, R. H. Kadlec and C. J. Richardson, eds.), pp. 257–268. The University of Michigan, Ann Arbor, Michigan.

Surakka, S., and Kamppi, A. (1971). Infiltration of wastewater into peat soils. *SUO* 22, 51–58. [Finnish-English summary.]

Tamm, C. O., and Holmen, H. (1967). Some remarks on soil organic matter turnover in Swedish podzol profiles. *Medd. Det. Nor. Skogsforsoksves vollebekk Norway.* 23, 69–88.

Technicon. (1975). "Manual of Methods for Auto-analyzer Analyses." Technicon, Tarrytown, New York.

Thorpe, V. A. (1967). Study of moisture, ash, and organic content of peat samples. *J. of the A.O.A.C.* 50, 394–397.

Thorpe, V. A. (1968). Determination of the volume weights, water-holding capacity, and air capacity of water-saturated peat materials. *J. of the A.O.A.C.* 51, 1296–1299.

Thorpe, V. A. (1973). Collaborative study of the cation exchange capacity of peat materials. *J. of the A.O.A.C.* 56, 154–157.

Tieszen, L. L. (1972). The seasonal course of aboveground production and chlorophyll distribution in a wet arctic tundra at Barrow, Alaska. *Arctic and Alpine Res.* 4, 307–324.

Tilton, D. L. (1975). "The growth and nutrition of tamarack (*Larix laricina*)." Ph.D. Dissertation. University of Minnesota, St. Paul, Minnesota.

Tilton, D. L. (1977). Seasonal growth and foliar nutrients of *Larix laricina* in three wetland ecosystems. *Can. J. Bot.* 55, 1291–1298.

Toth, A. (1968). Effect of trace elements on low bogs near Lake Balaton. *Second Int. Peat Congr., Leningrad, USSR.* 2, 721–724.

Tourbier, J., and Pierson, R. W., Jr. (1976). "Biological Control of Water Pollution." University of Pennsylvania Press, Philadelphia, Pa.

Tusneem, M. E., and Patrick, W. H., Jr. (1972). Nitrogen transformations in waterlogged soil. *Louisiana State Univ. Dept. of Agronomy. Agr. Exp. Sta. Bull.* 6, 75 pp.

van Cleemput, O., Patrick, W. H., Jr., and McIlhenng, R. C. (1975). Formation of chemical and biological denitrification products in flooded soil at controlled pH and redox potential. *Soil Biol. Biochem.* 7, 329–332.

Verry, E. W. (1975). Streamflow chemistry and nutrient yields from upland-peatland watersheds in Minnesota. *Ecology* 56, 1149–1157.

Walsh, T., and Barry, T. A. (1958). The chemical composition of some Irish peats. *Proc. Royal Irish Acad.* 59, 305–328.

Weetman, G. F. (1962). Mor humus; a problem in a black spruce stand at Iroquois Falls, Ontario. *Pulp Pap. Res. Inst., Montreal Tech. Rep. Ser.* 277, 18 pp.

Weetman, G. F. (1968). The relationship between feather moss growth and the nutrition of black spruce. *Third Int. Peat Congr. Proc., Quebec City, Canada,* 366–370.

Wentz, A. W. (1976). "The effects of simulated sewage effluents on the growth and productivity of peatland plants." Ph.D. Dissertation. The University of Michigan, Ann Arbor, Michigan.

Wolfe, J. A. (1962). "Analytical procedures for plant and soil samples used in vegetation studies related to the movement of radioactive wastes." U.S. Atomic Energy Comm. and the University of Tennessee, Contract Pub. No. At-40-1-2077.

Woodwell, G. M. (1974). Variation in the nutrient content of leaves of *Quercus alba, Quercus coccinea,* and *Pinus rigida* in the Brookhaven forest from bud-break to abscission. *Am. J. Bot.* **61,** 749–753.

Zoltai, S. C., Pollett, F. C., Jeglum, J. K., and Adams, G. D. (1975). Development of wetland classification for Canada. *Proc. 4th North Am. For. Soils Conf., Quebec City, Canada.* **4,** 497–511.

SEASONAL PATTERNS OF NUTRIENT MOVEMENT IN A FRESHWATER TIDAL MARSH

Robert L. Simpson and Dennis F. Whigham[1]

Biology Department
Rider College
Lawrenceville, New Jersey 08648

and

Raymond Walker

Biology Department
Rutgers University
Camden, New Jersey 08102

Abstract The distribution and movement of dissolved O_2, CO_2, NO_3-N, NH_3-N and PO_4-P in the surface waters of a freshwater tidal marsh were studied. Tidal action, particularly periodic inundation and flushing, resulted in distinctly different patterns of nutrient distribution in the major wetland habitats. Inorganic N and PO_4-P were accumulated in the marsh during summer with emergent vegetation appearing to play an important role in the uptake and retention of nutrients. Although evidence is accumulating that some N and P may be translocated into belowground parts by several perennial macrophytes, most is rapidly leached after death of vascular plants with up to 80% of the total N and even more of the P lost within 1 month. In pond-like areas where filamentous algal blooms develop following the fall dieback of vascular plants, inorganic N and PO_4-P levels remain depressed through the winter and spring. On the basis of presently available evidence, it appears almost all habitats of freshwater tidal marshes may be sinks for inorganic N and PO_4-P during the vascular plant growing season and that certain habitats may continually function as sinks.

Key words *Carbon dioxide, decomposition, dissolved oxygen, emergent aquatic macrophytes, filamentous algae, freshwater tidal marsh, New Jersey, nitrogen, nutrient distribution, phosphorus, tide cycle.*

[1] Present address: Chesapeake Bay Center for Environmental Studies, Smithsonian Institution, Route 4, Box 622, Edgewater, Maryland 21037.

INTRODUCTION

Only limited data exist on nutrient movements in freshwater tidal marshes (Stevenson et al., 1977), but it has been suggested that these wetlands may serve as sinks for certain nutrients, at least during part of the year (Good et al. [1975], Grant and Patrick [1970] and Whigham and Simpson [1975, 1976a, 1976b]). The purpose of this paper is to consider recent studies on nutrient movements in the Hamilton Marsh complex (the northernmost freshwater tidal marsh on the Delaware River) and their relationship to the annual cycle of production and decomposition known to occur in this and other New Jersey freshwater tidal marshes.

The Hamilton Marshes occupy 500 ha of tidal and nontidal land along an old meander adjacent to the Delaware River near Trenton, New Jersey (Fig. 1). Four distinct habitats occur in the marsh: (1) streams and streambanks include Crosswicks Creek, the main stream through the wetland and its tributary channels, (2) pond areas that are continually covered with water although the flow direction reverses with the tide, (3) areas that are pond-like during much of each tide cycle and are drained only at low tide, and (4) high marsh areas (the largest habitat in areal extent) that are covered by shallow water (usually <15 cm) at high tide only (Whigham and Simpson, 1976a).

The marsh is dominated by combinations of perennial (*Nuphar advena*, *Peltandra virginica*, *Lythrum salicaria*, *Sagittaria latifolia*, *Typha latifolia*, *Typha angustifolia*, and *Typha glauca*) and annual species (*Bidens laevis*, *Zizania aquatica* var. *aquatica*, *Polygonum arifolium*, *Polygonum sagittatum*, *Ambrosia trifida*, and *Impatiens capensis*) (Whigham and Simpson, 1975). Peak aboveground biomass for 6 vegetation types ranged from 650 to 2,100 g/m^2 and averaged 950 g/m^2 (Whigham and Simpson, 1976a). These data, however, underestimate net annual community production because Whigham et al. (1978) have shown that high marsh vegetation produces in excess of 4,000 g·m^{-2}·yr^{-1}. After the fall dieback, most of the marsh is devoid of aboveground vascular plants but certain sections, notably the pond-like areas, develop dense mats of filamentous algae. The only major nutrient loading to the marsh comes from the 28.4 × 10^6 litres of secondarily treated effluent discharged daily by the Hamilton Township Sewage Treatment plant into Crosswicks Creek (Fig. 1).

METHODS

For comparison of the patterns of nutrient distribution in the surface waters of the major wetland habitats throughout the year, water samples were collected for a 1-year period beginning in June 1974 at 4 stream sites (1, 2, 7, 8) on Crosswicks Creek (Fig. 1), 1 pond-like site (4B), 1 pond site (4C), 2 sites (4, 4A) on the tributary linking the pond-like area and Crosswicks Creek, and 2 sites (5, 5A) on a second tributary draining an extensive high marsh area (5A). Sites 4 and 5 were on the tributary streams near their entrances to Crosswicks Creek. Samples were collected at 2-week intervals

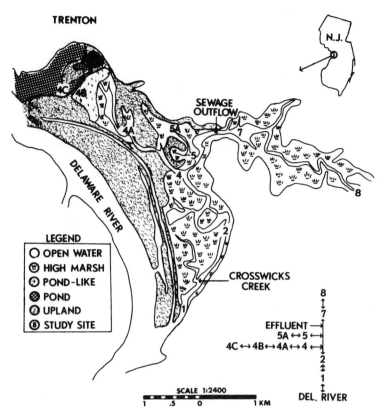

Fig. 1.   Map of the Hamilton Marshes showing the major wetland habitats and study sites. Tidal flow within the marsh is indicated by the diagram at the lower right of the map.

during the summer of 1974 and monthly thereafter at morning high slack water (hsw) and afternoon low slack water (lsw).

Based on the results of the 1974–75 study, investigations were conducted at high marsh Site 5A during the summers of 1975 and 1976 to trace the fate of selected nutrients through complete tide cycles. These studies were begun at either hsw or lsw and focused on both the water running off the marsh surface and water remaining on the marsh.

In the laboratory all samples were analyzed for dissolved O_2 (DO) using the azide modification of the Winkler method (American Public Health Association 1971), CO_2 by titration (American Public Health Association 1971), nitrate nitrogen (NO_3-N), nitrite nitrogen (NO_2-N), ammonia (plus amino acid) nitrogen (NH_3-N) and inorganic phosphate (PO_4-P) following Strickland and Parsons (1968) and total phosphate (total P) following Menzel and Corwin (1965). Because the sites were not gauged, all values are expressed in marshwater concentrations rather than total outputs from the wetland.

RESULTS

Dissolved Oxygen and Carbon Dioxide

Patterns of DO and CO_2 are presented in Fig. 2. The Crosswicks Creek sites (1, 2, 7, 8) and those downstream from the high marsh (Site 5) and pond-like areas (Site 4) show typical seasonal DO patterns. The high marsh Site 5A had depressed DO levels in the summer at both hsw and lsw. Morning high slack water values ranged from 5.5 mg/l to <4 mg/l in the early fall with lsw values usually 1–1.5 mg/l lower during this period. When compared to the main channel of Crosswicks Creek, late fall DO at lsw was noticeably depressed coinciding with the period of maximum decomposition of vascular plant material. Pond and pond-like areas (Sites 4B, 4C) were virtually depleted of DO (often <1 mg/l) in the summer and supersaturated to >16 mg/l at afternoon lsw in the winter and spring. The elevated winter values appeared with the development of dense mats of *Rhizoclonium* and other filamentous algae following the death of vascular plants in the fall at these sites. The channel draining the pond-like areas (Site 4A) reflected lsw patterns similar to the pond-like areas but with less intensity as Crosswicks Creek was approached (Site 4). Except for Sites 7 and 8 upstream from the sewage treatment plant outflow, and Sites 4B and 4C during the winter and spring months, all sites had generally lower DO levels at lsw.

High marsh Site 5A had raised CO_2 levels particularly at lsw. Maximum values of 28 mg/l occurred in the fall during the dieback of vegetation. Though less intense, this pattern was reflected downstream at Site 5. Pond and pond-like areas (Sites 4B, 4C) had elevated CO_2 levels during the summer months of DO depletion with pond Site 4C having values >25 mg/l throughout the summer. A noticeable decline in CO_2 to <3 mg/l occurred during the spring DO maximum. Crosswicks Creek (Sites 1, 2, 7, 8) had consistently lower CO_2 levels than the high marsh and pond-like areas with CO_2 levels usually <5 mg/l at hsw, and 10 mg/l at lsw. Afternoon low slack water values were always higher than hsw values except at Sites 7 and 8 upstream from the sewage outflow and Sites 4B and 4C in the pond-like and pond areas.

Inorganic Nitrogen

Nitrate and ammonia (plus amino acid) N patterns are shown in Fig. 3. Nitrate N levels were near the limits of detection during the summer in the pond-like and pond sites (4B, 4C), but rose dramatically in the early winter. Of particular interest is the fact that hsw values at Site 4B were consistently 10–40 μg–atoms N/litre higher than lsw values during the winter and early spring. This pattern was not seen at downstream Sites 4 and 4A. Site 5A had noticeably depressed NO_3-N levels of <40 μg–atoms N/l at lsw during the months of maximum growth of vascular plants. With the fall dieback, hsw and lsw values became about equal at 100 μg–atoms N/l. Mainstream channel

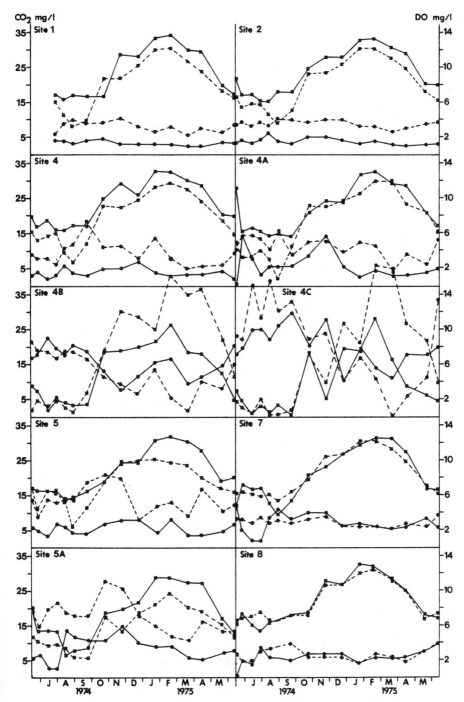

Fig. 2.   Changes in DO (squares) and CO_2 (circles) for 10 marsh sites between June 1974 and June 1975. Solid lines represent high slack water and dashed lines low slack water.

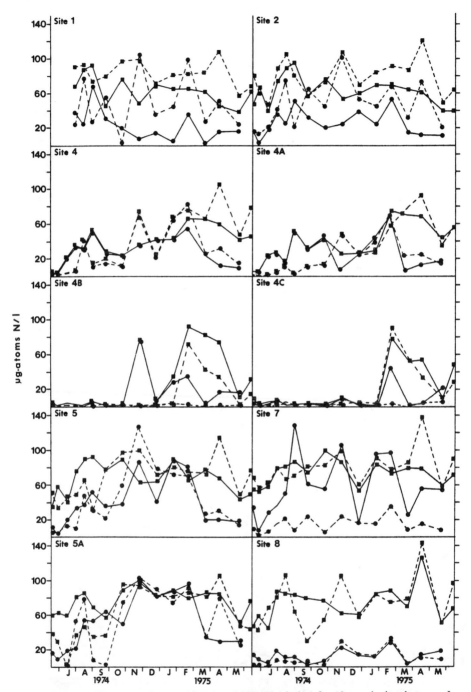

Fig. 3. Changes in NO_3-N (squares) and NH_3-N (circles) for 10 marsh sites between June 1974 and June 1975. Solid lines represent high slack water and dashed lines low slack water.

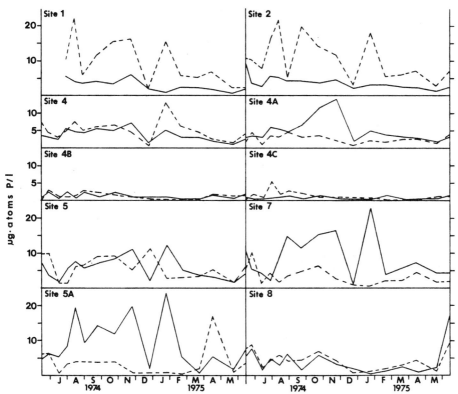

Fig. 4. Changes in PO_4-P for the 10 marsh sites between June 1974 and June 1975. Solid lines represent high slack water and dashed lines low slack water.

sites (1, 2, 7, 8) showed no particular patterns except that lsw NO_3-N values were generally 10–40 μg–atoms N/l higher than hsw levels downstream from the sewage outflow.

Ammonia N levels were extremely low ranging from 5 μg–atoms N/l to undetectable in the pond-like and pond areas (Sites 4B and 4C) during the summer. For the remainder of the year, values stayed <5 μg–atoms N/l at lsw, but typically ranged from 20–80 μg–atoms N/l at hsw. Much of the hsw NH_3-N present at downstream Sites 4 and 4A never reached these areas during the summer months. High marsh (Site 5A) levels were variable during the summer but on several occasions during this period, levels approaching zero were recorded at lsw. At other times during the year, levels were approximately the same at both hsw and lsw with maximum values of 100 μg–atoms N/l occurring during the winter. Ammonia N values at Site 8 at the head of the marshes on Crosswicks Creek were always low, generally being <20 μg–atoms N/l with little difference between hsw and lsw. The main stream channel Sites 1 and 2 normally had elevated values at lsw

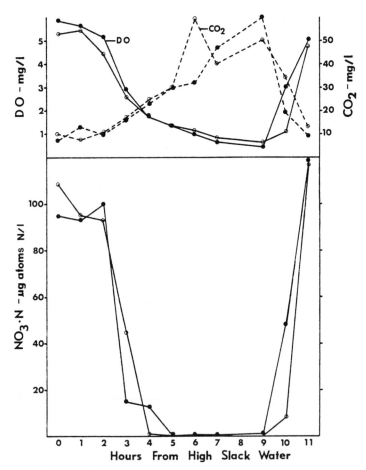

Fig. 5. Changes in DO, CO_2 and NO_3-N during 1 complete tide cycle (beginning with high slack water) at 2 locations on the high marsh surface near Site 5A on 13 June 1975.

while Site 7 upstream from the sewage outflow always had elevated NH_3-H levels at hsw.

Nitrite N during the summer months ranged from 4–8 μg–atoms N/l at high marsh Sites 5 and 5A and stream channel sites (1, 2, 7, 8). Pond-like areas had much lower values with hsw NO_2-N levels <2 μg–atoms N/l and lsw levels <0.5 μg–atoms N/l. Generally, values <2 μg–atoms N/l NO_2-N were found during the winter and spring at all sites, with <0.5 μg–atoms N/l present in the pond-like and pond Sites 4B and 4C.

Inorganic Phosphate

Inorganic phosphate changes during the study are presented in Fig. 4. Inorganic phosphate levels were always low usually being <5 μg–atoms P/l

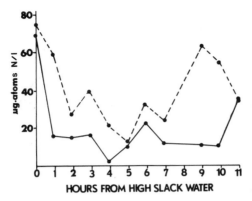

Fig. 6. Changes in NO_3-N leaving the high marsh surface (solid line) and in the adjacent stream channel (dashed line) during 1 complete tide cycle (beginning with high slack water) near Site 5A on 12 August 1976.

in pond-like and pond areas (Sites 4B and 4C). In these areas little difference was found between hsw and lsw. Inorganic phosphate was consistently higher at Site 4A, particularly at hsw, than at pond-like Site 4B. Except for April 1975, the high marsh Site 5A had substantially higher PO_4-P levels at hsw than at lsw. Indeed, levels were considerably higher than those at Site 5 which was a short distance downstream within the same channel. This extra PO_4-P loading most likely came from seepage from a nearby sludge lagoon. Substantially less PO_4-P was found at lsw at Site 5A. Except for Site 8 which is minimally influenced by the tide, all sites on Crosswicks Creek show elevated PO_4-P levels often exceeding 15 μg–atoms P/l that correspond to the direction of tidal movement.

Tide Cycle Studies

Figure 5 shows changes in DO, CO_2 and NO_3-N on the high marsh surface at 2 study areas near Site 5A through a complete tide cycle beginning at hsw. Dissolved O_2 levels were >5 mg/l until the tide ebbed when they dropped rapidly to <1 mg/l where they remained until the next flood tide covered the marsh surface. Then DO levels again rose to >5 mg/l. Carbon dioxide levels showed the reverse pattern, but rose more gradually from ≈10 mg/l to >50 mg/l during the 9 hours that the high marsh was not inundated, and then dropped rapidly to initial levels as the marsh flooded again. Nitrate N dropped dramatically from nearly 100 μg–atoms N/l while the marsh was flooded to undetectable levels during the hours when the marsh drained, and then rose quickly to nearly 120 μg–atoms N/l with the next high tide.

Studies of runoff from the high marsh surface into the tributary channel (Fig. 6) show similar NO_3-N patterns. Here it appears, however, that NO_3-N concentrations may decline somewhat more slowly. Nitrate N levels in the adjacent stream channel were almost always higher than levels leaving the

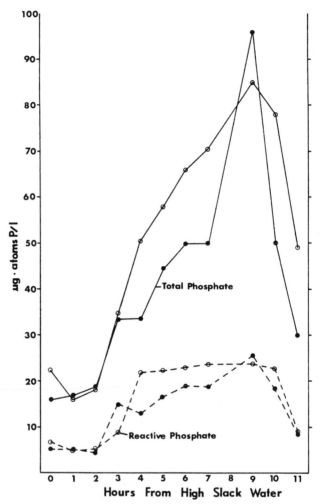

Fig. 7. Changes in PO_4-P and total-P during 1 complete tide cycle (beginning with high slack water) at 2 locations on the high marsh surface near Site 5A on 13 June 1975.

marsh surface. Ammonia N concentrations in this runoff water were always at or near the limits of detection.

Figure 7 shows PO_4-P and total-P levels on the high marsh surface for the same 2 high marsh study areas. Inorganic phosphate levels were at 5–7 μg–atoms, P/l while the marsh surface was flooded and then slowly rose to 20 μg–atoms P/l before falling rapidly back to 8 μg–atoms P/l as the next flood tide covered the high marsh surface. Total phosphate followed the same general pattern, but the magnitude of the increase was much larger going from a value of ≈20 μg–atoms P/l at hsw to nearly 100 μg–atoms P/l immediately before the marsh reflooded.

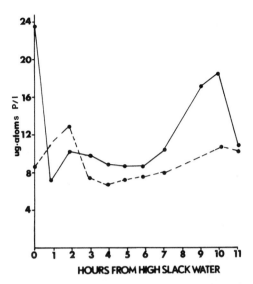

Fig. 8. Changes in PO_4-P leaving the high marsh surface (solid line) and in the adjacent stream channel (dashed line) during 1 complete tide cycle (beginning with high slack water) near Site 5A on 12 August 1976.

Surface waters leaving the high marsh show a somewhat different pattern for PO_4-P (Fig. 8). There was an initial rapid decline in PO_4-P just before water left the marsh surface, then a gradual increase in PO_4-P leaving the marsh through the remainder of the tide cycle. Unlike NO_3-N, PO_4-P concentrations in the adjacent stream channel were generally lower than those of water draining from the high marsh.

The patterns of change in all nutrients were basically the same regardless of when sampling was initiated in the tide cycle.

DISCUSSION

Different patterns of nutrient distribution were found in the major habitats of the marsh. Pond and pond-like sites (4B, 4C) were depleted of PO_4-P and NH_3-N throughout the year, NO_3-N from late spring to midwinter (especially at lsw) and DO in the summer and fall. Carbon dioxide values were elevated during the period of DO depletion with the pond site (4C), the site least affected by tidal action, exceeding 30 mg/l in the summer months. In winter and spring, CO_2 was virtually absent, particularly at afternoon lsw in close correspondence with the growth of dense *Rhizoclonium*-dominated mats of filamentous algae following the dieback of the emergent vegetation.

The major nutrient source for the pond and pond-like sites (4B and 4C) was flood tide Delaware River water that entered the area through tributary Sites 4 and 4A. High slack water inorganic N and PO_4-P concentrations at the tributary sites were always higher than those found at the pond and pond-like sites indicating that both inorganic N and PO_4-P were trapped in the pond-like area upstream from Site 4A. Low slack water inorganic N and PO_4-P concentrations were well below hsw levels at Sites 4 and 4A suggesting that trapped nutrients did not leave the pond-like area with the ebb tide.

Although there were differences in magnitude, marsh water nutrients as high marsh Site 5 were very similar to those in the pond and pond-like sites but only during the emergent macrophyte growing season. There were similar patterns of lsw DO depression, elevated CO_2, depletion of NO_3-N and usually NH_3-N, and depressed PO_4-P levels. Carbon dioxide levels were very elevated in October which corresponded with the period of most rapid decomposition of emergent plants. Similarities between high marsh and pond and pond-like sites disappeared following death and dieback of the emergent vegetation as little difference between hsw and lsw nutrient levels were seen at the high marsh sites during the winter and spring.

In contrast to high marsh, pond and pond-like areas, stream channel sites on Crosswicks Creek (Sites 1, 2, 7 and 8) did not show seasonal depletion of nutrients. Discharge from the Hamilton Township sewage treatment plant greatly influenced inorganic N and PO_4-P concentrations in the main stream channel. Sites 1 and 2 downstream from the discharge point had consistently lower DO and higher CO_2, inorganic N and PO_4-P levels at lsw and Site 7 upstream from the effluent discharge point displayed the reverse pattern. Site 8 at the head of the wetland was not perceptibly influenced by either Delaware River water or effluent. This site consistently displayed little difference in hsw and lsw DO and CO_2 levels, always had low NH_3-N and PO_4-P values, but did have variable NO_3-N levels in part due to runoff from the surrounding watershed following storm events.

These patterns of nutrient distribution were clearly influenced by the action of the tides within the wetland. The main stream channel which experiences a tidal fluctuation of ≈ 2 m (and strong tidal currents) was completely flushed with each tide cycle. The high marsh was likewise flushed with each tide cycle, but the water gently spreads to a depth of 10–15 cm over the wetland surface before receding. In contrast the pond and pond-like areas with a tidal amplitude of 30 cm or less were at best only partially flushed.

The impact of the completeness of flushing and rate of water movement was seen in the distribution of DO in the wetland with the well flushed stream channel sites being near saturation and the poorly flushed pond and pond-like sites constantly approaching complete depletion of DO especially during the summer and fall. The importance of this tidal action was also felt on the high marsh where the periodic inundation insured a renewed supply of DO that tide cycle studies show was virtually depleted within 2–4 h after the tide ebbs.

Periodic inundation with water from the Delaware River likewise insured a relatively constant supply of N and P to the high marsh and pond-like

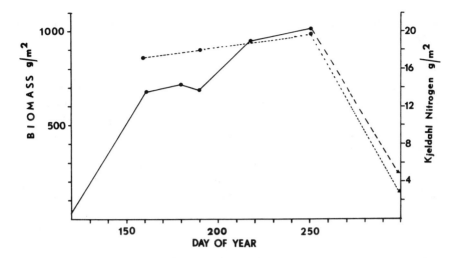

Fig. 9. Aboveground primary production (solid line) and Kjeldahl-N incorporated in that biomass (dashed line) for 1 high marsh study area near Site 5A during 1975. The values for year day 298 (triangles) represent litter mass and Kjeldahl-N in that litter ≈30 days after the end of the 1976 growing season in the same study area.

areas. The high marsh, where flood tide water gently flowed over the surface, apparently acted as a sink for inorganic N during the summer months and the pond-like and pond areas which are only partially flushed appeared to accumulate inorganic N throughout the year. Similar patterns of inorganic N uptake have been reported for the Tinicum Marsh on the Delaware River south of Philadelphia (Grant and Patrick, 1970). Likewise, brackish water marshes of Chesapeake Bay (Stevenson et al., 1977) and Hog Island and adjacent marshes on the Mullica River, New Jersey (Durand and Nadeau, 1972) seem to assimilate inorganic N during the late spring and summer growing period.

Nitrogen was rapidly incorporated by the emergent macrophytes early in the growing season as shown in Fig. 9. After this initial incorporation, N standing crop in the plants remained relatively constant. However, if leaf turnover which has been shown by Whigham et al. (1977) to be substantial is considered, the vegetation likely continues to need a N source to produce new tissue until growth ceases in late summer. It was precisely during this summer period of biomass accumulation that NO_3-N levels appeared lowest in the surface waters of the marsh. Likewise summer tide cycle studies showed NO_3-N was rapidly depleted on the high marsh surface once the flood tide waters recede. It appears, therefore, that at least part of this NO_3-N may have been assimilated by the plants. The fact that marsh vegetation exposed to sewage effluent high in inorganic N had substantially more tissue N than vegetation not exposed to sewage (Whigham and Simpson, 1976*b*) supports this conclusion.

Following death and dieback of the vascular plants at the first frost, there was a rapid loss of weight and even more dramatic loss of N from the standing dead plants. Within a month after death, 80% of the N incorporated in the litter standing on the marsh was lost. Similar losses of N have been found using litterbags placed on the marsh surface although there was only ≈ a 50% loss of N in the 1st month probably because of microbial colonization of the litter on the wetland surface (R. L. Simpson and D. F. Whigham, *personal observation*). Some N however, may be withdrawn from above-ground parts of marsh perennials including *Peltandra*, *Typha* and perhaps *Nuphar* (Good and Good, 1975; R. Walker and R. E. Good, *personal observation*) prior to the end of the growing season.

With death and dieback of vegetation in the fall, hsw and lsw inorganic N levels became roughly equal in the high marsh suggesting no net exchange of inorganic N between the marsh and Delaware River during the fall and winter. In contrast Stevenson et al. (1977) found a net export of inorganic N from a Chesapeake Bay low salinity marsh during the fall. This difference may be due to the fact that the Chesapeake Bay marshes are in agricultural areas whereas the Hamilton Marshes are more isolated from such activities. Although the fate of organic N is unknown for the Hamilton Marshes, studies of the Hog Island Marsh (R. E. Good, *personal communication*) and Chesapeake Bay low salinity marshes (Stevenson et al., 1977) suggest that freshwater and brackish tidal marshes may export organic nitrogen to the estuary.

In the Hamilton Marshes it appears that there was a net import of PO_4-P into pond-like areas. Differences in hsw and lsw surface water concentrations suggest that the high marsh also took up PO_4-P. However, unlike with inorganic N, substantial PO_4-P may be returned to the surface waters after the flood tide waters recede from the high marsh. Phosphorus appeared to be lost quickly when the high marsh vegetation decomposes in the fall with up to 95% of the P incorporated in living tissue lost within the 1st month after frost (R. L. Simpson and D. F. Whigham, *personal observation*). The net annual flux of P is unknown for the Hamilton Marshes, but Stevenson et al. (1977) have found that Chesapeake Bay low salinity marshes export P on an annual basis.

ACKNOWLEDGMENTS

This work was supported in part by the Hamilton Township Environmental Commission and Departments of Planning and Water Pollution Control, the Office of Water Research and Technology through Grant No. B-060-NJ, the National Geographic Society and a Sigma Xi Grant-in-Aid. Special thanks are extended to our many student assistants without whose assistance this work would have been impossible to complete.

LITERATURE CITED

American Public Health Association (1971). "Standard Methods for the examination of Water and Wastewater." 13th Ed. American Public Health Association, Inc., New York.

Durand, J. B., and Nadeau, R. J. (1972). "Water resources development in the Mullica River Basin. Part I. Biological evaluation of the Mullica River-Great Bay Estuary" New Jersey Water Resources Institute, Rutgers University, New Brunswick, N.J.

Good, R. E., and Good, N. F. (1975). Vegetation and production of the Woodbury Creek-Hessian Run freshwater tidal marsh. *Bartonia* **43**, 38–45.

Good, R. E., Hastings, R. W., and Denmark, R. E. (1975). "An environmental assessment of wetlands: A case study of Woodbury Creek and associated marshes." Rutgers University, Marine Sciences Center Technical Report 75-2. New Brunswick, New Jersey. 49 pp.

Grant, R. R. Jr., and Patrick, R. (1970). Tinicum Marsh as a water purifier. *In* "Two studies of Tinicum Marsh, Delaware and Philadelphia Counties, Pa." (J. McCormick, R. R. Grant and R. Patrick, eds.), pp. 105–123. The Conservation Foundation, Washington, D.C.

Menzel, D. W., and Corwin, N. (1965). The measurement of total phosphorus in seawater based on the liberation of organically bound fractions by persulfate oxidation. *Limnol. Oceanogr.* **10**, 280–282.

Stevenson, J. C., Heinle, D., Flemer, D., Small, R., Rowland, R., and Ustach, J. (1977). Nutrient exchanges between brackish water marshes and the estuary. *In* "Estuarine Processes. Vol. 2. Circulation, Sediments, Transfer of Material in the Estuary" (M. Wiley, ed.), pp. 219–240. Academic Press, New York.

Strickland, J. D. H., and Parsons, T. R. (1968). "A Practical Handbook of Seawater Analysis." Fish. Res. Bd. Canada, Ottawa.

Whigham, D. F., and Simpson, R. L. (1975). "Ecological studies of the Hamilton Marshes." Progress report for the period June 1974–January 1975. Rider College, Biology Department, Lawrenceville, New Jersey.

Whigham, D. F., and Simpson, R. L. (1976*a*). The potential use of freshwater tidal marshes in the management of water quality in the Delaware River. *In* "Biological control of water pollution" (J. Tourbier and R. W. Pierson, Jr., eds.), pp. 173–186. University of Pennsylvania Press, Philadelphia, Pa.

Whigham, D. F., and Simpson, R. L. (1976*b*). Sewage spray irrigation in a Delaware River freshwater tidal marsh. *In* "Freshwater wetlands and sewage effluent disposal" (D. L. Tilton, R. H. Kadlec and C. J. Richardson, eds.), pp. 119–144. The University of Michigan, Ann Arbor, Michigan.

Whigham, D. F., McCormick, J., Good, R. E., and Simpson, R. L. (1978). Biomass and primary production of freshwater tidal wetlands of the Middle Atlantic Coast. *In* "Freshwater Wetlands: Ecological Processes and Management Potential." (R. E. Good, D. F. Whigham and R. L. Simpson, eds.), pp. 3–20. Academic Press, New York.

NUTRIENT DYNAMICS:
SUMMARY AND RECOMMENDATIONS

Ivan Valiela

Boston University Marine Program
Marine Biological Laboratory
Woods Hole, Massachusetts 02543

and

John M. Teal

Woods Hole Oceanographic Institution
Woods Hole, Massachusetts 02543

The preceding chapters summarize our present understanding of nutrient cycling in freshwater wetlands dominated primarily by herbaceous vegetation. Boyd (1978) provides a useful review of the observed orders of magnitude for concentrations of a variety of nutrients in herbaceous vegetation. He emphasizes that there are no meaningful "average concentrations" of nutrients in wetland plants and that analysis should be performed in each investigation of nutrient cycling in freshwater wetlands.

Richardson et al. (1978) present the most comprehensive available comparison of nutrient concentrations in water, soils and vegetation of fens, bogs, swamps, and marshes. They show that incoming water is markedly transformed within the wetland, particularly by microbial transformation of N species and nutrient uptake by vascular plants. Their data suggest that sediments form the major nutrient pool, that low vascular plant productivity was associated with low N and P availability, and that Ca was most likely not limiting.

Klopatek (1978) shows that riverine wetlands retain N, and that P may not limit plant growth. He also documents distinct seasonality of nutrient standing crops with Na and K being quickly leached out of plants at the end of the growing season whereas other ions (Ca, Mg, Si) are released slowly. He suggests that the primary role of macrophytes is to pump nutrients from the sediments, thus making them available through leaching and decomposition. The "mining" of sediments by vegetation in a lakeshore wetland is further detailed by Prentki et al. (1978). They found that only 20% of the sediment P which was converted to aboveground biomass was ultimately returned to the sediment. Their calculations result from consideration of a very carefully obtained set of data on the interconversion of P through the various parts of a *Typha* stand. Whether detritus from *Typha* decays in situ and returns P to the sediment through leaching is not certain but net upward removal of P by plants no doubt takes place. It seems unlikely that wetland macrophytes are part of a temporary phenomenon that just mines deposited P. It is likely that there must be additional P inputs into the sediments. The source of additional P to replenish the sediments is not known but Prentki et al. (1978) demonstrate that groundwater, runoff, streamflow, rainfall and dry fall do not provide the input. An interesting aspect of their work indicates that the macrophytes must have evolved in high P ecosystems where there was not a selective advantage given to plants that exhibit a tight coupling of nutrient exchanges. Simpson et al. (1978) also provide data which suggest that freshwater wetlands may be leaky ecosystems. They found marked seasonal changes in N and P concentrations in the surface waters of a freshwater tidal wetland which suggests marked uptake of N and P during the growing season followed by leaching and downstream export during senescence.

Many fundamental questions are left unanswered in the ecology of freshwater wetlands. What are the limiting nutrients? Are these wetlands sinks or exporters of nutrients and how does this depend on nutrient level, if at all? It is especially clear that our knowledge of the dynamics of nutrients in freshwater wetlands is limited because of a lack of quantification of inputs and outputs as well as detailed understanding of the internal cycling of nutrients. The matter of inputs and outputs of nutrients is of special importance to potential use of wetlands for removing nutrients from sewage effluent, as discussed by Sloey et al. (1978), in the management potential section of this volume.

We suggest that several areas of research must be pursued to understand clearly the nutrient dynamics of freshwater wetlands. First, to solve problems of nutrient inputs and outputs, one needs increased emphasis on hydrology. Investigators must obtain precipitation data as well as data on flow rates of both surface and groundwater. Minimally, flow rates of surface water need to be determined.

Once flow rates are known and nutrient concentrations are determined, it will be possible to calculate inputs and outputs. Interpretation of the data may not, however, be straightforward. For example, dissolved organic nitrogen (DON) has been discussed in several papers. When attempts have been made to determine the composition of DON, the sum of such compounds as urea, amino acids and other biologically useful forms of N usually turns out to be a small fraction of the measured DON. Perhaps the rest is refractory and therefore of little use to consumers. Thus, the ecological meaning of exchange of DON is far less well defined than exchanges of, for example, NO_3. The same can be said of organic matter in general.

Both the general literature and the papers in this volume largely ignore the role of microorganisms in the input and output of nutrients. Microbial activity is particularly important when considering N and S. Rates of N fixation should be measured because both blue-green algae and bacteria associated with roots of aquatic plants can be significant contributors to the N budgets of a wetland. Denitrification can lead to major losses of N, as is pointed out by Richardson et al. (1978).

The second major area needing research is the exchange of nutrients among components of freshwater wetlands. The major pools (water, plants, interstitial water, sediments) have been identified and the papers of Richardson et al., Klopatek, Simpson et al., and Prentki et al. (1978) have provided useful information on the seasonal changes in pools, especially P and K. However, the authors themselves could not readily identify the causes of the seasonal changes in the pools or mechanisms that controlled the size of the pools.

To understand internal cycling it is necessary to measure rates of nutrient exchange. We need measurements of gross uptake rates of nutrients by plants and microorganisms. Several of the papers refer to "uptake rates" obtained by measuring the size of the pool of nutrients at peak standing crop and dividing by the time during which growth took place. This is a net measurement and a severe underestimate of the actual amount of nutrient processed by a plant. As reported in the decomposition section of this volume, plants may release materials through leaching while alive, and through decomposition and leaching after plant senescence and death. Prentki et al. (1978) showed good measurements of just such releases, and their methods can be adapted to the specific nutrient and situation. Here interaction with microbial ecologists would also be very helpful. Rates of translocation of nutrients from roots to aboveground parts and vice versa are needed to evaluate the ability of plants to transport nutrients out of sediments, a factor that is especially important to our understanding the fate of nutrients in anoxic sediments.

A third area which seems neglected is the role of animals in nutrient cycling. Muskrats, mice, snails, amphipods, midges and other insects are abundant and may be involved in the production of particulate matter and maintaining the stability of sediments. Excretion of NH_4 by animals must influence nutrients in water.

Fourth, we need to conduct experiments to understand limits and thresholds of nutrients and to predict the effects of changes in nutrient concentration. We need to know what would happen if groundwater nutrients were to increase or decrease. Only through experimentation can we readily identify the specific nutrients and mechanisms involved in regulation of nutrient cycles as well as predict the effect of alterations.

Experiments should be carried out at 2 scales. One is the manipulation of whole ecosystems. In the **Management Potential** section of the volume, Weller (1978) and Sloey et al. (1978) describe experiments involving drawdowns of water levels and fertilization, respectively. These types of experiments provide the key to understanding ecological processes at the ecosystem level. A second smaller scale experiment should also be conducted to test the effects of specific processes or factors. For example, there is some evidence that N and P, alone or in combination, may be limiting nutrients in freshwater wetlands. Further experimental work is needed on this and also the limiting roles of light, grazers and other factors.

In conclusion, we can make few generalizations about nutrient dynamics in freshwater wetlands at present except that they seem to be remarkably leaky systems in regard to what are probably their limiting factors. It is clear, however, that for answers to be forthcoming future research must emphasize the following areas: (1) wetland hydrology; (2) measurement of nutrient exchanges with emphasis on microbial aspects, especially in work concerning N and S; (3) attention to the role of animals, even if merely to demonstrate their relative lack of importance; and perhaps most importantly, (4) experiments on ecosystems to test new hypotheses and obtain answers most directly and efficiently.

LITERATURE CITED

Boyd, C. E. (1978). Chemical composition of wetland plants, In "Freshwater Wetlands: Ecological Processes and Management Potential" (R. E. Good, D. F. Whigham and R. L. Simpson, eds.), pp. 155–167. Academic Press, New York.

Klopatek, J. (1978). Nutrient dynamics of a freshwater riverine marsh and the role of emergent macrophytes. In "Freshwater Wetlands: Ecological Processes and Management Potential" (R. E. Good, D. F. Whigham and R. L. Simpson, eds.), pp. 195–216. Academic Press, New York.

Prenkti, R., Gustafson, R., and Adams, M. S. (1978). Nutrient movements in lakeshore marshes. In "Freshwater Wetlands: Ecological Processes and Management Potential" (R. E. Good, D. F. Whigham and R. L. Simpson, eds.), pp. 169–194. Academic Press, New York.

Richardson, C. J., Tilton, D. L., Kadlec, J. A., Chamie, J. P. M., and Wentz, W. A. (1978). Nutrient dynamics of northern wetland ecosystems. In "Freshwater Wetlands: Ecological Processes and Management Potential" (R. E. Good, D. F. Whigham and R. L. Simpson, eds.), pp. 217–241. Academic Press, New York.

Simpson, R. L., Whigham, D. F., and Walker, R. (1978). Seasonal patterns of nutrient movement in a freshwater tidal marsh. In "Freshwater Wetlands: Ecological Processes and Management Potential" (R. E. Good, D. F. Whigham and R. L. Simpson, eds.), pp. 243–257. Academic Press, New York.

Sloey, W. E., Spangler, F. L., and Fetter, C. W., Jr. (1968). Management of freshwater wetlands for nutrient assimilation. *In* "Freshwater Wetlands: Ecological Processes and Management Potential" (R. E. Good, D. F. Whigham and R. L. Simpson, eds.), pp. 321–340. Academic Press, New York.

Weller, M. (1978). Management of freshwater marshes for wildlife. *In* "Freshwater Wetlands: Ecological Processes and Management Potential" (R. E. Good, D. F. Whigham and R. L. Simpson, eds.), pp. 267–284. Academic Press, New York.

PART IV

Management Potential

MANAGEMENT OF FRESHWATER MARSHES FOR WILDLIFE

Milton W. Weller

Department of Entomology, Fisheries, and Wildlife
University of Minnesota
St. Paul, Minnesota 55108

Abstract   Although commonly practiced on wildlife management areas, marsh management is poorly founded in theory and as a predictive science. Major objectives have been to preserve marshes in a natural state and to maintain their productivity. System or community-oriented management techniques are encouraged as most likely to meet diverse public needs, whereas species-specific management is more difficult, costly and limited in application.

The structure of a marsh is a product of basin shape, water regimes, cover—water interspersion, and plant species diversity. Resultant vegetative patterns strongly influence species composition and size of bird populations. Food resources influence mammals as well as birds. Species richness (i.e., number of species) may be the most simple index to habitat quality, although various diversity indices need further evaluation.

Marshes are in constant change, and wildlife species have evolved adaptations of wide tolerance or mobility. Throughout the Midwest, water levels and muskrats (*Ondatra zibethicus*) induce most vegetative change, and the pattern of vegetation, muskrat and avian responses are predictable in a general way. This short-term successional pattern in marshes forms a usable management strategy. Various ramifications are discussed that may enhance or perpetuate the most beneficial stages.

Artificial management practices are discouraged as costly and of short-term value whereas systems based on natural successional patterns produce the most ecologically and economically sound results. Public pressures for single-purpose management often increase as management potential increases, but such problems often can be avoided by advance planning and public relations.

Marsh management projects for wildlife have rarely been adequately evaluated because of cost, manpower, and inadequate experimental study areas. Some high priority, management-oriented research goals are suggested.

Key words   *Freshwater marshes, management, marsh birds, marsh habitat cycles, muskrats.*

INTRODUCTION

The major objective of this paper should be to summarize the "state of the art" concerning the theory and practice of marsh management for wildlife. I hope this objective will be met but it is difficult because the theory has never been well-formulated, and the practice has developed by trial and error. Therefore, I will attempt (1) to summarize what is known of the marsh as a habitat for wildlife (especially birds), (2) examine the dynamics of marsh habitat that influence management potentials, and (3) outline evaluation efforts and research needs.

In addition to extensive use of the literature, my summary is based mainly on studies of the birds, muskrats, vegetation and invertebrates of glacial marshes in the Midwest. These marshes are characterized by dramatic seasonal as well as year-to-year changes, and with strong herbivore influences. The studies have focused on behavioral and population responses by wildlife to changes in the vegetative substrate produced by both natural and experimental manipulations. These studies lend support to some current practices and discourage the use of others. Above all, they dramatize the need for large-scale, experimental studies to develop a better understanding of marsh fauna and flora.

MANAGEMENT OBJECTIVES AND PRIORITIES

Most of the large individual marshes or marsh complexes preserved today are due to interests of state and federal wildlife conservation agencies or private groups interested in waterfowl hunting. Although these groups still may represent the majority of active managers, many additional agencies and private groups or individuals now are interested in maintaining marshes in a natural state, and in understanding their role in man-influenced as well as pristine systems. In establishing objectives, there is then a tendency to ask "Who is managing the marsh and for what purpose?" In most cases, this is a less significant question than "How can we maintain maximal productivity in this natural system?"

Because the best management may be merely protection via a fence, I include as the first objective of management the preservation of marshes in a natural and esthetically-pleasing state (with or without manipulation) as habitat for wildlife. A second objective is to maintain high productivity of characteristic flora and fauna in marsh units, whether for harvest, enjoyment, or natural biological processes.

In development of a management plan, a marsh should first be classified according to its mean condition or marsh type, and the program for it based on maintenance of the natural values for which it was protected. While this may sound obvious, it is often not the case. Commonly, marshes are viewed as basins that can be changed to a unit more productive of a single species or complex of species other than those found there at a given time. Subsequently, management of the marsh as a system or community should

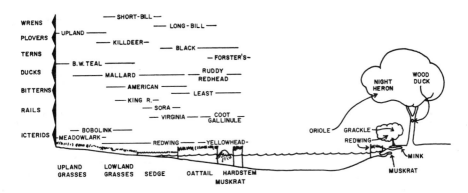

Fig. 1. A schematic drawing of habitat utilization by several families of marsh birds (from Weller and Spatcher, 1965).

produce a wetland attractive to botanists, ornithologists, hunters, photographers and others. Such system or community management requires an understanding of how wildlife species use a marsh, how dynamics in the marsh system affect wildlife, and how natural processes can be used for the benefits of wildlife.

STRUCTURE OF MARSHES AND THEIR
USE BY WILDLIFE

Marshes are known for their concentric zones of plants of various life forms surrounding a deeper, often open, water area. Several detailed studies of marsh vegetation have been conducted in the prairie pothole region (Millar, 1969; Stewart and Kantrud, 1972). The usual pattern is one of robust, deep-marsh emergents like cattail (*Typha* spp.) or bulrush (*Scirpus* spp.) surrounded by wet-meadow taxa like sedges (*Carex* spp.). In some marshes, taller *Phragmites* dominate the edge. Zonation is produced by varying water depths and results in a horizontal plant diversity attractive to different forms of wildlife. An example of wildlife use of a Midwestern marsh is shown in Fig. 1. How these species are ecologically separated is an exciting and complex subject but one outside the scope of this paper.

Beecher (1942) first demonstrated that the number of bird nests was positively correlated with the number of plant communities in marshes. His studies infer general benefits to wildlife by the presence of several plant zones rather than homogeneous stands. Steel et al. (1956) reported larger duck nesting populations in broken than in solid emergent vegetation, suggesting that edge was an important attractant to waterbirds. Patterson (1974) found wetland heterogeneity important in waterfowl productivity. Weller and Spatcher (1965) noted that many marsh bird species nested near water–cover interfaces or the meeting of 2 cover types. Moreover, most species favor marshes in a "hemimarsh" stage with a ratio of about 1:1

cover–water interspersion. Weller and Fredrickson (1974) noted a positive correlation between bird species and the percent open water or the number of open pools in the emergent cover. They also noted that good interspersion was important, but found that greatest species richness and greatest density of nests ranged from 1:1 to 1:2 cover–water interspersion for different species. The studies of MacArthur (1958) on warblers of forest areas focused on the role of layers as an influence on available niches—and, therefore, on species diversity. A marsh with a complex plant zonation also has several heights or layers of vegetation as shown in Fig. 1. Open water acts as another layer, attracting the swimming birds which feed in the open, but use the cover–water edge for nesting. One of these species, the Pied-billed Grebe (*Podilymbus podiceps*), builds a completely floating nest of debris.

Most studies (Beecher, 1942; Weller and Spatcher, 1965) indicate that it is the structure rather than the taxonomic composition of emergent marsh plants that is of greatest importance to nesting birds. Bird species that choose a particular habitat niche (tall and robust emergents, for example) tend to use that life form regardless of the species. Hence, Yellow-headed Blackbirds (*Xanthocephalus xanthocephalus*) may use cattail, hardstem bulrush (*Scirpus acutus*), *Phragmites* or small willows (*Salix* spp.) for nests when such stands are in water and adjacent to open water. They rarely use low sedges. Red-winged Blackbirds (*Agelaius phoeniceus*), however, are much more adaptable and will use low trees, cattail or sedges but clearly favor cattail-sized plants (Fig. 1). When Red-winged and Yellow-headed blackbirds are together, competition enters the picture to a limited degree, as may also be true of Short-billed (*Cistothorus platensis*) and Long-billed marsh wrens (*Cistothorus palustris*).

However, plant species are important when serving as food for wildlife. Cattail seems to be a better food plant for muskrats than is hardstem bulrush. Wet-meadow plants are better food for birds than is cattail. Whereas emergent vegetation may be crucial for nest sites for birds or food for muskrats, submergent plants are substrates for the invertebrates that serve as food for ducks (Krull, 1970). Biologists are just beginning to appreciate how vital certain food needs are to ducks during the reproductive phase (Krapu, 1974). Hence, each plant fills a different role depending upon the wildlife need and season. Therefore, various types of wetlands may serve wildlife needs that differ with season and reproductive cycle.

CLASSIFICATION OF WETLANDS IN
RELATION TO WILDLIFE USE

Various wetland classification systems have been developed, mostly by wildlife biologists, as an aid to quantitating wildlife uses of various wetland types to establish wildlife relationships (Shaw and Fredine, 1956; Stewart and Kantrud, 1971; Cowardin and Johnson, 1973; Golet and Larson, 1974; Jeglum et al., 1974; Bergman et al., 1977; and Millar, 1976). These systems are based mainly on water permanence and vegetation types. A new system is under development with the aid of specialists from many fields that should serve a still more diverse audience (Sather, 1976). The system formulated by Shaw and Fredine (1956) has been much used to classify purchased or leased areas, and is integrated into laws concerning the protection of wetlands. However, in actual fact, very little use has been made of such systems to evaluate habitat for wildlife or even to correlate use by birds in a quantitative way.

In addition to vegetation type, size of a wetland is vital to maintenance of a marsh fauna, especially when the marsh is a relict. In much-drained regions of northern Iowa, several isolated units suggest that a typical wildlife fauna can be preserved in wetlands ≈100 ha in size. Rush Lake near Ayreshire, Iowa is 162 ha and it attracted a typical complement of marsh birds when a 162-ha wetland complex serving as a control was totally dry and unattractive to most marsh wildlife (Weller and Fredrickson, 1974). Goose or Anderson Lake near Jewell, Iowa is 55 ha and is nearly as typical in bird fauna but a 110-ha lake and marsh nearby certainly serves as an attractant (Weller and Spatcher, 1965). Minimal functional sizes need to be established to aid in acquisition of typical marsh areas for wildlife conservation. Many marsh areas now are "habitat islands" due to drainage of surrounding areas, and the outlook for the future is ominous.

However, to preserve a typical marsh avifauna, it is best to have several wetland types, as well as upland areas, present. Heterogeneity of wetland types in a complex creates habitat diversity inducing high species richness. In general, high species richness can be equated to periods of high productivity in a single marsh or in an area (Weller and Fredrickson, 1974), and this index or other mathematical calculations of bird-species diversity may be simple, useful tools to measure habitat quality for some groups of marsh birds (Bezzel and Reichholf, 1974). Because of the need for rapid and inexpensive assessments of marsh bird habitats in evaluating marsh management, considerable work needs to be done on this topic.

THE MARSH AS A DYNAMIC WILDLIFE HABITAT

Marshes usually are in constant change, and those in unstable climates change most readily, and probably are more productive as a consequence. Stability seems deadly to a marsh system, at least where terrestrial or semiaquatic faunas are preferred to open marsh or lake faunas. The pattern of change from dense marsh to open lake is so regular that it may be regarded as a successional trend. This short-term "cycle" tends to be the reverse of the classical textbook concept of marsh succession which is a directional change from an eutrophic lake to a sedge meadow or forest bog in a geologic time frame. Few studies estimate the longevity of long-term wetland succession but one 10-ha pond studied by McAndrews et al. (1967) had a history of ≈11,000 years and was still extant. However, management strategies must

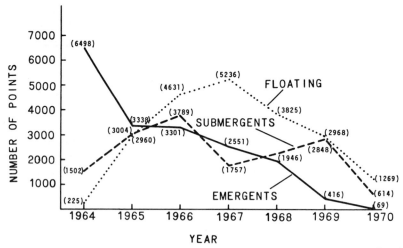

Fig. 2. Numerical changes in plant life-forms along point-count transects at Rush Lake, Iowa, 1964–1970 (from Weller and Fredrickson, 1974).

aim at exploiting short-term changes in marshes which are the products of water levels and herbivores. The chronology of such short-term successional changes usually involves the development of submergents and the elimination of many of the emergents over a period of 3 to 7 years (see example in Fig. 2), but there are natural reversals due to reduced water levels, and similar management strategies may lengthen the process.

Dramatic changes in vegetation are caused by plant–water relationships as demonstrated in both observational and experimental studies: Bourn and Cottam, 1939; Crail, 1951; Dane, 1959; Kadlec, 1960, 1962; Harris and Marshall, 1963; Weller and Spatcher, 1965; Harter, 1966; and Meeks, 1969. Many emergent plants (e.g., cattail) germinate only in shallow water or on mud flats, and revegetation of an open marsh occurs rapidly only when water levels are low. Even vegetative propagation of cattail is more rapid in low than in high water (Weller, 1975). Deep basins or those with stable water levels may remain open because vegetative propagation does not equal losses at such levels, and are open marshes or lakes rather than typical marshes.

In a marsh that has been naturally devegetated and is open, a natural or artificial dewatering ("drawdown") produces a subsequent "germination" phase. At first, sedges and other shallow-marsh plants dominate the plant community. These are gradually replaced by the more water-tolerant species such as cattail (Weller and Fredrickson, 1974), resulting in a shallow, "dense-marsh" stage dominated by robust aquatic emergents. A gradual reopening due to flotation or herbivores may result in excellent interspersion of emergent cover and water ("hemimarsh"). Continued devegetation re-

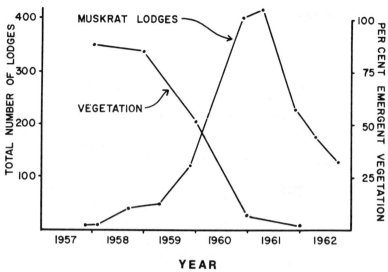

Fig. 3. The relationship between muskrat lodges and the percentage emergent cover at Goose Lake near Jewell, Iowa, 1957–1962 (from Weller and Spatcher, 1965).

sults in a "deep-water" or "open marsh" phase, which resembles a lake system. Variations in the pattern result from timing of drawdown, germination, and reflooding sequence.

Herbivores are the second-most important influence on the structure of Midwestern marshes and, although there are countless invertebrates consuming marsh plants, it is the cutting of vegetation by muskrats (or nutria, *Myocaster coypus*, in the South), that induces dramatic changes especially impacting upon birds. Much of this cutting by muskrats is for construction of "lodges" rather than for direct consumption (Fig. 3), although stored plant tubers later may be eaten. Errington et al. (1963) traced the population history of an especially dramatic muskrat "eat-out," and Weller and Spatcher (1965) recorded its impact on the vegetation (Fig. 3) and on bird populations. Lynch et al. (1947) observed the habitat consequences of herbivory by both geese and muskrats.

Marsh invertebrates clearly respond to water and vegetation as do other consumer-level taxa and seem to be a direct influence of bird species using the area. The general pattern of population change in some major invertebrate groups is related to a schematic pattern of marsh habitat phases in Fig. 4 (from Voigts, 1976). Although additional data are badly needed, especially from experimental manipulation, the patterns help to explain why the hemimarsh is so productive of many nesting bird species, and why many bird species concentrate to feed in semiopen marshes for both invertebrates and submergent plants.

Wildlife responses to dramatic changes in vegetation and water have been

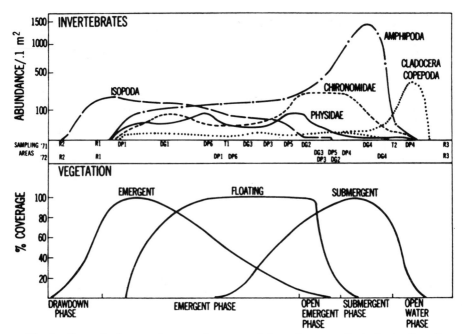

Fig. 4. Generalized vegetation macrofauna associations along a gradient of natural vegetation change in freshwater marshes (from Voigts, 1976).

reported by several workers based on observational studies: Wolf, 1955; Evans and Black, 1956; Johnsgard, 1956; Yeager and Swope, 1956; Weller et al., 1958; Boyer and Devitt, 1961; Bednarik, 1963; Rogers, 1964; Weller and Spatcher, 1965; Anderson and Glover, 1967; Krapu et al., 1970; Smith, 1970; and Stewart and Kantrud, 1974. Fewer experimental studies have successfully assessed wildlife responses to marsh changes (Schroeder et al., 1976; Weller and Fredrickson, 1974). However, these observations demonstrate the responsiveness and adaptiveness of wildlife species associated with dynamic habitats.

Species like Red-winged Blackbirds are adaptable to a wide range of marsh nest sites and are regularly recorded each year despite dramatic differences in water level and vegetation (Beecher, 1942). Niche-specific birds like certain ducks and Yellow-headed Blackbirds can be totally eliminated in some years. In abandoning unsuitable habitats, it is also obvious that birds invade newly available or suboptimal areas readily. The invasion rate of pioneering species in new habitats is an important component of a management strategy. Species with wide ranges in tolerance pioneer easily, but those with specific requirements are forced to pioneer to fill those needs. Information on the rates and causes of pioneering are sparse. Pioneering is a product of a complex evolutionary history involving marsh instability, interspecific competition, geography, and geologic influences.

Population shifts must occur regularly between marshes in a wetland cluster, but they also are seen in areas as widespread as the prairie potholes *versus* the Alaskan and Siberian tundra. Pintails (*Anas acuta*) and other ducks seem to shift to suboptimal and marginal ranges when the optimal habitats go dry (Hansen and McKnight, 1964; Henny, 1973; and Pospahala et al., 1974). However, one must be cautious not to infer that reproductive success in those areas is equal to that in optimal areas; it undoubtedly isn't, but it may be sufficient to warrant the energy required to shift as opposed to total failure or even lack of survival of breeding stock that would result by remaining in drought-stricken areas.

Another form of adaptability is in chronology of nesting. In my studies of waterbirds in northwest Iowa, I have seen 2 cases where Coots did not abandon a devegetated marsh, but remained in flocks until new vegetation appeared late in the season. Then, at least some of the population nested. Late migration and breeding by Short-billed Marsh Wrens also may be an adaptive response to suitable habitat conditions. There are cases, however, where ducks flock during a drought and seemingly never nest (Smith, 1971).

Muskrats also respond quickly to changing conditions. New habitats are invaded promptly, probably as a product of population shifts after the breeding season and prior to winter establishment (Errington, 1963). Reproductive rates vary markedly with habitat conditions with a density-dependent growth rate that influences population regulation. Very often, however, their numbers exceed estimates based on house counts made from shore, and destruction of emergent vegetation progresses so rapidly that population control comes after-the-fact. For this reason, regular surveys of lodge-building and feeding activities by muskrats requires travel by foot or canoe.

Although year-to-year water levels regulate major vegetation growth and wildlife responses, this does not mean that water levels must change constantly. Indeed, stable levels are important during breeding of birds that nest overwater (Low, 1945) and muskrats rearing young in lodges (Errington, 1937).

THE USE OF NATURAL PROCESSES IN MARSH MANAGEMENT

The short-term marsh habitat cycle based on water and herbivores as dominant causal factors, and the response of consumer-level organisms like invertebrates, muskrats and birds, is a natural dynamic pattern that is predictable and reproducible in a general way. It is, therefore, the theoretical basis on which we can establish management procedures. The resultant management practices may be very drastic to untrained eyes and they are commonly misinterpreted, but they are totally natural and, as such, are the most ecologically and economically sound methods. Some general principles of marsh management for wildlife are as follows:

1) System, rather than species management, results in widespread benefits

to all plants and wildlife. Although there is some evidence of competition in birds, losses in production of game species due to high species richness of nongame species have not been demonstrated.

2) Manipulation to produce early plant successional stages results in longer lasting benefits, and creates diverse habitat niches. A marsh then proceeds through various phases with productivity of any one species being a dynamic component of the system. Methods producing long-term results are less expensive and more natural. Usually, this means use of natural tools for management.

3) To maintain heterogeneity in wetland complexes, all marsh units in an area should not be managed in the same way at the same time, even with extreme climatic conditions. This permits local population shifts of wildlife to more optimal niches.

4) Tools such as remote sensing offer exciting opportunities to enhance and document marsh management (Olson, 1964; Cowardin and Myers, 1974) but there is no substitute for the manager getting into the marsh.

Water-level regulation is the ideal way to influence marsh vegetation, but many wetlands lack a regular supply of water or suitable control structures. Where water control is possible, the following steps outline procedures for management of the plant community based mainly on observation by Weller and Fredrickson (1974):

1) When a marsh has become so open that it no longer provides cover and food for wildlife, revegetation comes most readily via drawdown of water levels to produce the germination phase. Overwinter drawdowns seem to stimulate greater growth than single-growing season drawdowns. Timing of the drawdown influences plant species composition, as does soil moisture (Meeks, 1969).

2) Following germination, water levels must be increased slowly in late summer or fall to avoid excessive loss of plants due to flotation or shading.

3) Similar water-level management during the following growing season will reduce bird use by some groups but will foster the greatest vegetative response. By the fall of the second growing season, water level management will be related more to wildlife use than to plant growth.

4) Vegetative propagation of emergents can be stimulated secondarily by lowering water levels. Whether the rate of growth can compensate losses to herbivores is dependent on their population status at the time. Further study is essential, especially in reference to soil and water chemistry, nutrient cycling and invertebrate populations, because few marshes seem to remain productive with continuous inundation.

5) Establishment of submergents vital to invertebrate production usually requires several years of stable levels of moderate depths. Flooding during the growth phase should be avoided because of clouding of the water with silt and other particulate matter that reduces light penetration.

To enhance the growth or regulation of populations of major marsh wildlife, the timing and degree of water-level manipulations are very important:

1) Dramatic drawdowns should take place in early spring prior to bird territory establishment or in fall prior to muskrats lodging for winter. If properly timed, lowered water levels enhance harvest of muskrats and prevent wasteful mortality.
2) Low water levels essential to survival of plants may induce use mainly by wet-meadow birds or, in larger units, restrict deep-marsh birds and muskrats to deeper, central areas of the marsh.
3) Drastic manipulation of water levels following vegetation establishment may aid in regulating muskrat populations that threaten the vegetation. Habitat conditions influence reproductive rates and trapability.
4) As central portions of the marsh are opened by muskrats and flotation, raising water levels makes peripheral areas more suitable for use by wildlife.
5) Some small fish live in marshes but the consequences of drawdowns do not seem to have been documented. Marshes associated with lakes often are major spawning beds and are, therefore, of importance in planning marsh drawdowns. Nongame fish like carp are detrimental to survival of emergents and development of submergents (Robel, 1961) and may be controlled by drawdown procedures.

Several other natural processes have been adapted in marsh management. Grazing of a less intensive nature than performed by muskrats occurred naturally by deer (*Odocoileus virginianus*) and bison (*Bison bison*); now cattle may fill a similar role. Their indirect actions via the trails they make usually are more impacting than is their direct food consumption. There has been little experimental study of grazing impacts on freshwater marsh vegetation but careful studies of salt marsh have been made (Reimold et al., 1975). Some serious consequences of grazing on duck populations were reported during severe drought (Weller et al., 1958), and considerable effort has been devoted to measuring negative effects of grazing on ducks that nest in upland vegetation (Kirsch, 1969).

Burning in permanent marshes, especially those where *Phragmites* is dominant, has produced some benefits in nutrient recycling and regrowth attractive to nesting ducks (Ward, 1968), but careless burning of marsh vegetation during the nesting season produces direct losses (Cartwright, 1942). Burns in northern peat-bottomed marshes may produce deep potholes that prevent growth of emergents (Linde, 1969).

Modification of water salinity to achieve freshwater and freshwater vegetation is a common procedure along coastal areas (Palmisano, 1972). Natural drainages often are blocked in barrier wetlands and the accumulating freshwater leaches out salt. A resulting shift of freshwater vegetation attracts different wildlife species. Similar techniques have been used in marshes around Great Salt Lake where impoundment and leaching are necessary to produce freshwater marsh vegetation (Nelson, 1954; Christiansen and Low, 1970). A potential conflict thus exists between groups interested in the preservation and management of these 2 marsh types.

ARTIFICIAL PROCEDURES USED FOR MODIFYING MARSHES

It is possible to simulate advanced successional stages in marshes that lack water control structures. Openings can be created in dense stands of cattail and other emergents if they are cut during drought or on ice at low water levels and the cut ends are flooded during the next growing season (Nelson and Dietz, 1966). Herbicides are perhaps still easier to use on large areas (Martin et al., 1957) and water levels are insignificant, but side effects are uncertain and, due to time for natural decay of standing plants, openings are not evident until the following year. Obviously, the interspersion thus created may last a shorter time. Theoretically, a more permanent cover–water interspersion could be produced by creating a marsh basin of uneven depth during construction, but this is rarely done because contour data are gross and equipment operation is expensive. It often occurs by accident where streams are the water source and where stream meanders remain in the upper portion of the basin, but this usually affects only a small area. A common practice is to create islands for nesting birds in artificially constructed marshes (Uhler, 1956; Linde, 1969). Level ditching combines some benefits of both of these concepts (see Mathiak and Linde, 1956) but the product is esthetically questionable.

The use of dynamite to open dense marshes has been common (Provost, 1948) with some evaluation of longevity and bird use (Strohmeyer and Fredrickson, 1967). Blasting with ammonium nitrate or other less expensive explosives, and bulldozing have revitalized this effort (Mathiak, 1965). In most cases, too little of the bottom is modified to create cover–water interspersion comparable to hemimarsh. Such newly created ponds are attractive to certain birds at certain stages, but a more significant issue is whether such areas should be permanently modified. Sedge marsh areas, the wetland type most commonly blasted, may serve different wildlife than those for which many agencies tend to manage (e.g., Short-billed Marsh Wrens instead of ducks). Moreover, such marshes also may change marsh types from shallow to deeper marsh during high water levels, and attract species that have shifted from marshes then less attractive due to water levels and muskrats.

In marshes that tend to remain too long in the open phase, there have been efforts to plant perennial cover such as bulrushes and cattail (Steenis, 1939; Addy and MacNamara, 1948). This is expensive in manpower and time, is rarely successful, and opposes natural successional trends.

In wet meadows, dewatering, cutting and fertilizing have been used to stimulate growth of goose foods (Givens and Atkeson, 1957; Owen, 1975). Experimental fertilization of freshwater marshes seems to have been tried only rarely, but its value may be more tied to understanding nutrient cycles than for any feasible management procedure.

The most intensive, species-oriented form of management is the use of artificial nest sites for ducks. These have been used where natural sites are not present regularly or for increased protection against predators. Such platforms, baskets or boxes attract ducks and they are reasonably successful (Bishop and Barratt, 1970), and such birds return to these sites in subsequent years (Doty and Lee, 1974). This method appeals to sportsmen but is esthetically unsatisfying and is less sound either ecologically or economically when more natural management systems are possible. It is unlikely that a harvestable continental population of any duck can be retained on the basis of this type of management. The widespread practice of erecting nest boxes has successfully attracted Wood Ducks (*Aix sponsa*) and, although duck nesting success is higher than in natural cavities (Bellrose et al., 1964), the proportion of the total national production that occurs in such boxes must be modest.

SOCIOLOGICAL ASPECTS OF
MARSH MANAGEMENT

In spite of the potential of managing marshes for a diverse public, it is impossible to please all interest groups. Many individuals and some organized private groups object to management in any form and one has difficulty in arguing against a pursuit of natural values. Realistically, there are large areas where protection alone is the only available management option. However, many of the most sought-after marshes are relicts with far-from-natural watersheds, flora or fauna. In such cases, management is essential and can be accomplished with natural tools in an esthetically pleasing manner. Conflicts over the issue of manipulation are most likely to occur where artificial and esthetically questionable systems are used.

A still more common conflict is between management-oriented groups with different goals, or those attempting decisions on the basis of insufficient knowledge of a complex subject. In semipermanent marshes, fishing may be an important local activity. Dramatic drawdowns designed to revegetate the marsh produce a direct conflict that is not easily resolved if the wetland is in public ownership. Such problems often are worsened when a water control structure is added, when the water supply is stabilized, or when a dredge is readily available because public pressure by fishermen, boaters, waterskiers and cabin owners may be to change a marsh to an open marsh or lake. Advance public relations programs are essential to prevent the development of such local pressures. As greater appreciation of marshes develops, this may be less of a problem, but it is an issue worthy of consideration as management strategies are developed for specific areas.

EVALUATING MARSH MANAGEMENT
TECHNIQUES AND PROGRAMS

Because of the long period required for experimental work, and the lack of readily available controls for demonstrating cause and effect in a complex ecosystem, too little effort has been devoted to evaluation. An assessment of the size of a given wildlife population has been the major means of judging habitat quality or management results, but this is an expensive procedure for a large project of long duration. Qualitative judgements of before and after conditions, especially of marsh vegetation, often replace quantitative pre- and post-population data. While these procedures may have been adequate for gross management operations, they have not made marsh management a quantitative effort. More quantitative data are needed for modeling the system so that the consequences of manipulations can be predicted more precisely. Data on plant germination during drawdowns are fairly extensive but are not representative of a great variety of marsh types. The literature contains some data on muskrat densities but usually not with a quantitative and concurrent measurement of the food resource. Of the references reporting increased duck use after reflooding, few demonstrate increases in the foods that attracted them there. Quantitative data on songbird breeding populations in marshes are surprisingly uncommon. Such observational data gathered with clear-cut management objectives would be very valuable.

What is most needed to advance marsh management theory and practice are experimental data gathered concurrently by a team of specialists in marsh plants, limnology, hydrology, invertebrates, and vertebrates. If such a group also had an isolated, artificial, multi-unit marsh complex designed exclusively for experimentation on the ecology of marshes and wildlife, the potential for understanding marsh systems and devising proper marsh management procedures for marsh wildlife would be unsurpassed. Among the many topics that need to be addressed by careful experimental study are:

1) Habitat stimuli that attract wildlife to marshes
2) The development of indices to wildlife production in marshes
3) The size of isolated areas essential to the development and/or maintenance of marsh fauna (i.e., the marsh as a habitat "island")
4) The diversity or heterogeneity of wetland areas in a complex essential to attract and maintain marsh wildlife
5) Wetland:upland ratios conducive to preservation of typical prairie-wetland biotas
6) Germination conditions that make marsh drawdowns or other water manipulations more effective and predictable
7) A better understanding of water and soil chemistry of marsh systems to expand on the work of Cook and Powers (1958) and Harter (1966)
8) The role of siltation, fertilizers and other man-made products in modifying productivity of wetland areas

9) Objective experimentation on grazing, burning and other natural procedures to assess their role in marsh management for wildlife
10) The relationship of invertebrates to marsh dynamics
11) Detailed studies of the biology of dominant aquatic plants, such as the work of Linde et al., 1976.

Freshwater marshes often have been viewed as transient stages of either terrestrial or aquatic ecosystems. They are, however, unique, long-lived, and highly productive systems. They are a valuable resource and have not received the sophisticated and intensive study necessary to understand, manage and preserve them for posterity. Wildlife are among the most attractive components of the marsh, and their study in relation to their habitat will induce broadly applicable concepts and understanding of this complex system.

LITERATURE CITED

Addy, C. E., and MacNamara, L. G. (1948). Waterfowl management on small areas. Wildl. Manage. Inst., Washington, D.C. 80 pp.

Anderson, D. R., and Glover, F. A. (1967). Effects of water manipulation on waterfowl production and habitat. Trans. N. Am. Wildl. Nat. Resour. Conf. 32, 292–300.

Bednarik, K. E. (1963). Marsh management techniques, 1960. Ohio Dep. Nat. Resour. Game Res. Ohio 2, 132–144.

Beecher, W. J. (1942). Nesting birds and the vegetation substrates. Chicago Ornithol. Soc. Chicago, Illinois. 69 pp.

Bellrose, F. C., Johnson, K. L., and Meyers, T. V. (1964). Relative value of natural cavities and nesting boxes for wood ducks. J. Wildl. Manage. 28, 661–675.

Bergman, R. D., Howard, R. L., Abraham, K. F., and Weller, M. W. (1977). Waterbirds and their wetland resources in relation to oil development at Storkersen Point, Alaska. U.S. Fish Wildl. Serv. Resour. Publ. 129. 38 pp.

Bezzel, E., and Reichholf, J. (1974). Die Diversität als Kriterium zur Bewertung der Reichhaltigkeit von Wasservogel-Lebensräumen. J. Ornithol. 115, 50–61.

Bishop, R. A., and Barratt, R. (1970). Use of artificial nest baskets by mallards. J. Wildl. Manage. 34, 734–738.

Bourn, W. S., and Cottam, C. (1939). The effect of lowering water levels on marsh wildlife. Trans. N. Am. Wildl. Nat. Resour. Conf. 4, 343–350.

Boyer, G. F., and Devitt, O. E. (1961). A significant increase in the birds of Luther Marsh, Ontario, following fresh-water impoundment. Can. Field-Nat. 75, 225–237.

Cartwright, B. W. (1942). Regulated burning as a marsh management technique. Trans. N. Am. Wildl. Nat. Resour. Conf. 7, 257–263.

Christiansen, J. E., and Low, J. B. (1970). Water requirements of waterfowl marshlands in Northern Utah. Salt Lake City, Utah Div. Fish Game Publ. No. 69-12. 108 pp.

Cook, A., and Powers, C. F. (1958). Early biochemical changes in the soils and waters of artificially created marshes in New York. New York Fish and Game J. 5, 9–65.

Cowardin, L. M., and Johnson, D. H. (1973). A preliminary classification of wetland plant communities in north-central Minnesota. U.S. Fish Wildl. Serv. Spec. Sci. Rep.-Wildl. 168. 33 pp.

Cowardin, F., and Myers, V. (1974). Remote sensing for identification and classification of wetland vegetation. J. Wildl. Manage. 38, 308–314.

Crail, L. (1951). Viability of smartweed and millet seed in relation to marsh management in Missouri. Missouri Conserv. Comm., 16 pp.

Dane, C. W. (1959). Succession of aquatic plants in small artificial marshes in New York State. *New York Fish and Game J.* **6**, 57–76.

Doty, H. A., and Lee, F. B. (1974). Homing to nest baskets by wild female mallards. *J. Wildl. Manage.* **38**, 714–719.

Errington, P. L. (1937). Drowning as a cause of mortality in muskrats. *J. Mammal.* **18**, 497–500.

Errington, P. L. (1963). Muskrat populations. Ames, Iowa. Iowa State Univ. Press.

Errington, P. L., Siglin, R., and Clark, R. (1963). The decline of a muskrat population. *J. Wildl. Manage.* **27**, 1–8.

Evans, C. D., and Black, K. E. (1956). Duck production studies on the prairie potholes of South Dakota. U.S. Fish Wildl. Serv. Spec. Sci. Pre.-Wildl. No. 32. 59 pp.

Givens, L. S., and Atkeson, T. Z. (1957). The use of dewatered land in southeastern waterfowl management. *J. Wildl. Manage.* **21**, 465–467.

Golet, F. C., and Larson, J. S. (1974). Classification of freshwater wetlands in the glaciated northeast. U.S. Fish Wild. Serv. Resour. Publ. 116. 56 pp.

Hansen, H. A., and McKnight, D. E. (1964). Emigration of drought-displaced ducks to the Arctic. *Trans. N. Am. Wildl. Nat. Resour. Conf.* **29**, 119–127.

Harris, S. W., and Marshall, W. H. (1963). Ecology of water-level manipulations on a northern marsh. *Ecology* **44**, 331–343.

Harter, R. D. (1966). The effect of water levels on soil chemistry and plant growth of the Magee Marsh Wildlife Area. Ohio Dep. Nat. Res. Game Monogr. No. 2. 36 pp.

Henny, C. J. (1973). Drought displaced movement of North American pintails to Siberia. *J. Wildl. Manage.* **37**, 23–29.

Jeglum, J. K., Boissonneau, A. N., and Haavisto, V. F. (1974). Toward a wetland classification for Ontario. Can. For. Serv., Sault Saint Marie, Ont. Inf. Rep. O-X-215. 54 pp.

Johnsgard, P. A. (1956). Effects of water fluctuations and vegetation change on bird populations, particularly waterfowl. *Ecology* **37**, 689–701.

Kadlec, J. A. (1960). The effect of a drawdown on the ecology of a waterfowl impoundment. Mich. Dep. Cons. Game Div. Rep. 2276. 181 pp.

Kadlec, J. A. (1962). Effects of a drawdown on a waterfowl impoundment. *Ecology* **43**, 267–281.

Kirsch, L. M. (1969). Waterfowl production in relation to grazing. *J. Wildl. Manage.* **33**, 821–828.

Krapu, G. L. (1974). Feeding ecology of pintail hens during reproduction. *Auk* **91**, 278–290.

Krapu, G. L., Parsons, D. R., and Weller, M. W. (1970). Waterfowl in relation to land use and water levels on the Spring Run area. *Iowa State J. Sci.* **44**, 437–452.

Krull, J. N. (1970). Aquatic plant macroinvertebrate associations and waterfowl. *J. Wildl. Manage.* **34**, 707–718.

Linde, A. F. (1969). Techniques for wetland management. Wisconsin Dep. Nat. Resour. Res. Rep. 45. 156 pp.

Linde, A. F., Janisch, T., and Smith, D. (1976). Cattail—the significance of its growth, phenology and carbohydrate storage to its control and management. Wisconsin Dep. of Nat. Resour. Tech. Bull. No. 94. 27 pp.

Low, J. B. (1945). Ecology and management of the redhead, *Nyroca americana*, in Iowa. *Ecol. Monogr.* **15**, 35–69.

Lynch, J. J., O'Neil, T., and Lay, D. W. (1947). Management significance of damage by geese and muskrats to Gulf Coast marshes. *J. Wildl. Manage.* **11**, 50–76.

Martin, A. C., Erickson, R. C., and Steenis, J. H. (1957). Improving duck marshes by weed control. U.S. Fish Wildl. Serv. Circ. 19-Revised. 60 pp.

Mathiak, H. A. (1965). Pothole blasting for wildlife. Wisconsin Cons. Dep. Publ. 352. 31 pp.

Mathiak, H. A. and Linde, A. F. (1956). Studies on level ditching for marsh management. Wisconsin Cons. Dep. Tech. Wildl. Bull. No. 12. 48 pp.

MacArthur, R. H. (1958). Population ecology of some warblers of northeastern coniferous forests. *Ecology* **39**, 599–619.

McAndrews, J. H., Stewart, R. E. Jr., and Bright, R. C. (1967). Paleoecology of a prairie pothole; a preliminary report. Pp. 101–113. *In* "Mid-western Friends of the Pleistocene Guidebook" 185h Ann. Field Conf. (Clayton, Lee, and Freers, Eds.). North Dakota Geol. Surv. Misc. Ser. 30.

Meeks, R. L. (1969). The effect of drawdown date on wetland plant succession. *J. Wildl. Manage.* **33**, 817–821.

Millar, J. B. (1969). Observations on the ecology of wetland vegetation. pp. 49–56. *In*: Saskatoon Wetlands Seminar. Can. Wildl. Serv. Rep. Ser. 6. 262 pp.

Millar, J. B. (1976). Wetland classification in western Canada: a guide to marshes and shallow open water wetlands in the grasslands and parklands of the Prairie Provinces. Can. Wildl. Serv. Rep. Ser. No. 37. 38 pp.

Nelson, N. F. (1954). Factors in the development and restoration of waterfowl habitat at Ogden Bay Refuge, Weber County, Utah. Utah Dep. Fish Game Publ. No. 6. 87 pp.

Nelson, N. F. and Dietz, R. H. (1966). Cattail control methods in Utah. Utah Dep. Fish Game Publ. No. 66-2. 31 pp.

Olson, D. P. (1964). The use of aerial photographs in studies of marsh vegetation. Maine Agric. Exp. Stn. Bull. 13, Tech. Ser. 62 pp.

Owen, M. (1975). Cutting and fertilizing grassland for winter goose management. *J. Wildl. Manage.* **39**, 163–167.

Palmisano, A. W. Jr. (1972). The effect of salinity on the germination and growth of plants important to wildlife in the Gulf Coast marshes. *Proc. Ann. Conf. SE Assoc. Game and Fish Comm.* **25**, 215–223.

Patterson, J. H. (1974). The role of wetland heterogeneity in the reproduction of duck populations in eastern Ontario. Can. Wildl. Serv. Rep. Ser. No. 29, 31–32.

Pospahala, R. S., Anderson, D. R., and Henny, C. J. (1974). Population ecology of the Mallard II. Breeding habitat conditions, size of the breeding population, and production indices. U.S. Fish Wildl. Serv. Resour. Publ. No. 115, 73 pp.

Provost, M. W. (1948). Marsh-blasting as a wildlife management technique. *J. Wildl. Manage.* **12**, 350–387.

Reimold, R. J., Linthurst, R. A., and Wolf, P. L. (1975). Effects of grazing on a salt marsh. *Biol. Cons.* **8**, 105–125.

Robel, R. J. (1961). The effects of carp populations on the production of waterfowl food plants of a western waterfowl marsh. *Trans. N. Am. Wildl. Conf.* **26**, 147–159.

Rogers, J. P. (1964). Effect of drought on reproduction of the Lesser Scaup. *J. Wildl. Manage.* **28**, 213–220.

Sather, J. H., ed. (1976). Proceedings of the national wetland classification and inventory workshop. U.S. Fish. Wildl. Serv. 110 pp.

Schroeder, L. D., Anderson, D. R., Pospahala, R. S., Robinson, G. W., and Glover, F. A. (1976). Effects of early water application on waterfowl production. *J. Wildl. Manage.* **20**, 227–232.

Shaw, S. P., and Fredine, C. G. (1956). Wetlands of the United States. U.S. Fish Wildl. Serv. Circ. 39. 67 pp.

Smith, A. G. (1971). Ecological factors affecting waterfowl production in the Alberta Parklands. U.S. Fish Wildl. Serv. Resour. Publ. No. 98. 49 pp.

Smith, R. I. (1970). Response of pintail breeding populations to drought. *J. Wildl. Manage.* **34**, 934–946.

Steel, P. E., Dalke, P. D. and Bizeau, E. G. (1956). Duck production at Gray's Lake, Idaho, 1949–1951. *J. Wildl. Manage.* **20**, 279–285.

Steenis, J. H. (1939). Marsh management on the Great Plains waterfowl refuges. *Trans. N. Am. Wildl. Conf.* **4**, 400–405.

Stewart, R. E., and Kantrud, H. A. (1971). Classification of natural ponds and lakes in the glaciated prairie region. U.S. Bur. Sport Fish Wildl. Resour. Publ. 92. 57 pp.

Stewart, R. E., and Kantrud, H. A. (1972). Vegetation of prairie potholes, North Dakota in relation to quality of water and other environmental factors. U.S. Dep. Int. Geol. Survey Professional Paper. 585-D.

Stewart, R. E., and Kantrud, H. A. (1974). Breeding waterfowl populations in the prairie pothole region of North Dakota. *Condor* 76, 70–79.

Strohmeyer, D. L., and Fredrickson, L. H. (1967). An evaluation of dynamited potholes in northwest Iowa. *J. Wildl. Manage.* 31, 525–532.

Uhler, F. M. (1956). New habitats for waterfowl. *Trans. N. Am. Wildl. Conf.* 20, 453–469.

Voigts, D. K. (1976). Aquatic invertebrate abundance in relation to changing marsh vegetation. *Am. Midl. Nat.* 95, 313–322.

Ward, P. (1968). Fire in relation to waterfowl habitat of the Delta marshes. *Proc. Tall Timbers Fire Ecol. Conf.* 8, 254–267.

Weller, M. W. (1975). Studies of cattail in relation to management for marsh wildlife. *Iowa State J. Sci.* 49, 333–412.

Weller, M. W., and Fredrickson, L. H. (1974). Avian ecology of a managed glacial marsh. *Living Bird* 12, 269–291.

Weller, M. W., and Spatcher, C. E. (1965). Role of habitat in the distribution and abundance of marsh birds. Iowa Agric. Home Econ. Exp. Stn. Spec. Rep. No. 43. 31 pp.

Weller, M. W., Wingfield, B., and Low, J. B. (1958). Effects of habitat deterioration on bird populations of a small Utah marsh. *Condor* 60, 220–226.

Wolf, K. E. (1955). Some effects of fluctuating and falling water levels on waterfowl production. *J. Wildl. Manage.* 19, 13–23.

Yeager, L. E., and Swope, H. M. (1956). Waterfowl production during wet and dry years in North-Central Colorado. *J. Wildl. Manage.* 20, 442–446.

TIDAL FRESHWATER MARSH ESTABLISHMENT IN UPPER CHESAPEAKE BAY: PONTEDERIA CORDATA AND PELTANDRA VIRGINICA

E. W. Garbisch, Jr., and L. B. Coleman

Environmental Concern Inc.
P.O. Box P
St. Michaels, Maryland 21663

Abstract   The effects of tidal elevation, substrate type, and fertilization on the establishment of *Peltandra virginica* and *Pontederia cordata* by seeding and transplanting seedling stock has been determined at a freshwater location in the Upper Chesapeake Bay, Maryland.

Germination percentages ranged from 93% to 5% for *Peltandra virginica* and from 20% to 5% for *Pontederia cordata* with the higher percentages occurring at the high tidal elevations. The percentages of seedlings that survived the study period averaged ≈30% for both species, but the surviving seedlings developed poorly. The establishment of either *Peltandra virginica* or *Pontederia cordata* by seeding is not considered feasible in unsheltered tidal areas.

No transplanted 1.5-month-old seedlings of *Peltandra virginica* survived at the intermediate and low elevations because of wave stress, debris deposition, and animal depredation. Those surviving at the high elevations did not flower or develop much beyond their stage at the time of planting. Because of the low productivity of 1st-year *Peltandra virginica* seedlings, their satisfactory establishment in unprotected tidal environments is not promising. Planting 1st-year bulbs or 2nd-year seedling stock may yield better results.

The survival of the 3-month-old seedling transplants of *Pontederia cordata* was relatively high at all elevations and in all substrate types. Both the number of flowering stems and the aboveground standing crop values were significantly greater at the lower tidal elevations. Fertilization effected significant increases in productivity, particularly in sand at the high tidal elevation and peat at the low tidal elevation. Seedling transplants of *P. cordata* became satisfactorily established at all tidal elevations. It is estimated that a tidal *P. cordata* marsh exhibiting a 1st-year aboveground standing crop of 1×10^3 to 4×10^3 kg/ha can be established in the Chesapeake Bay region by planting single seedling transplants on 1- to 0.5-m centers, respectively.

Key words   *Arrow arum, marsh establishment, Maryland,* Peltandra virginica, Pontederia cordata, *tidal.*

INTRODUCTION

Tidal marshes are lands that are periodically flooded by tidal waters and that support emergent aquatic plants. They generally are associated with the protected salt, brackish, and freshwaters of estuaries. The values of tidal marshes for wildlife, fish, water quality control, shore stabilization, and flood control are now well recognized.

Testimony to the importance of the tidal marsh natural resource is that all coastal states now have wetland legislations that strive to protect existing tidal marshes and to regulate their loss through man's activities. In addition, Section 150 of the 1976 Water Resources Development Act encourages the establishment of new marshes in connection with federal dredging projects.

Marsh establishment is the development of a desired community of emergent aquatic plants on substrates having appropriate elevations with respect to the regional water table or tidal range. The technology of marsh establishment has applications in (1) the improvement of existing wildlife habitats, (2) the restoration of marshes that have been damaged or destroyed through construction activities, (3) shore erosion abatement, (4) new habitat development on dredged materials, (5) mitigation of marsh losses from construction activities, and (6) the biological treatment of wastewater.

Applied research on marsh establishment, primarily in fresh and brackish water, began in the United States around the turn of the century. Reviews of this work (McAtee, 1939; Martin and Uhler, 1939) still provide practical qualitative guidelines for the development, improvement, and management of wildlife habitats within federal and state wildlife management areas. Subsequent marsh establishment research has not been well documented until recently.

Most of this recent work has concentrated on the establishment of salt marsh (Soil Conservation Service, 1968, 1973, 1975; Sharp and Vaden, 1970; Kadlec and Wentz, 1974; Woodhouse et al., 1974, 1976; Garbisch and Woller, 1975; Garbisch et al., 1975a, 1975b, 1975c; Dunstan et al., 1975; San Francisco District Corps of Engineers, 1976). However, 2 of the 5 marsh creation projects that are currently being sponsored by the Dredged Material Research Program at the Waterways Experiment Station are at tidal freshwater sites (H. Smith, *personal communication*; Garbisch, 1977a), and several recent accounts of freshwater marsh establishment are available (Stanley and Hoffman, 1974, 1975; Garbisch et al., 1975c; Ristich et al., 1976).

Garbisch (1977b) recently has surveyed marsh establishment projects throughout the contiguous United States and general guidelines are now available for tidal marsh establishment (Environmental Concern Inc., 1977; Garbisch, 1977b; Knutson, 1977).

Two plant species that have a high potential for use in the establishment of both tidal and nontidal freshwater marshes and for which no establishment information is available are *Peltandra virginica* (L.) Schott & Endlicher [see Blackwell, 1972] and *Pontederia cordata* (L.), commonly referred to as "arrow arum" and "pickerelweed," respectively.

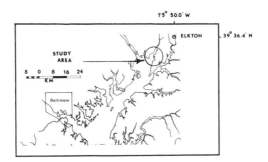

Fig. 1. Location of the study area in the uppermost freshwater region of the Chesapeake Bay, Maryland.

Arrow arum and pickerelweed are broadly distributed in freshwaters throughout the Atlantic and Gulf coasts, as well as inland, and may develop as monotypic stands, as an association, or in association with sedges (*Carex* spp.), smartweeds (*Polygonum* spp.), bulrushes (*Scirpus* spp.), ferns (*Osmunda* spp.), cattails (*Typha* spp.), and other emergent aquatics. Aboveground standing crop values for arrow arum and pickerelweed range from 4 × 10^3 to 9 × 10^3 kg/ha in Virginia. The aboveground parts of these plants decompose rapidly following the growing season, supplying detritus and nutrients to the upper and lower estuarine waters (Silberhorn et al., 1974). The seeds of both plants are valuable duck food; however, waterfowl do not utilize other aboveground parts or the rootstock (Martin and Uhler, 1939; Kadlec and Wentz, 1974). Additionally, these plants can grow throughout most of the tidal zone and are considered important in stabilizing shores and in abating shoreline erosion in freshwater areas. In ranking all marsh types in terms of combined values, Silberhorn et al. (1974) assigned the pickerelweed-arrow arum freshwater marsh to the highest ranked group.

This report describes the effects of tidal elevation, substrate type, and fertilization on the establishment of *Peltandra virginica* and *Pontederia cordata* by seeding and by transplanting seedling stock at a tidal freshwater location in the Upper Chesapeake Bay, Maryland.

Study Area

The study area (≈500 m²) shown in Fig. 1, is located in a small cove on the eastern shore of the Northeast River in the northernmost part of the Chesapeake Bay, Maryland. Studies were conducted throughout the southern shore of the cove and at the mouth of an adjoining creek. This shore has an open water fetch [= distance traversed by waves without obstruction] to the west-northwest of 2.8 km but is otherwise protected. The prevailing winds in the area are from the southwesterly quadrant during the summer and from the northwesterly quadrant during the balance of the year. Consequently, the study area is subject to frequent stress from wind-driven waves, particularly during the spring, fall and winter.

The annual mean tidal (semidiurnal) amplitude at Charlestown, Maryland (3 km north of the study area) is 0.58 m and the water salinity is <1‰ throughout the year.

The mineral sediments of the unvegetated shore are uniform, consisting of 22% gravel, 75% sand, and 3% silt plus clay. A small ($\approx$10 m^2) unvegetated peaty area is located at the lower elevations of the tidal zone and another low-elevation unvegetated area contains thick ($\approx$30 cm) deposits of unconsolidated and finely divided detrital materials. This latter area lies at the mouth of the adjoining creek and is sheltered from the west-northwest exposure to which the balance of the study area is subjected.

The natural freshwater emergent marsh plants growing throughout the study area are listed below in diminishing relative abundance:

High tidal elevation	Mean tidal elevation	Mean-Low tidal elevation
Eupatorium perfoliatum (boneset)	Scirpus americanus (American three-square)	Peltandra virginica (arrow arum)
	Acorus calamus (sweet flag)	Orontium aquaticum (goldenclub)
	Justicia americana (waterwillow)	Pontederia cordata (pickerelweed)
	Scirpus validus (softstem bulrush)	Iris prismatica (slender blue flag)
	Stachys hyssopifolia (hyssop hedge nettle)	Iris pseudacorus (yellow iris)
	Lobelia cardinalis (cardinal flower)	

METHODS AND MATERIALS

Experimental Design   The establishment of *Peltandra virginica* and *Pontederia cordata*, based upon the aerial standing crop values following a 109 day growth period at location, was evaluated as a function of combinations of the following treatments.

Elevation
　　1) HH (spring and storm tide above mean high water [MHW])
　　2) H (about MHW)
　　3) M (about mean tide)
　　4) L (about mean low water [MLW])
Substrate type
　　1) Sand (throughout all elevations)
　　2) Peaty (L elevations)
　　3) Organic (L elevations)

Plant stock
  1) Seed
  2) Peat-potted seedling transplant
Fertilization
  1) Unfertilized
  2) Fertilized (side-dressed at time of planting with 41 g of 3- to 4-month release Osmocote® 19-6-12)

The seed and seedling stock each were planted in replicates of 6 at each available elevation within each substrate type with fertilized and unfertilized treatments, for a total of 144 plots each of *P. virginica* and *P. cordata*. The plot size was 0.1 m²; each seeded plot contained 10 viable seeds and each transplant plot contained 1 peat-potted seedling.

Elevation   Elevations at the study area were empirically estimated. On 14 May 1976, the daytime high and low tides for the study area were predicted (NOAA, 1976) to be +0.21 m above MHW and at MLW, respectively, and to be separated in time by 7.0 h. No major weather systems had been in the area for several days and it was judged that the tides were not significantly influenced by winds and were normal.

At the time of high tide, polyvinyl chloride plastic stakes were positioned at the waterline. Such stake emplacements were repeated every 70 minutes, ending at the time of low tide. Another row of stakes was positioned at the high elevation litter (debris) line. The resulting 8 rows of stakes were then correlated with annual mean tide datum by assuming the tide change followed a standard sine function. Knowing the predicted tidal amplitude (0.79 m) and the time period (7.0 h) for the single tide change, Eq. 1 was derived.

$$y = 0.4 \sin[(0.4286°)(t)]$$

$$y = \text{height (m)} \qquad (1)$$
$$t = \text{time (min)}.$$

Equation (1) then was used to assign elevations relative to MLW to all rows of stakes. These elevations are given in Table 1.

Throughout the 109 day study period, the predicted tide elevations (NOAA, 1976) at the study area were significantly above the annual mean values. The 109 day MHW was predicted to average 0.082 m ($\sigma = 0.037$ m) above the annual value of 0.58 m and the 109-day MLW was predicted to average 0.098 m ($\sigma = 0.034$ m) above the annual low value of 0.0 m. In Table 1, the 2 low (L) elevations are shown to be slightly below the 109-day MLW, whereas the 109-day MHW is shown to be between the 2 high (H) elevations.

Plant Stock   Seeds of *Peltandra virginica* and *Pontederia cordata* were harvested in the James River near Hopewell, Virginia on 21 October 1975 and subsequently stored as collected in tap water at 4°C. After 3 months of

Table 1. Elevations of the 8 Rows of Stakes at the Study Area, Derived from Eq. (1)

Row	Elevation (m) above MLW[a]		Elevation designation[a]
1	litter line; ≈0.85[b]		HH
2		0.79	HH
3		0.73	H
	109-day MHW[a]	0.67	
4	MHW	0.59	H
5		0.40	M
6		0.20	M
	109-day MLW	0.098	
7		0.061	L
8	MLW	0.00	L

[a] MLW = mean low water, MHW = mean high water, HH = spring and storm tide above mean high water, H = about MHW, M = about mean tide, L = about mean low water.
[b] Not estimated from Eq. (1), but estimated in the field.

cold storage, germination tests (14 h at 32°C and 10 h at 15°C, seeds in standing water) afforded ≈35% germination of *P. cordata* within 2 weeks and ≈80% germination of *P. virginica* within 5 weeks. Comparable germination percentages were realized 9 months after harvest.

Although afterripening is not required of *P. virginica* seed (Riemer and MacMillan, 1972) and has not yet been fully assessed for seed of *P. cordata*, seedlings of both species that were obtained through incubator germinations in December, January, and February exhibited poor vigor after transplanting to the greenhouse, and gradually terminated growth and died. Seedlings derived from later germinations (March and April) developed well in the greenhouse and provided suitable stock for use at the study area.

Seedlings derived from incubator germinations were transplanted to sand-filled peat pots in plastic lined beds in the greenhouse. Hoagland's nutrient solution (Hoagland and Aron, 1938) was added to the beds until the solution levels reached the surface of the pots. Thereafter, tap water was periodically added to maintain the nutrient solution at its original level. *Peltandra virginica* and *P. cordata* seedlings were cultivated in this manner for 1.5 and 3 months, respectively, prior to transplanting to the study area. On the date of transplanting, dry weight determinations were made on 3 representative transplants of each species. These dry weights of the aerial and root portions of *P. virginica* were 0.30 g ($\sigma = 0.08$) and 0.70 g ($\sigma = 0.19$) respectively, and those for *P. cordata* were 2.81 g ($\sigma = 0.76$) and 2.21 g ($\sigma = 0.65$), respectively.

Planting, Monitoring, and Work-up   *Pontederia cordata* and *Peltandra virginica* were seeded to the study area on 14 May 1976 and transplanted to the study area on 25 May 1976 and 2 June 1976, respectively. All planting was accomplished in plots that were clear of any vegetation. The plots were seeded by broadcasting seed (10 viable seeds per plot) and fertilizer, when scheduled (41 g of 3- to 4-month-release Osmocote® 19-6-12), on the sediment surface and hand raking the seed or seed–fertilizer mixture to a subsurface depth of 1–3 cm. The peat-potted seedling stock was planted so the surface of the pot was 1 to 2 cm below the sediment surface. When scheduled, fertilizer was included on the surface of the pot in the amount cited above. Each plot was identified with a labelled plastic (polyvinyl chloride) pipe.

The study area was inspected on 6 and 11 June and every 2 weeks thereafter until terminating the study on 2 September 1976. During each inspection, foreign vegetation (if any) in the vicinity of the plots was removed and notations were maintained for each plot regarding: (1) germination of the seed, (2) effects of wave action, deposits of organic debris, and driftwood, (3) numbers of flowering plants, (4) extent of animal and insect damage, and (5) specific and general appearance of the plants.

On 2 September 1976, the aboveground parts of the plants in all plots were clipped at the sediment surface and returned to Environmental Concern Inc. for greenhouse pre-drying (6 to 8 weeks) and final ovendrying at 80°C ± 4° to constant weight.

Statistical testing for differences in standing crop and flowering stems between treatments was accomplished using analysis of variance (Sokal and Rohlf, 1969; Snedecor and Cochran, 1973). The following convention has been adopted to denote significance of all such statistical tests:

$$\alpha = .05 \equiv .95 \leqslant P < .99, \alpha = .01 \equiv .99 \leqslant P < .999,$$

$$\text{and } \alpha = .001 \equiv P \geqslant .999.$$

RESULTS AND DISCUSSION

Peltandra virginica

The results of the establishment of P. virginica at the study area are summarized in Table 2. Seed germinations of P. virginica in unfertilized and fertilized plots at the HH elevation were significantly greater ($\alpha = .01$ and $\alpha = .05$, respectively) than germinations at the H elevation which were, in turn, significantly greater ($\alpha = .05$) than germinations at all lower elevations, including those in the peat and organic substrates. Fertilizer application significantly ($\alpha = .01$) reduced seed germinations at the HH elevation from 93% to 60%; however, the similar reduction from 42% to 23% at the H elevation was not found to be significant. At the M and L elevations in sand, peaty, and organic substrates, seed germinations were low, averaging 5%, with no significant differences between elevation, substrate type, or fertilizer treatment.

Of the P. virginica seed that germinated, ≈45% and ≈15% of the resulting seedlings survived the study period at the HH and H elevations, respec-

Table 2. Results of the Attempted Establishment of *Peltandra virginica* at the Study Area[a]

		Survivors on 2 September 1976			
	Seed germination percentages per plot[c] ($\bar{x} \pm$ SD)	Seedlings		Transplants	
Elevation[b]		N	Aboveground dry wt ($\bar{x} \pm$ SD)	Percent[c]	Aboveground dry wt per transplant[f] ($\bar{x} \pm$ SD)
		Sand			
HH	93.3 ± 8.2	27	0.23 ± 0.18	67	0.19 ± 0.20
HH fertilized	60.0 ± 22.8	16	0.067 ± 0.048	33	0.77 ± 0.66
H	41.7 ± 33.7	4	0.11 ± 0.08	100	0.56 ± 0.39
H fertilized	23.3 ± 18.6	2	0.30 ± 0.18	33	2.64 ± 1.37
M	3.3 ± 5.2	0	. . .	0[d]	. . .
M fertilized	11.7 ± 14.7	0	. . .	0[d]	. . .
L	3.3 ± 5.2	0	. . .	0[d]	. . .
L fertilized	0	0	. . .	0[d]	. . .
		Peat			
L	0.33 ± 0.52	0	. . .	0[d]	. . .
L fertilized	0.83 ± 1.60	0	. . .	0[d]	. . .
		Organic			
L	0.33 ± 0.52	0	. . .	0[e]	. . .
L fertilized	0.17 ± 0.41	0	. . .	0[e]	. . .

[a] All weights are in grams. Seeds and transplant stock were planted on 14 May 1976 and 2 June 1976, respectively, and harvested on 2 September 1976.

[b] For elevation designation abbreviations, see footnote of Table 1.

[c] Values represent cumulative germination (out of 10 seeds in each of 6 plots) through 19 July 1976, whether or not seedlings survived through that date.

[d] All aboveground portions of plants were lost between 11 June 1976 and 13 July 1976.

[e] 50% of aboveground portions of plants were missing by 7 June 1976 and 100% of aboveground portions were missing by 13 July 1976.

[f] At the time of planting, the mean dry wt (g) for the aboveground portion of the transplant material was 0.30 g ($\sigma = 0.08$).

tively. No seedlings survived the study period at the M and L elevations. The aboveground dry weight values of surviving seedlings were either comparable to or significantly ($\alpha = .05$) less than those of the 1.5-month-old nursery stock transplant seedlings at the time of planting. Of the surviving seedlings, fertilizer significantly ($\alpha = .001$) retarded growth at the HH elevation, but had no significant influence at the H elevation. With the exception of those at the fertilized HH elevation, all surviving seedlings exhibited comparable growth.

All transplants of the 1.5-month-old potted seedlings of *P. virginica* below the H elevation suffered mortality in all substrates within 6 weeks after

Table 3. Results of the Attempted Establishment of *Pontederia cordata* at the Study Area[a]

	Seed germination percentages per plot[c] ($\bar{x} \pm$ SD)	N	Survivors on 2 September 1976			
				Transplants		
			Seedlings		Aboveground dry wt per transplant[d] ($\bar{x} \pm$ SD)	No. of flowering and post-flowering stems per transplant ($\bar{x} \pm$ SD)
Elevation[b]			Aboveground dry wt	Per-cent[c]		
Sand						
HH	0	0	. . .	100	10.8 ± 6.1[e]	1.00 ± 1.26
HH fertilized	0	0	. . .	100	24.6 ± 23.9	0.50 ± 0.55
H	5	5	7.0 (SD = 3.8)	50	6.6 ± 1.2[e]	0.67 ± 0.58
H fertilized	0		. . .	50	38.8 ± 17.6	1.00 ± 1.00
M	1		1.3	50	88.5 ± 47.7	3.33 ± 0.58
M fertilized	3		4.6	83	135.1 ± 64.9	3.60 ± 2.07
L	0		. . .	100	97.7 ± 78.8	2.50 ± 1.76
L fertilized	0		. . .	100	174.7 ± 96.5	3.50 ± 2.26
Peat						
L	1		0.22	100	22.9 ± 16.6[e]	0.67 ± 0.82
L fertilized	1		0.10	100	89.8 ± 52.5	2.33 ± 1.86
Organic						
L	0		. . .	67	188.5 ± 90.4[e]	2.50 ± 1.29
L fertilized	1		74.6	100	280.7 ± 81.4	4.17 ± 1.17

[a] All weights are in grams. Seeds and transplant stock were planted on 14 May 1976 and 25 May 1976, respectively, and harvested on 2 September 1976.

[b] For elevation designation abbreviations, see footnote[a] of Table 1.

[c] Values represent cumulative germination (out of 10 seeds in each of 6 plots) through 29 July 1976, whether or not seedlings survived through that date.

[d] At the time of planting the mean dry wt for the aboveground portion of the transplant material was 2.81 g ($\sigma = 0.76$).

[e] Japanese beetle damage to foliage of at least 33% of plants in these plots was observed.

planting. Similar results that were encountered in Virginia in 1976 (Garbisch, 1977a) were attributed to depredation by fish and wave damage. Whereas wave stress and the deposition of driftwood and other debris certainly contributed to the mortality, animal clipping of the aboveground portions of the plants probably contributed also, because the organic substrate portion of the study area is sheltered. Of the surviving transplants at the HH and H elevations, only those at the fertilized H elevation had aboveground dry weight values on 2 September 1976 that were significantly ($\alpha = .05$) greater than those at the time of planting. None of the seedlings or transplants flowered during the study period.

Fig. 2.   Three seedlings of *Pontederia cordata* in a seeded (fertilized) plot in sand substrate at the M elevation on 2 September 1976 just prior to harvesting.

Peltandra virginica seedlings have been cultivated in greenhouses and in outside impoundments under conditions of limited physical stress at Environmental Concern, Inc. for 2 years. Although no numerical results are available, it has been found, using seeds from Wisconsin and Virginia and using various nutrient solutions and fertilizers, that *P. virginica* seedlings have a low productivity and do not reach maturity (or flower) during their 1st year's growth. Because of this characteristic, establishing *P. virginica* by seeding or by transplanting seedlings has not been thought to offer much promise except in areas that are protected from wildlife depredation and physical stress. This conclusion is substantiated by the qualitative results of this study. Establishing *P. virginica* by planting 1-year-old dormant bulbs or 2nd-year seedling stock may be more promising. This approach is currently under investigation.

Pontederia cordata

The results of the establishment of *P. cordata* at the study area are collected in Table 3. The germination of *P. cordata* seed was poor (20% or less) at all elevations and in all substrate types. No seed germinated at the HH elevation. The generally lower seed germination in the fertilized plots as compared with the unfertilized plots at the same elevation was not found to be statistically significant. Seed germinations at the H and M elevations were statistically comparable. Germination at the unfertilized H elevation was significantly ($\alpha = .005$) greater than germinations at the L elevations in the sand and peaty substrates, but not in the organic substrate.

Fig. 3. Two fertilized seedling transplants of *Pontederia cordata* at the L elevation in sand substrate on 2 September 1976, just prior to harvesting. The measuring stick is 1 m tall.

Seedlings from only 27% of the germinated seed survived the study period and 75% of these were at the H and M elevations. The dry weights of the aerial parts of the surviving seedlings were comparable to (sand substrate) or less (peaty substrate) than those of the 3-month-old seedling transplants at the time of planting, with the exception of the sole surviving seedling in the fertilized organic substrate. This seedling dwarfed all of the others. Figure 2 shows the 3 surviving seedlings in one of the fertilized plots at the M elevation on 2 September 1976.

The survival of the 3-month-old seedling transplants of *P. cordata* was satisfactory at all elevations and in all substrate types. Observations made during the bimonthly surveys of the study area indicated that the lowest percentage survival at the H and M elevations was attributed to stresses inflicted by the tidal depositions of sizable pieces of driftwood and other organic litter. Without such stresses, the seedlings and transplant survival rates at these elevations probably would have been improved.

During the nursery cultivation of *P. cordata*, it was observed that seedlings initiated flowering ≈3 months following seed germination, regardless of the prevailing natural photoperiod. At the study area, seedlings and transplants flowered throughout the period of study. There were no significant differences between the numbers of flowering and flowered stems per transplant in the fertilized and unfertilized plots at each elevation. However, the mean numbers of flowering and flowered stems per transplant were found to be significantly ($\alpha = .05$) greater at the M and L elevations of the sand and organic (but not peaty) substrates than at the H and HH elevations of the sand substrate.

Transplant fertilization led to a significant ($\alpha = .05$) increase in the aboveground dry weight values at the H elevation (6× increase) and in the peaty substrate (4× increase). In all other instances, the 1.5–2× increases in aboveground dry weight values that were realized after fertilization were not found to be significant. Particularly for permeable sandy substrates, which exist at the study area, the fertilization rate may best be increased by 50% in order to maximize initial plant establishment.

In the sand substrate, the mean aboveground dry weight values for the *P. cordata* transplants were not significantly different at the HH and H and at the M and L elevations; however, those at the M and L elevations were significantly ($\alpha = .001$) greater than those at the HH and H elevations by an average factor of 8. At the L elevation, the mean aboveground dry weight values were greatest ($\alpha \leq .05$) in the organic sediment and least ($\alpha \leq .05$) in the peaty sediment. Figure 3 shows representative seedling transplants of *P. cordata* at the study area on 2 September 1976.

The creation of *P. cordata* marsh in suitable tidal freshwater areas of the Chesapeake Bay appears feasible. According to the results described, *P. cordata* freshwater tidal marsh exhibiting a 1st-year aboveground standing crop value between 1×10^3 and 4×10^3 kg/ha can be produced in sand and in sediments with high organic content using 3-month-old seedling stock planted on 1- to 0.5-m centers, respectively, and side-dressed with slow release 19-6-12 fertilizer at a rate of 410 kg to 1,640 kg/ha, respectively. The 1st year aboveground standing crop may increase in unconsolidated mud (silt and clay) sediments (qualitative observations at Environmental Concern Inc.) or by adopting a higher fertilization rate. The entire freshwater tidal zone can be vegetated with *P. cordata* in areas subject to tidal amplitudes of up to at least ⅔ metre.

Seeding does not appear feasible for the establishment of *Pontederia cordata* in areas subject to significant wave exposure. In such areas, seed germination and seedling survival are unacceptably low except, possibly, for the high elevation of the tidal zone.

ACKNOWLEDGMENTS

We thank Michael Riner for his assistance with the periodic evaluations of the study area, the harvesting and processing of the plants, and the numerical tabulations. This study was made possible by membership contribution to Environmental Concern Inc.

LITERATURE CITED

Blackwell, Jr., W. H. (1972). The combination *Peltandra virginica* (L.) Schott & Endlicher. *Rhodora* **74**, 516–518.

Dunstan, W. M., McIntire, G. L., and Windom, H. L. (1975). *Spartina* revegetation on dredge spoil in SE marshes. *J. Waterways, Harbors, and Coastal Engineering Div.* **101**, 269–276.

Environmental Concern Inc. (1977). Brochure: Scope of services available, plant material costs, vegetative establishment costs, specifications for plant establishment. St. Michaels, Maryland.

Garbisch, E. W., Jr. (1977a). The propagation of vascular plants at the James River habitat development site. Contract report to the Norfolk District Corps of Engineers. Waterways Exp. Station, Vicksburg, Mississippi. (*In press*).

Garbisch, E. W., Jr. (1977b). Recent and planned marsh establishment work throughout the contiguous United States. A survey and basic guidelines. Contract Report D-77-3. U.S. Army Engineer Waterways Experiment Station, Vicksburg, Mississippi.

Garbisch, E. W., Jr. and Woller, P. B. (1975). Marsh development on dredged materials at Slaughter Creek, Maryland. Contract DAC W31-74-C-0120. Report prepared for the Baltimore District Corps of Engineers, Baltimore, Maryland. 110 pp.

Garbisch, E. W., Jr., Woller, P. B., and McCallum, R. J. (1975a). Saltmarsh establishment and development. TM-52, U.S. Army Corps of Engineers Coastal Engineering Research Center, Fort Belvoir, Virginia.

Garbisch, E. W., Jr., Woller, P. B. and McCallum, R. J. (1975b). Saltmarsh establishment on intertidal dredged material areas on the coast of Virginia. Contract DAC W65-74-C-0062. Report prepared for the Norfolk District Corps of Engineers, Norfolk, Virginia. 48 pp.

Garbisch, E. W., Jr., Woller, P. B., and McCallum, R. J. (1975c). Biotic techniques for shore stabilization. In: "Estuarine Research, Vol. II" (L. E. Cronin, ed.), pp. 405–462. Academic Press, New York.

Hoagland, D. R., and Aron, D. I. (1938). The water-culture method for growing plants without soil. Circular 347, California Agriculture Station.

Kadlec, J. A., and Wentz, W. A. (1974). State-of-the-art-survey and evaluation of marsh plant establishment: induced and natural. Contract report D-74-9 prepared for the U.S. Army Coastal Engineering Research Center. Waterways Exp. Station, Vicksburg, Mississippi.

Knutson, P. L. (1977). Planting guidelines for marsh development and bank stabilization. CDM77-1, U.S. Army Corps of Engineers, Coastal Research Center, Fort Belvoir, Virginia. (In press).

Martin, A. C., and Uhler, F. M. (1939). Food of game ducks in the United States and Canada. USDA Tech. Bull. 634.

McAtee, W. L. (1939). "Wildlife Food Plants, Their Value, Propagation and Management." Collegiate Press, Inc., Ames, Iowa.

NOAA. (1976). Tide Tables: East Coast of North and South America. U.S. Department of Commerce, Washington, D.C.

Reimer, D. N., and MacMillan, W. W. (1972). Seed germination in Arrow Arum (Peltandra virginica). Northeast Weed Control Conf. Proc. 26, 183–188.

Ristick, S. S., Fredrick, S. W., and Buckley, E. H. (1976). Transplantation of Typha and the distribution of vegetation and algae in a reclaimed estuarine marsh. Bull. Torrey Bot. Club 103, 157–164.

San Francisco District Corps of Engineers. (1976). Dredge disposal study San Francisco Bay and Estuary, Appendix K, Marshland Development, San Francisco, California.

Sharp, W. C., and Vaden, J. (1970). 10-year report on sloping techniques used to stabilize eroding tidal river banks. Shore and Beach, April, pp. 31–35.

Silberhorn, G. M., Dawes, G. M., and Barnard, T. A. Jr. (1974). Coastal Wetlands of Virginia. Guidelines for Activities Affecting Virginia Wetlands. Special Report No. 46, Virginia Institute of Marine Sciences, Gloucester Point, Virginia.

Snedecor, G. W., and Cochran, W. G. (1973). "Statistical Methods." The Iowa State University Press. Ames, Iowa.

Soil Conservation Service. (1968). Vegetative Tidal Bank Stabilization. College Park, Maryland. September 4, 7 pp.

Soil Conservation Service. (1973). Shore Erosion Protection. Tech. Guide Section III-H. Raleigh, North Carolina. August.

Soil Conservation Service. (1975). A Partial Summary on Soil Conservation Service Plant Materials Activities for Shoreline Erosion Control. Draft. March. Northeast Technical Center, Broomall, Pennsylvania, (brochure), 16 pp.

Sokal, R. R., and Rohlf, F. J. (1969). "Biometry." W. H. Freeman and Co., San Francisco.

Stanley, L. D., and Hoffman, G. R. (1974). The natural and experimental establishment of vegetation along the shorelines of Lake Oahe and Lake Sakakawea, Mainstem Missouri River Reservoir. Annual report submitted to the Omaha District Corps of Engineers, Omaha, Nebraska.

Stanley, L. D., and Hoffman, G. R. (1975). Further studies on the natural and experimental establishment of vegetation along the shorelines of Lake Oahe and Lake Sakakawea, lakes of Mainstem Missouri River. Annual report submitted to the Omaha District Corps of Engineers, Omaha, Nebraska.

Woodhouse, W. W., Jr., Seneca, E. D., and Broome, S. W. (1974). Propagation of *Spartina alterniflora* for substrate stabilization and salt marsh development. TM-46, U.S. Army Corps of Engineers Coastal Engineering Research Center, Fort Belvoir, Virginia.

Woodhouse, W. W., Jr., Seneca, E. D., and Broome, S. W. (1976). Propagation and use of *Spartina alterniflora* for shoreline erosion abatement. TR76-2. U.S. Army Corps of Engineers Coastal Engineering Research Center, Fort Belvoir, Virginia.

EFFECTS OF CANALS ON FRESHWATER MARSHES IN COASTAL LOUISIANA AND IMPLICATIONS FOR MANAGEMENT

James H. Stone, Leonard M. Bahr, Jr., and John W. Day, Jr.

Department of Marine Sciences
Center for Wetland Resources
Louisiana State University
Baton Rouge, Louisiana 70803

Abstract Water flow and quality determine and control species composition and function in the freshwater marshes of coastal Louisiana. Man's activities alter this when he disrupts or removes the marsh. For example, man-made canals can change the hydrologic regime, depending on its alignment and local elevations, from -1% to -35% of normal flow. This in turn likely accelerates land loss from increased wave action. It is estimated that perhaps 172 hectares per year of freshwater marsh in coastal Louisiana is being lost due to man's activities. Canals also tend to divert runoff water away from the marsh (where it would be purged of pollutants) to open water bodies, thereby probably causing eutrophication. For example, the P loading rate of several freshwater lakes in the upper Barataria Basin is estimated between 1.5 and 4.3 $g \cdot m^{-2} \cdot yr^{-1}$, which may be from 4 to $11\times$ greater than the loading limit for eutrophication. We do not know how these effects are quantitatively coupled to primary production, consumers in general, and fishery harvest in particular. But on the basis of preliminary data, we suggest that decision makers make full use of computer simulation models, of energy cost accounting, and of other more specific recommendations for marsh management.

Key words *Canals, coastal zone, eutrophication, hydrologic changes, impoundment, land loss, Louisiana management implications, spoil banks.*

INTRODUCTION

The Louisiana coastal ecosystem comprises a broad zone made up of wetlands and water bodies; it is characterized by low elevation and little natural relief. The coastal zone of Louisiana has recently been defined as extending landward from the Gulf of Mexico to the Pleistocene terrace, a distance ranging from about 30 to 150 km (McIntire et al., 1975) and encompassing an area of $\approx 31 \times 10^3$ km^2. Within the coastal ecosystem, 4 contiguous zones of emergent wetland have been distinguished, based on plant community composition (Chabreck, 1972): saline marsh, brackish marsh, intermediate marsh, and freshwater marsh. In addition, there are extensive areas of forested wetland that generally occur inland from the freshwater marsh zone. Figure 1 shows the full extent of the coastal ecosystem and the approximate boundaries between wetland types. The inclusion of all wetland types in a single coastal ecosystem is based on an array of data on exchanges of matter and energy across the entire zone. For example, tidal effects are sometimes detectable far inland from the point at which salinity effects are seen.

Because all wetland zones are interconnected by water bodies, water flow transports dissolved and particulate matter and organisms between different zones. This flow seems to be the primary regulator of the entire ecosystem, because the hydrologic and salinity regimes determine local plant community composition. Emergent plants in any given area integrate chemical and physical variations over time and reflect long-term average conditions. Water flow and alternating water levels perturb the wetland zones and maintain them in a successional state that allows net community production. Thus the hydrologic regime serves as a natural energy subsidy of significant importance. For these reasons, we normally do not view freshwater marshes as a separate entity in coastal Louisiana.

Man has imposed his activities on the Louisiana coastal ecosystem in a variety of ways, which usually affect the natural function adversely. We believe that management actions should reflect and complement, as much as possible, the natural function. We will restrict our discussion to the freshwater marshes of coastal Louisiana but we will indicate other aspects of the coastal wetlands when they relate to the freshwater marshes.

In this discussion our objectives are to:

1) Describe briefly the natural function of freshwater marshes in the Louisiana coastal ecosystem
2) Describe potential and actual changes from man's activities—such as canalling—to freshwater marshes
3) Illustrate how these changes affect and relate to the natural function
4) Draw implications and make recommendations for management and mitigation of these adverse effects in freshwater marshes.

We use the word "management" in its broadest sense—namely, those policy and day-to-day decisions affecting the use of air, land, and water in and about freshwater marshes.

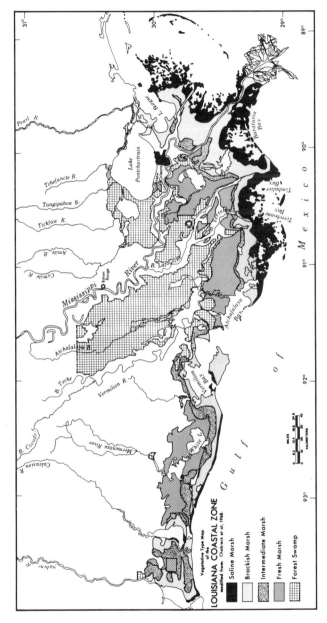

Fig. 1.   Coastal zone of Louisiana divided into 5 vegetative types.

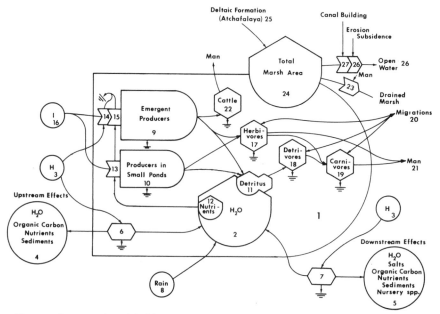

Fig. 2.   Conceptual model of freshwater marshes in coastal Louisiana. H = hydrology which includes all hydrodynamic processes influenced by natural forces and man's activities.

Natural Function of Freshwater Marshes in Coastal Louisiana

The function of any complex ecosystem is most clearly and concisely described through the use of a conceptual model in which the major components, processes, and forcing functions are indicated schematically. Such a model for the freshwater marsh ecosystem in Louisiana is shown in Fig. 2. Symbols used in the model were developed by Odum (1971) as a way of depicting flows, storages, and interactions of energy and matter. **Each component in Fig. 2 is numbered, and these numbers are referred to in the following discussion of the model.**

The overall freshwater marsh ecosystem (1) is depicted with the symbol for a producer, because there is a net export of organic matter in the form of detritus from the marsh into adjacent open water bodies.

Water associated directly with the marsh is indicated with a storage symbol (2). Water (particularly water flow) has already been discussed as the primary integrator and forcing function of the entire coastal marsh ecosystem of Louisiana. The local hydrologic regime (3) is a function of water inputs from rainfall and terrestrial sources, slope, marsh "friction," the ratio of wetland to open water, and all natural and man-made relief features. Hydrology and man's effect on it are discussed further in the next section. In addition, Table 1 lists selected physical and biological characteristics for marsh types in coastal Louisiana.

Table 1. Selected Physical and Biological Characteristics of Coastal Marsh Types in Louisiana[a]

Marsh type	Salinity range[b] (‰)	Organic proportion of substrate (%)	Marsh:water ratio	Frequency of inundation (times/yr)	Net primary production rate (kJ m⁻²yr⁻¹)	Plant species (n)	Dominant producers[c]	Vertebrate species (n)			
								Amphibia	Reptilia	Aves (yr-round)	Mammalia
Saline	5–35 (12.4)	5	0.5:1	160	56,480	17	(1), (2)	0	4	15	8
Brackish	0.5–16 (5.6)	23	3:1	40	82,840	40	(2), (3)	5	16	16	11
Intermediate	0.5–8 (2.1)	≈25	20:1	...	69,660	54	(3)	6	16	14	11
Freshwater	0–0.5 (<0.1)	usually >50	13:1	20	50,840	93	(4), (5), (6)	18	24	11	14

[a] Extracted from Bahr and Hebrard, 1976; C. B. Rainey, *personal communication*.

[b] Means in parentheses.

[c] Dominant producer species: (1) *Spartina alterniflora*; (2) *Distichlis spicata*; (3) *Spartina patens*; (4) *Sagittaria falcata*; (5) *Panicum hemitomon*; and (6) *Alternanthera philoxeroides*.

Two ecosystems normally border freshwater marshes in Louisiana. One is the inland upstream areas (4), either swamp forest with associated water bodies, or high land (Pleistocene terrace) and river basins; the other ecosystem is the downstream brackish and intermediate salinity marshes with associated estuaries (5). Both of these adjacent areas exchange matter and energy with the freshwater marsh, and these exchanges are indicated by bidirectional work gates (6 and 7).

Water (4) brings with it organic matter, nutrients, toxins, and sediments. Nursery species and salts (5) are also introduced to the freshwater marsh ecosystem via downstream areas; and some freshwater areas in Louisiana appear to serve as primary nursery areas. Of special importance are the export (via 7) of organic carbon downstream from freshwater marshes and the migration of aquatic nursery species along the same pathway. Rainfall (8) represents a major source of water and a minor source of nutrients to the ecosystem.

Marsh producers are divided into the predominant emergent vascular plants (9) and less important pond algae and floating aquatic plants in small marsh ponds (10). Because the marsh trophic structure is primarily based on a detrital system, the pathway from (9) to a detritus—microbial storage module (11) is very important. Nutrients (12) stimulate primary production (13 and 15). The nutrient pool and man's effect on it are discussed in a following section. Hydrology (3) is a major determinant of community composition and primary production in marsh plants (9) as shown by work gate (14). Insolation (16), of course, provides the major energy requirement of all primary producers.

Natural consumers in the freshwater marsh ecosystem have been lumped in the model into 3 groups: herbivores (17), detritivores (18), and carnivores (19). Herbivores and detritivores comprise terrestrial and aquatic organisms including insects, gastropods, crustaceans, finfish, birds, and mammals. Carnivores are represented by a diverse assemblage of birds, insects, mammals, and reptiles.

It should be noted that many marsh consumers migrate either daily or seasonally into the marsh to feed and find shelter. This activity is indicated in the model by number 20.

Man harvests various organisms that use the freshwater marshes and water bodies (21). These animals include waterfowl, furbearers, alligators, finfish and crayfish. This recreational and/or commercial harvest represents a major resource to Louisiana.

Cattle (22) are often allowed to graze in freshwater marshes in Louisiana. Natural marsh areas are usually marginal for cattle, however, and are therefore often impounded and drained to increase their agricultural value (23). Cattle grazing also probably disturbs natural marsh areas, although this is not documented for coastal Louisiana.

Total freshwater marsh area in Louisiana (24) is ≈282,000 hectares, but this area is changing because of the dynamic nature of coastal wetland areas and man's alterations. A newly emerging major delta is rapidly accreting wetlands in the Atchafalaya Bay region (Fig. 1). This sedimentation process is indicated in Fig. 2 (25). Loss of wetland (including freshwater marsh) occurs via natural processes of erosion and subsidence (26) and man-caused changes such as canal building (27) and draining (23). Some of these processes and the management implications they represent are the subject of the following section.

The conceptual model, represented by Fig. 2, allows us to identify the most important features of the freshwater marsh ecosystem, to view these features in relation to each other and to assign research priorities. Thus, we believe that changes in water flow and quality are 2 of the most important problems in coastal Louisiana, especially in the freshwater marshes. In the following section we illustrate this thesis by considering the effect of selected activities of man on hydrology, which in turn causes land loss and changes to the nutrient cycle.

Our conceptual model is derived from existing data on freshwater marshes and from the many years of our combined field experience. The model is only one of many possibilities but it illustrates and identifies the principal factors of how a freshwater marsh functions, what the control features are, and what the possible interrelationships and interactions are.

MAN'S ACTIVITIES AND THEIR IMPACTS IN COASTAL LOUISIANA

Eight use-issue categories of man's activities in coastal Louisiana are given in Table 2. A variety of activities are identified under each of these categories. This list is probably incomplete, but it is not our purpose to describe the various economic activities and their environmental impacts in the coastal zone (see Byrne et al., 1976; Conner, 1976; Grimes and Pinhey, 1976; and Van Sickle et al., 1976). However, Table 2 does allow us, to identify the most important environmental impacts of these activities. All of these activities require the use of land, air, and water. For example, oil and gas exploration entails several activities. Dredging of marshlands is required to clear right-of-ways, lay pipelines, and to prepare plant sites. Workmen and their families need places to live. Agricultural production requires the clearing of land, pest control, fertilization and a variety of harvesting techniques. As a result of these land uses, various chemical pollutants are often discharged into surrounding water bodies. All these activities occur in each of the vegetative zones of coastal Louisiana, including the freshwater marshes.

Below we discuss some of the environmental impacts in the freshwater marshes of coastal Louisiana due to the disruption of the marsh by man's activities, specifically as a result of digging canals, disposal of spoil, and the discharge of agricultural and urban runoff through these canals. These ac-

Table 2. Selected Use-Issue Categories and Man's Activities in Coastal Louisiana[a]

Use-Issue category	Activity
Mineral and energy extraction	Exploration Dredging Drilling Casing and cementing Treating oil field emulsions Pipe laying Brine disposal Drilling mud disposal Facility abandonment Oil spills
Navigation and transportation	Canal construction and maintenance Spoil disposal Dock construction Waterweed control Boat traffic Harbor-port development and use Airport construction and use Highway construction and use Railroad construction and use

Flood control and hurricane protection	Levee construction
	  a) dredging
	  b) spoil disposal
	  c) right-of-way
	Channel improvement
	  a) dredging (cutoff dredging, improvement dredging)
	  b) revetments
	  c) dikes
	Water control construction
	  a) spillways
	  b) pumping

Recreation and tourism	Sportfishing Beach, river and lake activities Camping Boating Outdoor games Hiking Hunting Tourism
Fishing and trapping	Harvesting (commercial and sportfishing) Boat operation Commercial processing Aquaculture Trainasse building (Trapper's Canal)
Wetlands maintenance	Weir construction Pesticide and herbicide application Mechanical tilling Marsh burning

Table 2.   Continued

Use-Issue category	Activity
	Pothole, plug and ditch construction
	Cattle grazing
	Impoundment construction
Agriculture and forestry	Commercial harvesting
	Soil preparation
	Pest control
	Cultivation
	Irrigation
	Fertilization
	Land use conversion for agriculture
	Various management practices (flood and saltwater intrusion control)
Urban development	Sundry economic activities

ᵃ Extracted from Conner, 1976; V. R. Bennett, and W. W. Burke *personal communication*.

tivities result in 3 important and interrelated environmental impacts: hydrologic changes, land loss, and changes in nutrient cycling.

Changes to the Hydrologic Regime

Our model illustrates the importance of water to the function of the freshwater marsh. When a marsh is disturbed or removed, local hydrology changes. Because there is so much dredging and canalling activity in coastal Louisiana, we are repeatedly asked: how much hydrologic change occurs when a canal is dug? where does the change occur? how can we mitigate its effects? As a result of these questions, one of our workers helped us to model the water circulation within a small piece of marsh. This approach should give us—at least initially—a means for making an order-of-magnitude estimate on the effects and for planning mitigative procedures.

McHugh (1976) has created a two-dimensional hydrodynamic model for predicting water levels and current vectors within a small area of given size, shape and boundary conditions. It simulates water flow over a small piece of marsh by using a variable-size grid system. For each block of the grid, current speed and direction, water volumes, and water height are given for each time step. The time steps can be made to duplicate a tidal cycle. For this discussion we will present data only on water volumes. However, the model will eventually enable us to study the effects of winds, exchange rates, and dissolved or suspended materials. Basic assumptions of the model are:

1) There is a constant horizontal velocity at all depths
2) There is no vertical velocity

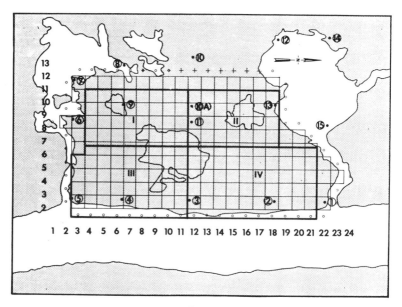

Fig. 3. Marsh layout and grid system used for testing, via McHugh's model, various physical disturbances to marsh. Black dots indicate elevation locations taken in field. Large grid (I, II, III, and IV) superimposed over smaller grids was used to calculate total water flow over a 20-h tidal cycle.

3) There is no vertical shear due to horizontal velocity gradients

4) There are no pressure and buoyancy forces due to variations in salinity.

The model uses an implicit method of solving the equations of water motion in alternating directions. The type of marsh used for testing the model is immaterial because it takes into account only the elevations or topography and the changes in water levels and velocities. For details of the model the reader is directed to McHugh (1976). The model has been tested on a small area ($\approx 185 \times 304$ m) of brackish marsh located in Caminada Bay, Louisiana (see Fig. 3). We believe that we can generalize from the model because coastal wetlands geometry in Louisiana tends to be very similar—namely, dendrite or branching streambeds with a very gentle slope; however, field verification of the model is still underway. In addition, the model can be used to study almost any area of marsh with open and closed boundaries. However, it should be emphasized that our results are only preliminary and deal presently only with water volumes.

Figure 3 shows the general layout of the study area and the grid system used for calculating current vectors. The larger grid system (namely, the 4 sectors) superimposed on the small grids is used for calculating flow rates over 4 larger areas so that average and total values could be derived. We

Table 3. Results of Computer Simulation Studies of Water Flow over a Piece of Louisiana Brackish Marsh 185 m × 304 m for a Variety of Test Conditions[a]

Test condition	Total water flow ($m^3 \times 10^4$) per 20-h tidal cycle	Percent of normal
Normal circulation (Fig. 3)	6.0752	100.0
1) North to south canal and spoil bank (complete)	5.5385	91.2
2) No. 1 with openings in spoil bank		95.9
3) East to west canal and spoil bank (complete; Fig. 4)	5.7178	94.1
4) No. 3 with openings in spoil bank	5.8962	97.0
5) North to south canal and spoil bank (blind-ended)	5.6105	92.3
6) No. 5 with openings in spoil bank	5.8602	96.5
7) East to west canal and spoil bank (blind-ended)	5.7889	95.3
8) No. 7 with openings in spoil bank	5.9322	97.6
9) Levee surrounding on 3 sides	5.1109	84.1
10) No. 9 with vertical canal and spoil bank (partial, but not extending from levee)	4.8608	80.0
11) No. 10 (partial but extending from levee)	4.8962	80.6
12) No. 10 with openings in spoil bank	4.9682	81.8
13) No. 11 with openings in spoil bank	5.0039	82.4
14) Impoundment (31 m × 274 m)	3.9308	64.7

[a] A mean tide height of 0.14 m was used.

discuss here only the change in the water volume flowing over the marsh, before and after simulating physical modifications to the marsh.

Table 3 presents the preliminary results of computer simulations. We will discuss 3 types of physical disturbances: (1) canals and their spoil banks, (2) surrounding levees, and (3) impoundments. These are common to a variety of economic activities; however, for the actual alignment of each there exist many possibilities so they should be considered only as illustrative. The 3 types of physical disturbances and their variations are as follows:

Canal and Spoil Bank
  North-south alignment (cutting completely across marsh)
  North-south alignment with openings in spoil bank
  East-west alignment (cutting completely across marsh)
  East-west alignment with openings in spoil bank
  North-south alignment (not cutting completely across marsh)
  North-south alignment with openings in spoil bank
  East-west alignment (not cutting completely across marsh)
  East-west alignment with openings in spoil bank
Surrounding Level (three-sided)
  Surrounding levels (by itself)
  Surrounding levels with vertical canal and spoil bank (not next to levee)

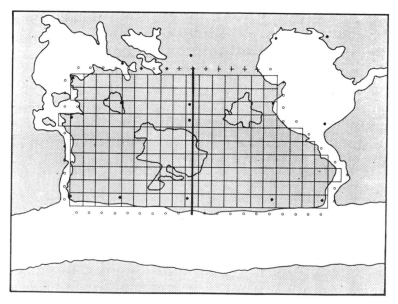

Fig. 4. An example of an east-to-west canal with spoil bank (barrier) cutting completely across marsh area.

Surrounding levels with vertical canal and spoil bank (not adjacent to
   levee) with openings in spoil bank
Surrounding levels with vertical canal and spoil bank (adjacent to levee)
Surrounding levels with vertical canal and spoil bank (adjacent to levee)
   with openings in spoil bank
Impoundment (four-sided)
Impoundment (by itself)

The canal was assumed to be 2.4 m deep and $\approx$30.5 m wide and the spoil bank assumed to be $\approx$30.5 m wide. In the north-south and east-west alignments, the canal cut completely across the marsh. A variation of these 2 alignments was made by simulating that the canal did not cut completely across the marsh; rather it extended about 80% of the way into the marsh then terminated. For all alignments, an additional experiment involved simulating 15.2-m openings every 15.2 metres. This allowed water to flow through the spoil bank or barrier. Figure 4 illustrates an east-west canal used for one of the simulations.

The surrounding levee was assumed to be a barrier on 3 sides of the marsh area (i.e., north, south and west). An additional simulation was made by placing a canal and spoil bank in the middle of the area surrounded by the 3 levees. In one instance, the canal and spoil bank abutted the levee while in another instance it did not; in neither case did the canal and spoil bank cut completely across the levee or marsh. The effects of creating breaks or openings in the spoil bank of the canal were also simulated.

The configuration assumed for the impoundment was a four-sided levee or barrier in the middle of the marsh, the area being ≈8.5 km².

Among the various canal alignments simulated, north-to-south canals produced the greatest hydrologic changes; water flow (cases 1 and 5) was decreased by approximately 9 to 8% below normal conditions. However, it should be emphasized that the values in Table 3 represent water flow over the entire marsh and some sectors (I, II, III or IV, as shown in Fig. 3) had larger hydrologic changes, but the change in total flow was not great. For example, the canal alignment illustrated in Fig. 4 reduced water flow to 94% of normal, but flow in sector II was 86% of normal and in sector IV, 90% of normal.

The alignments used for testing the effects of a three-sided levee showed more of a hydrologic change than did the canals. Water flow over a 20-hour tidal cycle was ≈80% of normal flow.

The impoundment showed the largest hydrologic changes as it reduced water flow 35% of normal.

Openings in the spoil bank (or barrier) were simulated for some of the alignments; in general, they increased the water flow closer to normal. For example, in Table 3, compare water flow in cases 1 and 2, 3 and 4, 5 and 6, 7 and 8, 10 and 12, and 11 and 13.

The implications of this preliminary work with a model that simulates water flow over a marsh are: first, hydrologic changes (at this stage only in terms of water volumes) do occur as a result of canalling and some of these changes appear to be quite significant; second, data on only 1 parameter are presented—data are needed on water heights, frequency of inundation, current speeds and directions, and distribution and abundance of dissolved and suspended material; third, this type of work should be coupled more closely with biological research such as nutrient dispersion, primary production, and consumers in general.

We believe that decision makers should use models such as the one above even in its rather incomplete state. Decision makers will be more able to identify the major impacts of an economic activity, to form initial rough estimates of these impacts, and, possibly to design mitigative procedures. For example, in the case of canals, they could select the direction of a canal that would produce the least amount of change or disruption.

Land Loss

The amount of available marshland is of considerable importance, as indicated in Fig. 2, because on it hinges the amount of primary production by the emergent producers. Because marshlands are under continued deterioration by both natural and man-made processes, and as the processes for forming new marshlands to augment these losses are now under man's control—with the exception of the new Atchafalaya delta—prevention of land loss is especially important. For example, Gagliano and van Beek (1970) estimate the total land loss for coastal Louisiana is 42.7 km²/yr, and man's activities probably account for ≈35% of the total. Land loss for the coastal zone of Louisiana can be estimated from Barrett (1970), Gagliano and van Beek (1970), Chabreck (1972), Adams et al. (1976), and Day et al. (1976).

Regional subsidence and erosion of marshes and barrier islands are probably the most important natural causes of wetland loss; however, substrate type greatly influences the rate of loss. Other factors being equal, wetlands characterize by peaty, organic soils will erode faster than those with more mineral soils (Day et al., 1976).

Land loss due to man is associated with a number of activities; these include flood control, canal construction with attendant spoil banks, and land reclamation. Flood control levees along the lower Mississippi River have eliminated most sediment input from the river which historically offset natural erosion. Canal construction (for navigation, oil rig access, pipelines) eliminates wetlands directly (the canal itself as well as where spoil is placed). These canals widen over time leading to additional land loss. Both canals and spoil banks lead to hydrological changes as discussed in the previous section. These changes, in turn, can lead to salinity encroachment, which can cause death of marsh grass. Land reclamation has taken place principally as the result of urban and agricultural expansion.

Estimates on total land loss for each vegetative zone in coastal Louisiana are given in Table 4. About 491 hectares of freshwater marsh is lost annually (≈35% man and 65% natural) in Louisiana (≈18% of total wetland lost in the state). Brackish and saline marshes have experienced even higher rates of land loss, being 49% and 25% of total wetland lost, respectively. There are at least 3 possible reasons for the higher rates in brackish and saline marshes. First, tidal currents are more pronounced in these areas than in freshwater marsh. Second, there is much more open water in the brackish and saline zones. These conditions of tidal currents and fetch probably result in strong wave action. Finally, salinity intrusion is more serious in the brackish zone, which often results in the death of vegetation and, in turn, more rapid erosion.

The land loss data for each vegetative zone (Table 4) have not yet been broken down into the percentage of causal factors. However, data exist on some of man's effects, particularly canal widening in a freshwater marsh (i.e., the Rockefeller Wildlife Refuge in southwestern Louisiana). L. G. Nichols (*personal communication*) studied the erosion and widening of canals mainly used by petroleum interests in the refuge. One canal system (Humble Oil Co.) was dredged ≈20 metres wide for a length of 8.5 km in 1954 and has widened at the average rate of 0.36 m/month (1.2 ft/month) since. The total land loss for this canal, including the levee, amounts to 204 hectares of freshwater marsh. Another canal system (Superior) was also constructed

Table 4. Land Loss (hectares/year) per Vegetative Type and Percent of Total Land Loss for Management Units of Louisiana Coastal Zone[a]

| Management unit | Marsh type | | | Swamp forest |
	Saline	Brackish	Fresh	
Pontchartrain/St. Bernard	122.0	430.0	3.0	72.0
	24%	64%	1%	11%
Mississippi River	1.5	94.0	116.0	...
	1%	37%	62%	...
Barataria Basin	331.0	365.0	76.0	58.0
	40%	44%	9%	7%
Terrebonne Basin	168.0	184.0	166.0	68.0
	30%	31%	28%	11%
Atchafalaya River[b]	0.7	4.7	43.0	6.6
	1%	9%	78%	12%
Vermilion Basin	5.2	161.0	11.1	15.1
	3%	84%	6%	7%
Chenier Plain	11.3	106.0	76.0	...
	6%	55%	39%	...
Totals	640.0	1344.0	491.0	220.0
	25%	49%	18%	8%

[a] From Craig and Day, 1976.
[b] This does not include current delta building in the Atchafalaya Bay.

≈20 metres wide for a length of 24.6 km in 1952. By 1958 the average canal width increased by 23 m. Total land loss of freshwater marsh was 262 ha by 1958.

The data in Table 4 (and shown above on canals) indicate that land loss has several cumulative impacts, some of which are interrelated. First, land loss due to the construction of canals often results in rapid salinity changes. This causes a change in the species composition of the area. One example of this is a result of the construction of the Mississippi River-Gulf Outlet. Since its construction 20 years ago it has changed the biotic composition of at least 18,000 ha from that of a freshwater marsh to that of a brackish marsh. Freshwater flora and fauna have been eliminated from the immediate area (van Lopik and Stone, 1974).

Another negative result of land loss is due to the construction of straight canals, which creates a condition whereby runoff water, such as from agricultural and urban sources, is shunted through the marsh into open water bodies. Eutrophication probably results, which we discuss in the next section. In addition, straight canals result in less buffering of chemical discharges and a direct loss of habitat necessary for consumer and especially fishery species.

It is possible to see two implications in the data presented on land loss, especially loss due to canals: (1) canals cause a direct loss of land by removal of the marshland and they cause a cumulative loss of land because erosion at the banks of canals is quite rapid; (2) there is a considerable amount of natural land loss in freshwater marshes of coastal Louisiana, but man's activities can significantly amplify and increase this.

We strongly suggest that managers and decision makers require that all canals not used for navigation be refilled and refurbished to their prior or natural condition.

Nutrient Changes

Disruption or digging in any type of marsh produces local changes in the sediments and associated chemical species. For example, it has been estimated (Stone et al., 1976) that the construction of a particular pipeline ditch through 21.6 km of freshwater marsh in coastal Louisiana would produce $\approx 1 \times 10^6$ m^3 of spoil and potentially could add to the surrounding aquatic environment a total inorganic nitrogen component of 2,400 kg, a PO_4^{-3} component of 610 kg, a dissolved SiO_2 component of 4,000 kg, a BOD load of 4.1×10^6 kg, and a H_2SO_4 component of 2.0×10^5 kg.

To provide some relative measure of the importance of some activities (such as digging the above pipeline), we made extensive, basic field studies (during 1973–1974) on the hydrology, C, and nutrients of the upper drainage basin of the Barataria estuary (Craig and Day, 1976). This is a freshwater area consisting of bayous, lakes, swamp forest, and marsh (Fig. 1). Export of materials from this wetland watershed was calculated by dividing total output by total watershed area. Final estimates were 19.3 g C per square metre per year, 2.7 g N per square metre per year, and 0.4 g P per square metre per year. (By comparison litter fall in the swamp amounted to 600 dry weight grams per square metre per year.) These figures are not highly precise and should be treated cautiously. From these data we estimated that the annual export to the lower estuary calculated from water discharge and materials concentrations is 8,016 metric tons of organic C, 1,047 metric tons of N, and 154 metric tons of P. Thus, the pipeline, cited above, could account for 0.25% of the total N export and approximately 0.45% of the total P export. In 1974 alone, 128 dredging permits were issued for the Barataria Basin, which suggests that the total impact on nutrient cycling due to dredging and/or pipeline construction could be quite significant.

The annual export data on nutrients clearly indicate that the fresh waters in the upper Barataria Basin are rich in N and P, but we believe that some of these nutrients are derived primarily from terrestrial sources such as agriculture and urban runoff. If these waters could flood over wetlands, the nutrients would be taken up by plants and incorporated into detritus. (There are, of course, other important transformations that would also occur in wetland soils, such as the adsorption of P and denitrification.) Many canals allow

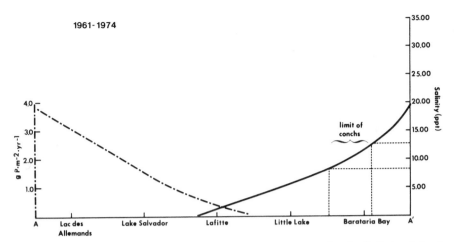

Fig. 5. Phosphorus loading rate (grams per square metre per year) and salinity levels (parts per thousand) at various locations in Barataria Basin, 1961–1974. Conch range illustrates southern limit of oyster production (Craig and Day, 1976).

runoff water to bypass wetlands and flow directly into water bodies. This eliminates the purging value of wetlands and may lead to eutrophication of water bodies.

Export of these materials represents an important source of raw materials to the lower estuary. And we know that there is a general trend of decreasing concentrations of total N, P, and organic C toward the brackish and saline marsh areas and the gulf (Happ et al., *In press*). This may indicate that part of the material supplied by the upper watershed is being consumed by the lower system. Water discharge from the basin is significant year-round, except during the summer when evaporation equals precipitation.

Research by Craig and Day (1976) indicates that the upper basin is eutrophicating. They estimate that the P loading rates into 2 of the largest freshwater lakes of the upper basin are 0.5 and 4.3 g per square metre per year, respectively, and that this may exceed the loading limit by about 4 and 11×, respectively (Fig. 5). Craig and Day believe that the primary sources for this P are natural background levels, plus runoff from agricultural and urban lands, and domestic sewage.

The freshwater system and marshes of the Barataria Basin serve as a major freshwater reservoir for maintenance of favorable salinities in the brackish and saline zones. The upper area thus contributes substantially to the productivity of the lower estuary. The major pulse of materials to the lower system begins in February and coincides with the time of high detrital formation and the arrival of migrant species which enter the estuarine area for growth and spawning purposes. An advantage of detrital export in spring is that it allows for overwinter enrichment of detritus by microbial action on the swamp floor. This probably produces a food of high quality for the

Table 5. Selected Recommendations for Management of Freshwater Marshes in Coastal
Louisiana[a]

Management recommendations	Reference and/or derivation sources
A. Policy recommendations	
Managers should know the general features of the natural function of freshwater marshes. Decisions should reflect, complement, and reinforce natural function.	Fig. 2; Clark, 1974; Odum, 1971.
Managers should make full use of simulation models for estimating environmental impacts at order-of-magnitude level.	Fig. 3 and Table 3; Eckenrod et al. 1977; R. A. Hinchee (*personal communication*); Hall and Day, 1977.
Managers should evaluate environmental impacts by means of the energy-cost accounting technique or other suitable optimalization techniques.	Odum, 1972; Odum and Odum, 1976; Gilliland, 1975; Day et al., 1976; Stone et al., 1976; Zucchetto, 1975.
Managers should support research that will quantify impact assessment.	McHugh, 1976; Craig and Day, 1976; Day et al., 1976.
Managers should recognize the limits of technology and require those who propose environmental changes to demonstrate no effect or to refurbish.	Happ et al., 1976; Schumacher, 1973.
B. Operational recommendations	
Preventative procedures:	
Use existing canals (whenever possible).	CM No. 26; L. G. Nichols (*personal communication*); Gagliano, 1973; Craig and Day, 1976.
Restrict new canals to natural corridors or levees.	Gagliano, 1973.
Stop construction of canals that connect the edge of the hydrological basin to the middle.	Gosselink et al., 1976.
Stop construction of blind-end canals or finger-fill developments.	Barada and Partington, 1972.
Limit construction of impoundments in marsh areas.	MM No. 14.
Limit canals between different vegetative types.	Stone and McHugh, 1977.
Mitigative procedures:	
Perform dredging operations as fast as possible.	Gosselink et al., 1976.
Place periodic openings in spoil banks so water circulation is not impeded.	MM No. 2, 4, 5, 8, 12 and 13; Day et al., 1976.
Place water control structure on all existing waterways.	St. Amant, 1971, 1972.
Plug pipeline canals on seaward side until construction is finished.	St. Amant, 1971, 1972.

Table 5. Continued

Management recommendations	Reference and/or derivation sources
Dispose of spoil with special care; if possible, place in nonwetland areas.	Craig and Day, 1976.
Use special care during times of wildlife migrations, spawning, and nesting.	Gosselink et al., 1976; Bahr and Hebrard, 1976.
Shunt agricultural and urban runoff waters into marsh areas rather than into open water bodies.	Turner et al., 1976.
Construct spoil banks lower than tidal amplitude.	Cite coastal elevations.
Dig transportation canals only as deep as the euphotic zone and not deeper than water body where canal ends.	Barada and Partington, 1972.
Time application of fertilizer to time of maximum uptake (eliminate fall and winter applications).	R. A. Hinchee (*personal communication*).
Use manure to condition soil and to replace chemical fertilizers.	R. A. Hinchee (*personal communication*).
Plow fertilizers (if used) as deeply as possible.	R. A. Hinchee (*personal communication*).
Refill and refurbish canals.	Happ et al., 1976.

[a] Recommendations are divided into policy (A) and operational levels (B). Reference cited as CM No. 1 through 27 refer to the conceptual model and its various parts as illustrated in Fig. 1 and described in the text; references cited as MM No. 1 through 14 refer to the marsh model and the various simulation tests described in Table 3.

detrital-based food chains both in the swamp and the area below it. Enrichment of detritus over time has been well demonstrated (Odum and de la Cruz, 1967).

The implications of these data are that canals or dredging has several impacts, all of which are interrelated. For example, by digging a canal the water is often diverted away from the marsh proper and nutrients in these waters are not removed by the marsh grass; the consequence of this is unfavorable conditions for most fishery species, namely eutrophication.

Decision makers should consider the use of marshlands for removing excessive nutrient loads from agricultural and urban runoff. Pilot programs with the Louisiana Sea Grant program demonstrate the validity and effectiveness of this approach (Turner et al., 1976).

RECOMMENDATIONS FOR MANAGEMENT OF FRESHWATER MARSHES

Table 5 summarizes some of our recommendations for management of freshwater marshes in regard to canalling. These are segregated in terms of policy and operational recommendations, the difference between the two levels being that policy deals more with general items while operations deal more with specific, or day-to-day items. Operational recommendations are subdivided into preventative and mitigative measures. For each recommendation we have cited a reference and/or derivation source.

We believe that the policy recommendations, perhaps the most obvious, are the most important. For example, a decision maker should know the rudiments of how a wetland system operates; thus, for example, if he is aware of the importance and role of water in the natural functioning of the freshwater marsh, he may be cautious in deciding to change or modify the marsh. He should realize that the marsh is a product of many years of evolution, and not easily duplicated by man and that it provides many natural services to man. If a development cannot be stopped, then he should use a variety of simulation models to estimate the impacts and he should use some form of energy-cost accounting (Odum, 1972; Gilliland, 1975; Stone et al., 1976; Zucchetto, 1975) in order to derive representative benefit:cost ratios. If research is not supported, then almost any opinion will be valid in regard to proposed economic activities and the discussion is apt to be clouded by emotion rather than data. Technology does not have all the answers, but those who propose economic-return activities in freshwater marshes should demonstrate that there will be no adverse effects on the marsh or how any potential adverse effects may be mitigated.

Table 5 lists recommendations to develop guidelines for coastal zone management in Louisiana. The reference or derivation sources are not complete; others could be added to the table but our knowledge of freshwater marshes is incomplete and much work remains to be done in order to perfect these guidelines. We are not certain that as more data and knowledge become available that eventually we will reach performance standards, comparable to engineering specifications. It is more probable that something in between our recommendations and specific performance standards will evolve.

Perhaps a more serious problem is how to implement or transfer these recommendations into action. The outlook is not particularly sanguine, but as with many things it will probably be done by a combination of laws, education, and moral persuasion. The frustration of this approach is that it is inefficient, slow, and very tortuous in its workings. Many of our problems in the management of marshes are more immediate and demand resolution faster than we can do the necessary research. We must try, nonetheless.

ACKNOWLEDGMENTS

Part of the research reported herein was supported by the Louisiana Sea Grant Program, a part of the National Sea Grant Program maintained by the National Oceanic and Atmospheric Administration of the U.S. Department of Commerce and by the Louisiana State Planning Office under the Coastal Zone Management program.

LITERATURE CITED

Adams, R. D., Barrett, B. B., Blackmon, J. H., Gane, B. W., and McIntire, W. G. (1976). Barataria Basin: Geologic processes and framework. Sea Grant Publication No. LSU-T-76-008, Louisiana State University Center for Wetland Resources, Baton Rouge, La.

Bahr, L. M., Jr., and Hebrard, J. J. (1976). Barataria Basin: Bilogical characterization. Sea Grant Publ. No. LSU-T-76-005. Louisiana State University, Center for Wetland Resources, Baton Rouge, La.

Barada, N., and Partington, W. M., Jr. (1972). Report of investigation of the environmental effects of private waterfront canals. Environmental Information Center of the Florida Conservation, Inc., Winter Park, Fla. 63 pp.

Barrett, B. B. (1970). Water measurements of coastal Louisiana. Louisiana Wildlife and Fisheries Comm., U.S. Dept. of the Interior Fish and Wildl. Serv., Bureau of Commercial Fisheries Proj. No. 2-22-T of P.L., 88–309.

Byrne, P., Borengasser, M., Drew, G., Muller, R. A., Smith, B. L., Jr., and Wax, C. (1976). Barataria Basin: Hydrologic and climatologic processes. Louisiana State Printing Office. Baton Rouge, La.

Chabreck, R. H. (1972). Vegetation, water, and soil characteristics of the Louisiana coastal region. Louisiana State University Agri. Expt. Sta., Baton Rouge, La. Bull. No. 664.

Clark, J. (1974). Coastal ecosystems: Ecological considerations for management of coastal zone. The Conservation Foundation, Wash., D.C. 178 pp.

Conner, W. H. (1976). Wetland use practices: Mineral and energy production. Louisiana State Planning Office, Baton Rouge, La.

Craig, N. J., and Day, J. W., Jr. (1976). Barataria Basin: Eutrophication case history. Louisiana State Planning Office, Baton Rouge, La.

Day, J. W., Jr., Craig, N. J., and Turner, R. E. (1976). Cumulative impact studies in Louisiana coastal zone: Land loss. Louisiana State Planning Office. Baton Rouge, La.

Eckenrod, R. M., Day, J. W., Jr., and Bahr, L. M., Jr. (1977). A simulation model of urban estuarine interactions. *In* "Proc. Marsh-Estuarine Systems Simulation Symposium" (R. F. Dame, ed.), Univ. of South Carolina *(In press)*.

Gagliano, S. M. (1973). Canals, dredging and land reclamation in the Louisiana coastal zone. Hydrologic and Geologic Studies of Coastal Louisiana, Report No. 14. Louisiana State University Center for Wetland Resources, Baton Rouge, La.

Gagliano, S. M., and van Beek, J. L. (1970). Geologic and geomorphic aspects of deltaic processes, Mississippi River system. Hydrologic and Geologic Studies of Coastal Louisiana, Report No. 1. Louisiana State University Center for Wetland Resources, Baton Rouge, La.

Gilliland, M. W. (1975). Energy analysis and public policy. *Science* 189, 1051–1056.

Gosselink, J. G., Miller, R. R., Hood, M., and Bahr, L. M., Jr., eds. (1976). Louisiana offshore oil port (LOOP): Environmental baseline study. 4 vols. LOOP, Inc., Harvey, La.

Grimes, M. D., and Pinney, T. K. (1976). Recreation potential of private lands in Louisiana Coastal Zone. Louisiana Agricultural Experiment Station. Sea Grant Publication No. LSU-T-76-010. Center for Wetland Resources, Louisiana State University, Baton Rouge, La.

Hall, C. A. S., and Day, J. W., Jr., eds. (1977). Ecosystem Modeling in Theory and Practice: An Introduction with Case Histories. John Wiley and Sons, New York.

Happ, G., Bennett, V. R., Burke, W. M., III, Conner, W. H., Craig, N. J., Hinchee, R. E., Bahr, L. M., Jr., and Stone, J. H. (1976). Impacts of outer continental shelf activities; Lafourche Parish, La. Louisiana State Planning Office, Baton Rouge, La.

Happ, G., Gosselink, J. G., and Day, J. W., Jr. *(In press)*. The seasonal distribution of organic carbon in a Louisiana estuary. *J. Coastal and Estuarine Mar. Sci.*

McHugh, G. F. (1976). Development of a hydrodynamic numerical model for use in a shallow well-mixed estuary. Sea Grant Publication No. LSU-T-76-008. Center for Wetland Resources, Louisiana State University, Baton Rouge, La.

McIntire, W. G., Hershman, M. J., Adams, R. D., Midboe, K. D., and Barrett, B. B. (1975). A
    rationale for determining the Louisiana coastal zone. Sea Grant Publ. No. LSU-T-75-006.
    Louisiana State University Center for Wetland Resources, Baton Rouge, La.
Newsom, J. D., ed. (1968). Proc. of the Marsh and Estuary Management Symposium held at
    Louisiana State University, July 19–20, 1967. Louisiana State University Division of Con-
    tinuing Education. Baton Rouge, La.
Odum, E. P., and de la Cruz, A. A. (1967). Particulate organic detritus in a Georgia salt marsh
    estuarine ecosystem. In "Estuaries" (G. H. Lauff, ed.), pp. 383–388. AAAS Publ. No. 83.
Odum, H. T. (1971). Environment, power, and society. John Wiley and Sons, Inc., New York.
Odum, H. T. (1972). Use of energy diagrams for environmental impact statements. In "Tools
    for Coastal Management," pp. 197–213. Proc. of the Conf. Marine Technology Society,
    Washington, D.C.
Odum, H. T., and Odum, E. (1976). Energy Basis for Man and Nature. McGraw-Hill, New
    York.
Schumacher, E. F. (1973). Small is Beautiful: Economics As If People Mattered. Perennial
    Library, Harper & Row, New York.
St. Amant, L. S. (1971). Impacts of oil on the Gulf Coast. Trans. of 36th N. Amer. Wildlife and
    Nat. Res. Conf. 36, 206–219.
St. Amant, L. S. (1972). The petroleum industry as it affects marine and estuarine ecology.
    Journal of Petroleum Technology. 46, 385–392.
Stone, J. H., Byrne, P. A., and Fannaly, M. T., III (1976). Selected environmental impacts of
    proposed Louisiana oil port project. In "Louisiana Offshore Oil Port: Environmental
    Baseline Study" (J. G. Gosselink, R. R. Miller, M. Hood, and L. M. Bahr, Jr., eds.), Vol. I,
    Chaps. 5–9. LOOP, Inc., Harvey, La.
Stone, J. H., and McHugh, G. F. (1977). Simulated hydrologic effects of canals in Barataria
    Basin: A preliminary study of cumulative impacts. Final Report to Louisiana State Planning
    Office, Baton Route, La.
Turner, R. E., Day, J. W., Jr., Meo, M., Payonk, P. M., Stone, J. H., Ford, T. B., and Smith,
    W. G. (1976). Aspects of land-treated waste applications in Louisiana wetlands. In
    "Freshwater Wetlands and Sewage Effluent Disposal" (D. L. Tilton, R. H. Kadlec, and C.
    J. Richardson, eds.), pp. 145–167. The University of Michigan, Ann Arbor, Michigan.
van Lopik, J. R., and Stone, J. H. (1974). Environmental Planning for Future Port Develop-
    ment. In "Port Planning and Development as Related to Problems of U.S. Ports and the
    U.S. Coastal Environment" (E. Schenker and H. C. Brockel, eds.), pp. 154–174. Cornell
    Maritime Press, Inc., Cambridge, Maryland.
Van Sickle, V. R., Barrett, B. B., Gulick, L. J., and Ford, T. B. (1976). Barataria Basin:
    Salinity changes and oyster distribution. Louisiana State University Center for Wetlands
    Resources, Baton Rouge, La. Sea Grant Publ. No. LSU-T-76-002.
Zucchetto, J. J. (1975). Energy basis for Miami, Florida, and other urban systems. Ph.D. Dis-
    sertation, University of Florida. 248 pp.

MANAGEMENT OF FRESHWATER WETLANDS FOR NUTRIENT ASSIMILATION

William E. Sloey,[1] Frederic L. Spangler,[1] and C. W. Fetter, Jr.[2]

[1] *Biology Department*
[2] *Geology Department*
University of Wisconsin–Oshkosh
Oshkosh, Wisconsin 54901

Abstract Nutrient transformation processes such as sorption, coprecipitation, active uptake, nitrification, and denitrification remove P and N from the free-flowing water of a wetland and transfer them to the substrate and biota for storage. Advantage is being taken of these processes by using wetlands to treat sewage in Germany, Holland, Finland and other European countries. In the USA, experimental application of cultural water (wastewater) to peatlands in Michigan, tidal marshes in Louisiana and New Jersey, cattail marshes in Wisconsin, cypress domes and sawgrass in Florida and many more have shown promising results. Denitrification may remove up to 3.5 kg $N \cdot ha^{-1} \cdot day^{-1}$ and as much as 20 g P/m^2 may be detained in a growing season. Natural release of nutrients between growing seasons, however, either restricts application periods, or demands management of the wetland systems to regulate such releases. The most obvious management tool, plant harvesting, removes only a few grams of P per square metre per year (<5) which is usually $<20\%$ of that detained. Most of the remainder is in the substrate-microbial compartment and is subject to between-season washout.

Other management techniques used and proposed, such as dikes, drains and intermittent application, relate to manipulating the hydraulics of the system to optimize conditions for assimilative biogeochemical processes and to prevent washout from reaching surface waters. Chemical treatment to reduce the surplus of P has also been considered.

Management options for a specific wetland will depend to a large extent upon that particular hydrologic regime. Riverine systems have different hydraulic patterns than lacustrine or palustrine systems. Flow-through systems cannot be managed like influent or seepage systems.

Already, changes in the biota of some experimentally treated wetlands indicate undesirable disturbances of these valuable natural resources. Caution in widespread use of natural wetlands to treat waste at this time is advised and careful monitoring of all biotic communities in experimental programs is essential.

Artificial marshes and peat filters offer feasible alternatives to other treatment methods for small systems and do not endanger natural wetlands.

Key words   *Artificial marshes, hydrology, nitrogen, nutrient assimilation, phosphorus, treatment, wastewater, wetland management, wetlands, Wisconsin.*

INTRODUCTION

Recently, considerable attention in North America and Europe has been directed toward the use of wetlands as water purification systems and nutrient traps to reduce the impact of man's activities upon ground and open surface waters (Tourbier and Pierson, 1976; Tilton et al., 1976). Nevertheless, the intentional use of wetlands to treat wastewater is still a very new "art" and scientific data on capabilities and impacts are meager. Most of the available data relate to nutrient inputs and outputs and to production and nutrient assimilation by the plants. Little is yet known about rates of transport to or assimilative capacities of the microbial and substrate compartments.

It is almost certain that long-term application of some wastewaters to wetlands will result in cases of nutrient toxicity, heavy metal accumulations or even public health problems. Few data are yet available on such matters, however, so this discussion is limited to considerations of nutrient assimilation, especially of N and P.

The goals of this chapter are to conceptualize for the reader the hydrologic variations and nutrient handling systems present in various types of wetlands, to review some of the recent findings on the use of wetlands in treating wastewater (cultural water), and to propose some possible management techniques which might optimize conditions for nutrient assimilation.

CHARACTERISTICS OF MAJOR FRESHWATER WETLAND TYPES

A wetland has been defined as ". . . land where the water table is at, or above the land surface for long enough each year to promote the formation of hydric soils and to support the growth of hydrophytes as long as other environmental conditions are favorable" (Cowardin et al., 1976).

Because hydrogeologic factors have a great influence on movement of water through and on the earth, they are a logical basis upon which to classify wetlands. From the standpoint of water quality management, the hydrogeology of a wetland is the thing that is most amenable to manipulation. The hydrogeologic categories of wetlands set up by Cowardin et al. (1976) are described below. Artificial marshes, previously not important in the literature, may become important in waste treatment and will constitute a special case.

Table 1. Surface and Groundwater Interactions of Palustrine Wetland Types

Surface water (type)	Groundwater interaction(s)
Flow-through	Influent, effluent, combination[a], none[b]
Inflow	Effluent, combination
Outflow	Influent, combination
Seepage	Combination

[a] Combination = groundwater is both entering (influent) and leaving (effluent) the wetland.
[b] None = perched bog.

Riverine Systems

The principal source of inflow to most riverine wetlands is surface water in the stream channel. In humid climates rivers are usually fed by groundwater. Such rivers often have adjacent wetlands into which freshwater seepage occurs. In arid climates, rivers may originate in mountain catchments and then flow across reaches where the water table is below the river channel. Such rivers lose water to the ground. The wetlands which fringe such rivers may be areas where seepage of river water into the ground takes place. Similar wetlands may be created by discharging of cultural water down a channel, some of which will infiltrate into the ground and recharge the water table.

Lacustrine Systems

The hydrogeology of a lacustrine wetland is a function of the morphology of the lake basin. Stands of emergent vegetation frequently occur along the margins of lakes, although they may be present in any shallow spots. The overall water balance of the lake regulates the water levels in the lacustrine wetlands. Lakes may be fed by both surface and groundwater inflow, overland runoff and direct precipitation. Losses may be due to evaporation, transpiration, surface water outflow and seepage into the groundwater. Water can generally exchange freely between open water and the lacustrine wetlands. Waves and currents in the lake promote such exchange.

Palustrine Systems

Palustrine systems are nontidal wetlands which are not confined by channels and are not marginal to lakes. They can contain wide varieties of vegetative types such as submergents (e.g., *Vallisneria*), floating-leaved aquatics (e.g., *Potamogeton*), emergent aquatics (e.g., *Typha*), mosses (e.g., *Sphagnum*), shrubs (e.g., *Cornus*), and trees (e.g., *Picea*). The palustrine wetlands are hydraulically isolated from open surface water so that there is not a ready exchange of water. Hydraulic residence times are high. These 2 factors make palustrine wetlands more amenable to management for wastewater treatment than other major wetland groups. Palustrine wetlands can be classified as flow-through, inflow, outflow or seepage on the basis of surface water movement. Possible interactions of surface and groundwater with palustrine wetlands are presented in Table 1.

NUTRIENT TRANSFORMATIONS IN WETLANDS

Any system such as a lake, forest, or wetland has a characteristic set of nutrient handling processes which may be quite different in relative importance if not in basic nature from the processes of other systems.

Nitrogen

Nitrogen is present in domestic wastewater in several organic and inorganic forms. The principal means by which N can be removed from water in a wetland is denitrification, with loss of N_2 to the atmosphere.

Denitrification occurs anaerobically. It is the reduction of nitrate to N_2 by organisms which oxidize organic matter and use nitrate as an electron acceptor in the process. In order for denitrification to be a significant factor in removing N from a wetland, the N_2 resulting must escape to the atmosphere before it can be fixed or converted to nitrate again. The N must not be allowed to accumulate or cycle in the wetland system. Instead, it should be leaked out as N_2 gas. If excessive aerobic water lies over anaerobic soil, the N_2 moving upward may not escape. Manipulation of water levels would be essential to achievement of good N removal. It is known that denitrification occurs mainly in the soil and not in the overlying water even if the water is anaerobic (Patrick et al., 1976). The flow of (diluted) cultural water in a wetland should, if possible, be regulated so that N_2 in the reduced zone should have the least possible resistance to its escape to the atmosphere.

Other possible fates of N should not be overlooked as they may be significant, depending upon local circumstances. For instance, loss of organic nitrogen via refractory compounds in sediment occurs (Godshalk and Wetzel, 1978). In fact, accumulation of undecomposed material is a dominant feature of some wetlands. This process is not very amenable to influence by human activity. Much interest centers on N uptake by higher plants. A logical expectation is that high growth rates during certain times of the year would lead to formation of a large pool of the element temporarily sidetracked on its journey. However, most of the work on the subject is directed toward understanding primary production.

Phosphorus

Municipal wastewater can be expected to contain 10–15 mg total P per litre (Rohlich and Uttermark, 1972). The enrichment with P is proportionally greater than for C or N. That becomes particularly important in view of the fact that P is suspected in many, if not most freshwater systems of being limiting. The P in domestic wastewater exists in a variety of forms but most of it can be readily converted to soluble, inorganic forms useable by plants.

Examination of the phosphorus cycle as it exists in wetlands, reveals no portion or step which leads to escape of P from the system (Prentki et al., 1978). There is no counterpart to denitrification. Permanent removal of P from a system can only be achieved by physically removing plants containing P or by allowing a natural surface water flush-out from the system between growing seasons (see below). Loss to the cycle (but not to the system) occurs when refractory or insoluble compounds enter the substrate. Unfortunately the substrate may be a source of P as well as a sink under some conditions.

In much of the literature, attention has been given to exchange rates between lake water and underlying sediments. For the time being, it is necessary to assume that the sediment–water exchange is governed by about the same factors in both lakes and marshes. The phosphorus exchange in lakes is largely a matter of absorption-desorption and is influenced most by pH, redox potential, and Ca (Kramer et al., 1972). Redox potential is a measure of the ratio of reduced to oxidized substances and depends greatly on the presence or absence of dissolved O_2. Mortimer (1971) believes that if O_2 concentration does not drop below 2 mg O_2/l sediments will not contribute P to the overlying water. The stratification of oxidized water over reduced substrate materials influences P conditions. Low redox potential (low dissolved O_2) promotes solubilization of P because of reactions similar to the one in which ferric iron is reduced with release of PO_4^{-3} from ferric phosphate. However, the chemical and physical nature of the substrate particles are as important as redox potential. Adsorption and ion exchange are very important also and depend upon particle size and pH (Hwang et al., 1976).

At pH 5–7, P is less likely to go into solution from sediments than it is at higher or lower pH values (Kramer et al., 1972). Since different kinds of wetlands vary widely in pH, the importance of this factor must not be overlooked. In Wisconsin, acid peat wetlands with soft water are more often in the northern region and alkaline muck wetlands with hard water are more often in the southern region (Phillips, 1970). Rate and extent of decomposition are related to pH, temperature and availability of N with less rapid and complete decomposition occurring in the northern peat areas. Decomposition produces organic colloids which offer good potential for adsorption.

The availability of Ca influences P transformations. Two mechanisms are involved. One is adsorption of P on $CaCO_3$ which precipitates in colloidal form in hard water lakes. The other, is formation of calcium phosphate. Wetzel (1975) discussed relationships between P and other materials including Ca and Fe in lake sediments. He concluded that iron–phosphate complexes are more important than $CaCO_3$ sorption in controlling P concentration in the sediment. A matter of great importance is the rate of transfer of P (or other material) across the sediment–water interface.

The pattern of contact of water with the substrate is different in wetlands than in lakes. Wetland substrates may routinely be exposed to the air. Also, the presence of vascular plants introduces complexity not found in lake sediments. Roots are locations of biological activity and in many cases carry O_2 into sediments immediately surrounding the roots which otherwise would be anaerobic (Hutchinson, 1975). This activity certainly influences redox potential, pH, and decomposition.

WETLAND STORAGE COMPARTMENTS AND THEIR SEASONAL BEHAVIOR

The objective of managing wetlands to permit the addition of cultural water is to divert nutrients from the free-flowing water into storage compartments such as biota and substrata in order to minimize the impact upon the receiving ground or surface water (Fig. 1). The free-flowing water includes the surface water and any groundwater with a reasonably short residence time. Trapped interstitial water is that groundwater not moving through the system, such as in perched bogs, or where the substrate has a low permeability. Nutrients dissolved in this water are, in effect, stored or "trapped." Inasmuch as we are concerned here only with water quality, any diversion of N into the atmosphere can also be called storage. The rate at which cultural (wastewater) water can be applied to a wetland (R_2) is dependent upon 3 factors:

1) The rates of input from the other sources such as rain and surface water, groundwater, and nitrogen fixation (R_1, R_3, R_N).
2) The rates of transfer to and from storage compartments (R_5 through R_{13}).
3) The rate at which nutrients may be permitted to enter the receiving waters (R_{14}, i.e., water quality standards).

The rate of discharge of nutrients from a wetland to the receiving stream can be expressed as in Eq (1):

$$R_{14} = R_4 - [(R_5 + R_7 + R_9 + R_{13}) - (R_6 + R_8 + R_{10} + R_{12})] \qquad (1)$$

That is: the nutrient outflow rate (R_{14}) is the inflow rate (R_4) minus the difference between the sums of the rates of transfer into and out of storage compartments. Denitrification (R_D) represents a transfer of N out of the microorganism compartment. The atmosphere, thus, merely expands the capacity of the microorganism compartment. There are, undoubtedly, also some nutrient exchanges between the trapped interstitial water and the free-flowing water, as well as with each of the storage compartments. The rates of transfer between the trapped interstitial water and the storage compartments are, however, not of importance in determining the quality of the outflow water. It is not possible at present to assign numbers to more than a

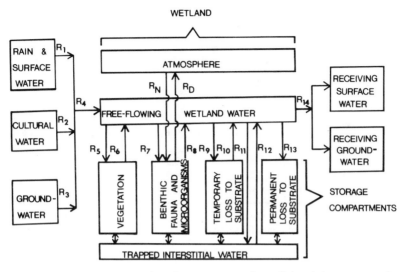

Fig. 1. Diagrammatic representation of sources, rates of transfer, and storage compartments of nutrients in a wetland ecosystem receiving cultural water.

few of the transfer rates. Perhaps we could now make some generalizations about the vegetation compartment but specific rates of transfer to and from the other storage compartments, especially under various loading conditions, are still not available. Such information is urgently needed.

Like other natural ecosystems, wetlands are subject to climatic variations which greatly affect the function of the various storage compartments. The vegetation is obviously the most affected compartment. While the growing season may vary in length depending upon latitude, the apparent nutrient storage period may be much longer than the actual storage period. Nutrient concentrations in aquatic vegetation have been shown to decrease as the growing season progresses (Boyd and Blackburn, 1970). This decrease is more than just dilution within the maturing tissues, but rather, includes an actual loss from the plants to the surrounding free-water by secretion (Boyd and Blackburn, 1970; Wetzel, 1969). Some species may be nutrient conser-

Table 2. Seasonal Aspects of Storage Compartment Activity in Wetland Ecosystems[a]

Storage compartment	Spring	Summer	Fall	Winter
Vegetation	+	±	−	−
Benthic fauna—microorganisms	±	+	±	−
Temporary loss to substrate	±	+	±	−
Permanent loss to substrate	+	+	+	+

[a] + = compartment functioning as a sink; − = compartment functioning as a source or is nonfunctional.

Table 3. Seasonal Distribution of Phosphorus in an Artificial Marsh Receiving Wastewater
in 1974

	17 June 1974		4 December 1974	
Plant part	g P/m^2	%	g P/m^2	%
Harvested shoots[a]	1.53	8.1	0.00	0.0
Unharvested shoots	1.83	9.7	0.75	12.3
Rhizomes	3.04	16.1	2.32	38.0
Roots	1.25	6.6	1.86	30.5
Total biomass	7.65	40.5	4.93	80.8
Gravel substrate	11.21	59.9	1.17	19.2
TOTAL	18.86	100.4	6.10	100.0

[a] *Scirpus validus*, planted in a plastic-lined basin containing 15 cm of pea-sized gravel and receiving secondarily treated domestic sewage.

vative and translocate some of the nutrients back into the underground storage organs between growing seasons, while others act as nutrient pumps and sacrifice nutrients in the aboveground tissue (Klopatek, 1975, 1978; Prentki et al., 1978). Only in the case of large trees can the nutrient storage capacity be considered permanent.

Each of the other nutrient storage compartments is also subject to seasonal variations in activity (Table 2). Even the nitrifying and denitrifying bacteria are limited in activity during droughts or low winter temperatures. Denitrification is, at best, intermittent in peak operation. Benthic fauna and other microorganisms also become inactive and may undergo severe population declines in winter and release large amounts of nutrients into the freeflowing water where they are lost to the receiving water (Whigham and Simpson, 1976). In a study of artificial marshes containing a gravel substrate, we found that almost all of the 11.2 g P/m^2 detained in the gravel during the growing season was eluted from the system between September and early December (Table 3; Spangler et al., 1976). The P was in microbial biomass, benthic fauna and temporary storage.

Dry periods which lower the water table in the wetland and expose the organic substrate can also induce loss of nutrients to the receiving water. Aeration accelerates the decomposition process. Should the marsh then be flooded before the next growing season, the wetland would discharge a slug of nutrients to the receiving water. All natural wetlands probably have compartments which function as nutrient sinks during some seasons and as sources during others. Northern cattail marshes are probably typical of temperate wetlands in that autumn and spring flushings wash out much of the P assimilated during the growing season (Spangler et al., 1977; Lee et al., 1975; Whigham and Simpson, 1976). Loucks et al. (1977), however, estimated the detention of storm water P by cattail marshes surrounding Lake

Wingra at Madison, Wisconsin. They found an 83% P detention in the summer, but only a 10% annual average. Unlike the other reports cited above, there were no net discharge periods, but winter and early spring detentions were only 1% and 8% respectively. Peatlands may have a greater permanent nutrient storage due to the high phosphorus sorption capacity of the litter component (Richardson et al., 1976).

EFFICIENCY OF NUTRIENT STORAGE BY WETLANDS

Most of the investigations to date on nutrient storage by wetlands have been directed toward application of cultural water (primarily domestic sewage) either to unmanaged natural systems or to artificial systems. Artificial systems are discussed under "USE OF ARTIFICIAL SYSTEMS CONTAINING MARSH PLANTS" below. Several studies on natural systems are discussed here.

Northern Peatlands

Studies in Michigan of a perched northern peatland in which additions of low volumes ($6.3–12.5$ litres$\cdot$m$^{-2}\cdot$wk^{-1}) for the growing season of "simulated sewage" (NH_4, NO_3, HPO_4-$7H_2O$, etc.) were applied, demonstrated that the nutrients were quickly bound to the litter in the upper layer. Lateral movement of P through the litter was restricted in spite of a 92 cm/s hydraulic conductivity. A deeper layer of more compact peat prevented vertical movement of groundwater (Richardson et al., 1976). The authors suggest that the P was apparently bound to the gel-like reduced ferrous compounds in the litter and the N was lost by denitrification. No P was apparently lost during winter or with the spring flushing. Application of much higher levels of nutrients (up to 506 g$\cdot$m$^2\cdot$wk^{-1} for the growing season of combined nutrients) increased plant growth and increased the concentration of both N and P in the aboveground tissues of willow (*Salix*), bog birch (*Betula*) and leatherleaf (*Chamaedaphne*). In spite of high concentrations of NH_4 in the interstitial water, there was little loss of N to the receiving stream.

In Finland, ditched peat systems are being used to treat domestic sewage. Farnham and Boelter (1976) found, in a review of Finnish literature, that P was reduced an average of 39% and total nitrogen by 62%. One project, however, demonstrated 82 and 90% reduction of P and N, respectively. The Finns were reported to have found it necessary to drain the water off the peat before application of sewage in order to force the sewage through the less porous decomposed peat for effective treatment. Tilton et al. (1976) and Lee et al. (1975) warn, however, that draining converts the surface layers to an aerobic condition which accelerates decomposition and could increase the loss of minerals from the system.

Cattail Marshes

We studied Brillion Marsh in Wisconsin which has been receiving domestic sewage since 1923 (P. O'Leary, McMahon and Assoc., *personal communication*) and agricultural runoff. Nutrient loadings were extremely high: the 1.6 km^2 cattail (*Typha*) marsh was receiving 148 kg P$\cdot$ha$^{-1}\cdot$yr^{-1}, primarily from the Brillion sewage treatment plant (Fetter et al., 1976). During a summertime intensive study period, P was reduced from about 2 mg P/litre (0.5–3.75) to a very consistent 1 mg P/litre. Ammonia values in the influent to the marsh ranged to 8 mg/l, but none was detected below the marsh. Nitrate originated primarily from agricultural sources and values above the marsh ranged from 0.4–2.1 mg/l. Below the marsh, nitrates remained at 0.1–0.2 mg/l. Other nitrogen species were not monitored. On an annual basis, the marsh retained 48 kg P/ha, but only 10 kg P/ha could have been removed by harvesting of the plants (0.49–1.0 mg P/m²). It is possible that this large retention of P was not "typical" for this or other temperate wetlands (Lee et al., 1975). There was, however, a considerable accumulation of P in the sediments of the influent stream channel, indicating a permanent sediment storage compartment in the organic muck.

Freshwater Tidal Marshes

Whigham and Simpson (1976) studied a freshwater tidal marsh near Trenton, New Jersey and found that the system behaved in a fashion similar to other temperate systems in the winter release of nutrients. The alga *Rhizoclonium*, however, developed on the mud in the winter and acted as a somewhat smaller N sink, thus dampening the impact from the loss of emergents. Nitrate-nitrogen levels decreased from 100 μg N/l at high standing water to only trace levels at low standing water. Reactive P, however, increased during low standing water.

Turner et al. (1976) applied fish processing waste both as a direct application to a Louisiana marsh and via an upland application. The overland flow and seepage from the upland application reduced N and P by 50%, but there was a high nitrate concentration entering the marsh. In neither the direct application nor the overland application was there a great increase of nutrients at the limits of their marsh sampling areas. The soil–water interface was identified as the major site of sorption, deactivation and denitrification. Engler and Patrick (1974) have reported 3.5 kg $N \cdot ha^{-1} \cdot day^{-1}$ denitrified from Louisiana marshes.

Everglades Sawgrass Marshes

Steward (1976) applied wastewater to a sawgrass marsh (*Cladium jamaicense*) in Florida at the rate of 2.2 kg $P \cdot ha^{-1} \cdot wk^{-1}$ and recorded a 95% reduction. Forty-three precent of the P removed was found in the sediments, but only 12% of the amount applied could be accounted for in the plant tissue. The capacity of the system was overloaded after only 8 weeks.

Cypress Domes

Odum et al. (1975) applied domestic sewage to Florida cypress domes (*Taxodium* sp.) which had seepage type hydrologic systems. Nitrogen was decreased from 15.5 mg/litre in the influent to 0.32–0.35 mg/l in the groundwater. Phosphorus was reduced from 8.2 mg/l to 0.21–0.44 mg/l.

Lacustrine Marshes

Tóth (1972) studied the effects of reed stands (*Phragmites*) on the impact of sewage effluent entering Lake Balaton in Hungary. He found that there was $\approx$98% reduction (5.57–0.082 mg/litre) of total P and $\approx$95% reduction in total N (21.6–1.024 mg/l). No such reduction occurred in a similar situation where the reeds were absent. Tóth attributes most of the action to precipitation and to uptake by flora (epiphytic algae and bacteria) and fauna (Mollusca, etc.) on the reed stems.

POTENTIAL MANAGEMENT TECHNIQUES TO OPTIMIZE CONDITIONS FOR NUTRIENT ASSIMILATION

Seasonal Application

It is possible to use any unmanaged wetland as a temporary nutrient detention basin providing that the nutrient uptake and hydraulic capacity are not exceeded, and providing that cultural water application is restricted to those seasons permitted by natural cycles (Table 4). Such use is restricted to seasonal activities such as recreation (de Jong, 1975) or food processing (Turner et al., 1976). If the volume is sufficiently small, it may be possible to store domestic wastewater in holding basins for seasonal application. Such use will at least keep the nutrients from entering the surface waters during the summer aquatic weed and algal bloom periods. It must be remembered, of course, that while N may have been removed via denitrification, much of the P may be discharged from the wetland if the wetland is flushed between growing seasons. Spring flushing could be especially detrimental for the receiving water as P concentrations at that time contribute to the magnitude of the summer algal blooms (Vollenweider, 1971). In marine situations, where incidence of red tides has been shown to be related to flushing from freshwater marshes (Prakash and Rashid, 1968), increased productivity due to nutrient additions might further aggravate the problem. Increased decomposition of humic compounds when wastewater is present (Vollenweider, 1971) could also exert an additional BOD loading on the receiving water. Application of wastewater to palustrine inflow and seepage wetlands may be continuous if soil conditions are suitable for P adsorption or nitrate reduction. Care must be taken that over-winter accumulations of wastes do not alter the natural integrity of the wetland. In tidal wetlands, it would appear advisable to restrict application of wastewater to periods of low standing water when more immediate contact with biota and substrate could be made (Whigham and Simpson, 1976).

Outflow Regulation

In some types of wetlands, it may be possible to prevent washout of nutrients to the receiving waters at undesirable periods by diking all or part of the wetland and regulating the outflow. For instance, if spring hydraulic

Table 4. Applicability of Potential Management Techniques to Major Wetland Groups[a]

				Palustrine[b]			
Management technique	Riverine	Lacus-trine	Tidal	Flow-through	Inflow	Out-flow	Seepage
Seasonal application	+	+	+	+	Not needed	+	Not needed
Outflow regulation	0	0	+0	+	0	+	0
Flushing	Natural	+0	Natural	+	0	+	0
Diversion (to prevent flushing)	0	0	0	+	Not needed	0	0
Upland application	+	+	+	+	0	+	0
Underground application	+0	+0	+0	+	0	+	+
Chemical pretreatment	+	+	+	+	+	+	+
In situ chemical treatment	0	0	0	+0	+	+0	+
Plant harvesting	+	+	+	+	+	+	+

[a] + = applicable; 0 = not applicable.
[b] Palustrine wetlands examples used: Flow-through = groundwater influent; Inflow = groundwater effluent; Outflow = groundwater influent; Seepage = combination.

loadings were not excessive, water could be held in the wetland until marsh communities were functioning at high uptake rates. It might also be possible to divert natural hydraulic surges around the wetland. Outflow regulation by dikes or drainage canals could be used to maintain sufficient water in the system to prevent aerobic decomposition of the substrate and maintain a proper balance of conditions for sequential nitrification and denitrification.

Flushing

A wetland system could be used to strip nutrients from dilute sources during the growing season. If the wetland were quite small, or if an artificial wetland were being employed, the concentrated nutrients could be flushed from the system after they had become mobilized in late autumn. This would require some energy expenditure and would demand that the concentrated flushing be captured for additional treatment. One approach might be to divert or pump the flushed materials onto agricultural land for an overland-flow type operation. This approach might be especially suitable where soils will not accept large hydraulic loads. It is unclear yet whether such flushed materials would constitute a human health hazard but there is some evidence to indicate that wetlands are effective at reducing pathogens (Seidel, 1971,

1976; de Jong, 1976; Wellings, 1976), toxic compounds (Seidel and Kickuth, 1967) and immobilizing heavy metals (Seidel, 1976).

Upland Application

On an areal basis most of the natural wetlands are groundwater influent. The flow of groundwater from an elevated water table under nearby uplands may be used to some advantage. Wastewater might be applied to drywells or as overland flow to the upland area. As the water percolates through the aerated zone above the water table, P could be adsorbed onto fine sand or clay, or precipitated as calcium phosphate (Fetter and Holzmacher, 1974; Fetter, 1977). The nitrate would pass freely through the soil and be transported via the groundwater to the wetland for denitrification. An attempt was made to utilize at least one form of this technique in treating highly nitrogenous fish processing wastes in Louisiana (Turner et al., 1976). The wastewater ("Stickwater") was applied as overland flow to dredge spoil banks from which it percolated into the groundwater and then to the adjacent marsh. Phosphorus was removed and there was a high level of nitrate in the groundwater at the downslope end of the upland, but, unfortunately, its ultimate fate in the marsh was not reported.

Underground Application

Underground application might be desirable where conditions permit and where aesthetics demand. High nitrate wastewater (i.e., animal wastes) could be piped via a manifold system directly into the substrate of a groundwater influent system. Under the anaerobic conditions present, the nitrate would be dentrified and at least some P would be adsorbed onto the substrate. If the flow rate was appropriate, no wastewater would be visible.

Harvesting Vegetation

Wetlands would be much more attractive as nutrient sinks if the sinks were permanent, or if the nutrients could be removed from the system at the end of the growing season. It seems logical that this could be accomplished by harvesting the large biomass of vegetation present. Klopatek (1975) found 37 kg P and 169 kg N per hectare in a Wisconsin *Typha, Carex, Scirpus* marsh. Retention capacity of the vegetation varies widely, however, as Steward (1976) found only 1.8 kg P and 48 kg N per hectare in the shoots of the low-nutrient-demanding Everglades sawgrass (*Cladium jamaicense*). It has been shown that addition of sewage to various types of wetlands not only increases the productivity, but also results in two to five-fold increases in concentrations of nutrients in plant tissue (Sloey et al., 1976; Turner et al., 1976; Steward, 1976). In order to force nutrient translocation and to minimize secretion losses, harvesting is best done when tissues are young. This means harvesting several times during the growing season (Sloey et al., 1976). By monthly harvesting, we were able to remove 30–50 kg P/ha in

Scirpus shoots. Unfortunately, only 5 to 20% of the nutrients detained by a wetland are generally stored in harvestable plant tissue (Steward, 1976; Turner et al., 1976; Sloey et al., 1976; also see Table 3). If 35 kg P/ha were harvested in plant tissue, and a per capita loading of 1.36 kg per annum assumed (Steward, 1970), the wastes from only 27 persons per hectare could be treated. Objection to wildlife habitat destruction and changes in plant species compositions resulting from large scale operations could further discourage the harvesting of plants in wetlands.

Chemical Treatment

Domestic wastewater is typically higher in proportion of P to N than would be optimum for the organisms in a wetland. The denitrification process further increases this proportion, and P is likely to be in surplus. It has been suggested that it may be necessary to strip wastewater chemically of most of the P before application to a wetland system (Farnham and Boelter, 1976). This could be accomplished with traditional lime, alum or ferric chloride precipitation as a pretreatment technique. In a few instances, it may also be possible to make direct application of chemicals and thereby precipitate the P *in situ*. This must be considered very hazardous to the biological integrity of the wetland, however, and should be considered only after careful testing and evaluation.

IMPACT OF NUTRIENT OVERLOADING AND MANAGEMENT ON WETLANDS

While wetland biota are adapted to wide ranges in water levels and perhaps also in nutrients, it can be expected that some species will be less tolerant of human activities than others (Burk et al., 1973).

The most obvious changes can be expected in the flora. Whigham and Simpson (1976) found that after only 1 season of sewage application to a tidal marsh, touch-me-not (*Impatiens capensis*) was eliminated from all enclosures receiving effluent. Bur marigold (*Bidens laevis*) was sensitive to aerial spraying when the spray was continuous. Perennials, wild rice (*Zizania aquatica* var. *aquatica*) and waterhemp (*Acnida cannabina*), however, were apparently not affected. Halberd tearthumb (*Polygonum arifolium*) increased in all treated enclosures, but general vegetation heights decreased. Ewell (1976) found a considerable increase in small floating plants such as duck-meat (*Spirodela oligorhiza*), water fern (*Azolla caroliniana*) and duckweed (*Lemna perpusilla*) in cypress domes receiving sewage. There was also a decrease in diversity of dominant species. Dog fennel (*Eupatorium compositifolium*) and fireweed (*Erechtites hieracifolia*) replaced fetterbush (*Lyonia lucida*) and staggerbush (*Lyonia mariana*). Water lilies (*Nymphaea odorata*) and bladderwort (*Utricularia* sp.) decreased. Richardson et al. (1976) found increased growth of most peatland species receiving nutrient loadings, but grasses and aster decreased.

Hanseter (1975) studied the impact of harvesting on Wisconsin marshland species, and found hardstem bulrush (*Scirpus acutus*) and softstem bulrush (*Scirpus validus*) recovered well and sustained high yields even when harvested every 2 weeks. Macereed (*Sparganium eurycarpum*), *Typha*, *Iris* and river bulrush (*Scirpus fluviatilis*) all decreased in shoot numbers and size after harvesting. Dykyjova and Husák (1973), however, found that harvesting of littoral reed stands (*Phragmites*) induced an increase in size and density of new shoots. Time of harvest may be of critical importance. Linde (1976) reports that *Typha* is most sensitive to harvest when the pistillate sheath leaf is shedding. This coincides with low sugar reserves in the rhizomes. Husák (1973) found *Phragmites* most sensitive to harvest before the emergence of the flowering "ear."

The impact of nutrient loading on insects in wetlands is as yet mostly unreported because so little is known about the ecology of wetland species (Witter and Croson, 1976). Jetter (1975) did report, however, an increase in dipterous detritivores in cypress domes receiving sewage. Witter and Croson feel that there is an urgent need to study the changes in nuisance biting and vectoring insects.

The impact of nutrient loading on wildlife is uncertain. Artificial systems do attract waterfowl, amphibians and shorebirds (Small, 1976), but this is probably not because of the nutrients *per se*. Changes in water levels by management practices will most certainly affect wildlife.

In the past, we caused the deterioration of the quality of our surface waters by using them to treat our wastes. When the practice was initiated, we marvelled at the remarkable ability of water to "self-purify." We based our decisions on short-term observations and immediate economics. Years later, the results of long-term overloading became evident. Lest we make the same mistake in handling our valuable and diminishing wetlands, it is mandatory that we carry out long-term, carefully monitored experiments at a severely limited number of sites. It is also important that those conducting the experiments document changes very carefully in the natural system that could signal future problems.

USE OF ARTIFICIAL SYSTEMS CONTAINING MARSH PLANTS

Artificial marshes offer the potential of capitalizing on nutrient transformation processes present in natural wetlands without disturbing or endangering valuable resources. The concept of using marsh plants in artificial systems, especially *Scirpus lacustris*, was most vigorously developed by Dr. Kathé Seidel at the Max Planck Institute. She and her co-workers have investigated the use of a wide variety of plants on many types of wastes. More than 90 publications have resulted from these studies. Primarily because of these studies, artificial marshes are being used to treat wastewater in several European countries (de Jong, 1976; Seidel, 1976) and on an exper-

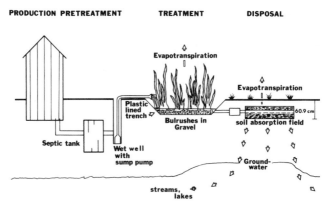

Fig. 2. Proposed design of artificial marsh treatment system for small-scale waste treatment. (Courtesy of *Groundwater*; C. W. Fetter, W. E. Sloey and F. L. Spangler, November–December 1976.)

imental basis in America. The Max Planck Institute has obtained a U.S. patent (3,770,623, 11/1973) to market complete package systems.

For limited scale use, artificial marshes have been found to be less energy intensive and less expensive than activated sludge plants, aesthetically pleasing, and they can have management capability designed in (de Jong, 1976; Farnham and Boelter, 1976; Small, 1976).

Use of Peat Filters

In Minnesota, partially decomposed peat was placed on top of a sand filter in an artificial system which demonstrated an amazing nutrient removal capacity (Stanlick, 1976; Farnham and Boelter, 1976). The system removed 99.3% of the P (3.4 g/m^2) and 49–80% of the N. Also surprising was the translocation of ⅔ of the phosphorus and 90% of the N detained into the harvestable portions of grasses cultivated on the peat. Biological oxygen demand reduction and fecal coliform elimination was almost total. Loss of nutrients in early spring may be occurring, but has not been documented (Stanlick, 1976).

Use of Artificial Marshes

Artificial marshes may be especially attractive where natural wetlands are not available, or for small systems where soil conditions are not suitable for septic tanks (Fetter et al., 1976). On Long Island, Woodwell et al. (1976) found artificial marsh–pond systems nearly 3× as effective as terrestrial forests in removing nutrients and particulates from sewage. In a small artificial marsh containing bulrush (*Scirpus*) planted in gravel and receiving primary and secondary treated effluent from a central Wisconsin community, we recorded very effective BOD and coliform reduction and up to 20 g P/m^2

per growing season detention. Phosphorus detention apparently is the factor which determines the size of system required for a given loading. This is because domestic wastewater is already high in P; and denitrification further accentuates the relative surplus in the system (see "NUTRIENT TRANS-FORMATIONS IN WETLANDS").

Thus, a system such as the one proposed in Fig. 2 for a single-family (4 persons) summer home would require ≈ 150 m² of marsh system. Harvesting of shoots at least once during midsummer would be desirable in order to prevent lodging, but it would accomplish little in terms of P removal. For, if 3.5 g $P \cdot m^{-2} \cdot yr^{-1}$ (the maximum that we were able to remove experimentally in Wisconsin) were removed from the 150-m² system, only about 0.45 kg P (⅓ person equivalent) would be harvested. If, however, P in the influent were reduced by not using high phosphate detergents or by chemically pre-cipitating much of the P in the septic tank, the size of the system could probably be reduced by ⅔. If the soil absorption field contained some clay or fine sand, P could be immobilized there (Dudley and Stephenson, 1973) while BOD, coliform and nitrogen reduction could take place in the artificial marsh. In Holland, artificial marshes are being used for new housing devel-opments (de Jong, 1976) and public recreation and campgrounds (de Jong, 1975; Kok, 1974). In Germany, artificial marshes are also being used to pretreat municipal drinking water (Czerwenda and Seidel, 1965). The United States Army Corps of Engineers has investigated the use of natural or artifi-cially developed marshes to filter, dewater and remove nutrients from dredge slurries (Lee et al., 1976) and has found the results very promising.

It seems logical that artificial marshes would be ideal for handling wastes such as feedlot runoff, other animal wastes, fish processing wastes, etc. where there is a high N:P ratio. More widespread experimental application of artificial wetland systems for wastewater treatment of various types of waste in this country warrants greater state and federal support and partici-pation.

ACKNOWLEDGMENTS

We thank Kathy Garfinkel for assistance in manuscript preparation and Diane Kromm for typing the final copy. We also thank the Water Well Journal Publishing Company for use of Fig. 2 and the following authors for permission to use unpublished data: K. Ewel, F. M. Wellings, H. Stanlick, R. E. Turner, M. Small, D. Tilton, C. Richardson, K. Steward, R. Simpson, J. Witter and R. Farnham.

LITERATURE CITED

Boyd, C. E., and Blackburn, R. D. (1970). Seasonal changes in the proximate composition of some common aquatic weeds. *Hyacinth Control Journal* 8, 42–44.

Burk, J. P., Hosier, P., Lawry, A., Lenz, A., and Mesrobian, A. (1973). "Partial recovery of vegetation in a pollution-damaged marsh." Project Completion Report. Water Resources Research Center, Univ. Mass., Amherst, Mass.

Cowardin, L., Carter, V., Golet, F., and La Roe, E. (1976). "Interim classification of wetlands and aquatic habitats of the United States." U.S. Dept. Interior, Fish and Wildlife Service.

Czerwenda, W., and Seidel, K. (1965). New methods of groundwater enrichment in Krefeld. *Das Gas-und Wasserfach*. **106, 30,** 828–833.

de Jong, J. (1975). Bulrushes and reed ponds. *Rijksdienst Voor de Ijsselmeerpolders. Flevobricht N.R.* **104.**

de Jong, J. (1976). The purification of wastewater with the aid of rush or reed ponds. *In* "Biological Control of Water Pollution" (J. Tourbier and R. Pierson, eds.), pp. 123–132. University of Pennsylvania Press, Philadelphia, Pa.

Dudley, J. G., and Stephenson, D. A. (1973). "Nutrient enrichment of groundwater from septic tank disposal systems." Univ. Wis. Inland Lake Renewal and Shoreline Management Demonstration Project Report. Oshkosh, Wisconsin.

Dykyjova, D., and Husák, S. (1973). The influence of summer cutting on regeneration of reed. *In* "Ecosystem Study on Wetland Biome in Czechoslovakia" (S. Hejny, ed.), IBP/PT-PP Report No. 3, Trebon, Czechoslovakia.

Engler, R., and Patrick, W. H., Jr. (1974). Nitrate removal from floodwater overlying flooded soils and sediments. *Jour. Environ. Qual.* **3** (4), 409–413.

Ewel, K. (1976). Effects of sewage effluent on ecosystem dynamics in cypress domes. *In* "Freshwater Wetlands and Sewage Effluent Disposal" (D. L. Tilton, R. H. Kadlec and C. J. Richardson, eds.), pp. 169–195. The University of Michigan, Ann Arbor, Michigan.

Farnham, R. S., and Boelter, D. H. (1976). Minnesota's peat resources: their characteristics and use in sewage treatment, agriculture and energy. *In* "Freshwater Wetlands and Sewage Effluent Disposal" (D. L. Tilton, R. H. Kadlec and C. J. Richardson, eds.), pp. 241–255. The University of Michigan, Ann Arbor, Michigan.

Fetter, C. W. (1977). Attenuation of wastewater elutriated through glacial outwash. *Groundwater*. **15,** (*In press*).

Fetter, C. W., and Holzmacher, R. G. (1974). Groundwater recharge with treated sewage. *Water Poll. Contr. Fed. Journ.,* **46** (2), 260–270.

Fetter. C. W., Sloey, W. E., and Spangler, F. L. (1976). Potential replacement of septic tank drain fields by artificial marsh wastewater treatment systems. *Groundwater*, **14** (6), 396–402.

Godshalk, G., and Wetzel, R. (1978). Decomposition in the littoral zone of lakes. *In* "Freshwater Wetlands: Ecological Processes and Management Potential" (R. E. Good, D. F. Whigham and R. L. Simpson, eds.), pp. 131–143. Academic Press, New York.

Hanseter, R. (1975). "Recovery, Productivity and Phosphorus Content of Selected Marsh Plants After Repeated Cuttings." MST Thesis, Univ. Wisconsin–Oshkosh. Oshkosh, Wisconsin. 81 pp.

Husák, S. (1973). Destructive control of stands of *Phragmites communis* and *Typha angustifolia* and its effects on shoot production followed for three seasons. *In* "Littoral of the Nesyt Fishpond. Ecological Studies." (J. Květ, ed.). Studdie CSAU 15, Academia. Praha, Czechoslovakia.

Hutchinson, G. E. (1975). "A Treatise on Limnology. Vol. III. Limnological Botany." Wiley-Interscience, New York.

Hwang, C. P., Lackie, T. H., and Huang, P. M. (1976). Adsorption of inorganic phosphorus by lake sediment. *Journ. Water Pollution Contr. Fed.* **48** (12), 2754–2760.

Jetter, W. A. W. (1975). Effects of treated sewage on the structure and function of cypress dome consumer communities. *In* "Cypress, Dome Wetlands for Water Management, Recycling and Conservation" (H. T. Odum, K. Ewel and J. Ordway, eds.), pp. 588–610. 2nd Ann. Rep., Center for Wetlands. Univ. Florida, Gainesville.

Klopatek, J. (1975). The role of emergent macrophytes in mineral cycling in a freshwater marsh. *In* "Mineral Cycling in Southeastern Ecosystems" (F. G. Howel, J. B. Gentry and H. M. Smith, eds.), pp. 367–393. ERDA Symposium Series (CONF-740513).

Klopatek, J. (1978). Nutrient dynamics of a freshwater riverine marsh and the role of emergent macrophytes. *In* "Freshwater Wetlands: Ecological Processes and Management Potential" (R. E. Good, D. F. Whigham and R. L. Simpson, eds.), pp. 195–216. Academic Press, New York.

Kok, T. (1974). The purification of sewage from a camping site with the aid of a bulrush pond. *H_2O* 7 (24), 536–544.

Kramer, J. R., Herbes, S. E., and Allen, H. E. (1972). Phosphorus analysis of water, biomass and sediment. *In* "Nutrients in Natural Waters" (H. E. Allen and J. R. Kramer, eds.), pp. 51–100. John Wiley and Sons, New York.

Lee, C., Hoeppel, R., Hunt, G., and Carlson, C. (1976). "Feasibility of the Functional Use of Vegetation to Filter, Dewater and Remove Contaminants from Dredged Material." Tech. Rep. D-76-4. U.S. Army Eng. Waterways Exp. Sta., Vicksburg, Miss.

Lee, G. F., Bentley, E., and Amundson, R. (1975). Effects of marshes on water quality systems. *In* "Coupling of Land and Water Systems" (A. D. Hasler, ed.), pp. 105–127. Springer-Verlag, New York.

Linde, A. (1976). Cattail: the significance of its growth, phenology and carbohydrate storage to its control and management. *Tech. Bull. No. 94*. Dept. Natural Resources. Madison, Wisconsin.

Loucks, O., Prentki, R., Watson, U., Reynolds, B., Weiler, P., Bartell, S., and D'Allessio, A. (1977). "Studies of the Lake Wingra Watershed: An Interim Report. Center for Biotic Systems, Inst. for Environmental Studies, Univ. Wisconsin-Madison, Report 78.

Mortimer, C. H. (1971). Chemical exchanges between sediments and water in the Great Lake—speculation of probable regulatory mechanisms. *Limnol. Oceanogr.* **16**, 387–404.

Odum, H. T., Ewel, K., Mitsch, W., and Ordway, J. (1975). Recycling treated sewage through cypress wetlands in Florida. *Occasional Publ. No. 1*. Center for Wetlands. Univ. Florida, Gainesville.

Patrick, W. H., Jr., Delune, R. D., Engler, R. M., and Gotoh, S. (1976). "Nitrate Removal from Water at the Water-mud Interface in Wetlands." EPA-600/3-76-042. Ecological Research Series.

Phillips, J. (1970). Wisconsin wetland soils, a review. *Research Report 57*. Wis. Dept. Natural Resources. Madison, Wisconsin.

Prakash, A., and Rashid, M. A. (1968). Influence of humic substances on the growth of marine phytoplankton. *Limnol. Oceanogr.* **13** (4), 598–606.

Prentki, R., Gustafson, T., and Adams, M. (1978). Nutrient movement in lakeshore marshes. *In* "Freshwater Wetlands: Ecological Processes and Management Potential" (R. E. Good, D. F. Whigham and R. L. Simpson, eds.), pp. 169–194. Academic Press, New York.

Richardson, C. J., Wentz, W. A., Chamie, J. P., Kadlec, J. A., and Tilton, D. L. (1976). Plant growth, nutrient accumulation and decomposition in a central Michigan peatland used for effluent treatment. *In* "Freshwater Wetlands and Sewage Effluent Disposal" (D. L. Tilton, R. H. Kadlec and C. J. Richardson, eds.), pp. 77–117. The University of Michigan, Ann Arbor, Michigan.

Rohlich, G. A. and Uttermark, D. P. (1972). Wastewater treatment and eutrophication. *In* "Nutrients and Eutrophication" (G. E. Likens, ed.), pp. 231–245. Amer. Society of Limnol. Oceanogr., Special Symposia, Vol. I.

Seidel, K. (1971). Wirkung hoherer pflanzen auf pathogene keime in gewassern. *Naturwissenschaften* **58** (4), 150–151.

Seidel, K. (1976). Macrophytes and water purification. *In* "Biological Control of Water Pollution" (J. Tourbier and R. Pierson, eds.), pp. 109–122. University of Pennsylvania Press, Philadelphia, Pa.

Seidel, K. and Kickuth, R. (1967). Biological treatment of phenol-containing wastewater with bulrush (*Scirpus lacustris* L.). *Wassersirtschaft-Wassertechnik.* **17** (6), 209–210.

Sloey, W. E., Spangler, F. L. and Fetter, C. W. (1976). Productivity and phosphorus distribution in natural and artificial marshes in Wisconsin. *Aquatic Ecology Newsletter* **9**, 20–21 (abstract).

Small, M. (1976). Marsh/pond sewage treatment plants. *In* "Freshwater Wetlands and Sewage Effluent Disposal" (D. L. Tilton, R. H. Kadlec, and C. J. Richardson, eds.), pp. 197–213. The University of Michigan, Ann Arbor, Michigan.

Spangler, F. L., Fetter, C. W., and Sloey, W. E. (1977). Phosphorus accumulation—discharge cycles in marshes. *Water Resources Bulletin (In press)*.

Spangler, F. L., Sloey, W. E., and Fetter, C. W. (1976). Experimental use of emergent vegetation for the biological treatment of municipal wastewater in Wisconsin. *In* "Biological Control of Water Pollution" (J. Tourbier and R. Pierson, eds.), pp. 161–171. University of Pennsylvania Press, Philadelphia, Pa.

Stanlick, H. T. (1976). Treatment of secondary effluent using a peat bed. *In* "Wetlands and Sewage Effluent Disposal" (D. L. Tilton, R. H. Kadlec, and C. J. Richardson, eds.), pp. 257–268. The University of Michigan, Ann Arbor, Michigan.

Steward, K. K. (1976). Physiological, edaphic and environmental characteristics of typical stands of sawgrass. *Aquatic Ecology Newsletter* **9**, 22–23 (abstract).

Steward, K. K. (1970). Nutrient removal capacity of various aquatic plants. *Hyacinth Control Journal* **8**, 34–35.

Tilton, D. L., Kadlec, R. H., and Richardson, C. J., eds. (1976). "Freshwater Wetlands and Sewage Effluent Disposal" The University of Michigan, Ann Arbor, Michigan.

Tóth, L. (1972). Reeds control eutrophication of Balaton Lake. *Journ. Internat. Assoc. Water Pollution Res.* **6** (12), 1533–1539.

Tourbier, J. and Pierson, R. W., Jr., eds. (1976). "Biological Control of Water Pollution." University of Pennsylvania Press, Philadelphia, Pa.

Turner, R. E., Day, J. W., Jr., Meo, M., Payonk, P. M., Ford, T. B., and Smith, W. G. (1976). Aspects of land-treated waste application in Louisiana wetlands. *In* "Freshwater Wetlands and Sewage Effluent Disposal" (D. L. Tilton, R. H. Kadlec and C. J. Richardson, eds.), pp. 145–167. The University of Michigan, Ann Arbor, Michigan.

Vollenweider, R. A. (1971). "Scientific Fundamentals of Eutrophication of Lakes and Flowing Waters, with Particular Reference to Nitrogen and Phosphorus as Factors in Eutrophication." Tech. Report OCED, Paris.

Wellings, F. M. (1976). Viral aspects of wetland disposal of effluent. *In* "Freshwater Wetlands and Sewage Effluent Disposal" (D. L. Tilton, R. H. Kadlec, and C. J. Richardson, eds.), pp. 297–305. The University of Michigan, Ann Arbor, Michigan.

Wetzel, R. G. (1969). Factors influencing photosynthesis and excretion of dissolved organic matter by aquatic macrophytes in hardwater lakes. *Verh. Internat. Verein. Limnol.* **17**, 72–85.

Wetzel, R. G. (1975). "Limnology." Saunders, Philadelphia, Pa.

Whigham, D. F., and Simpson, R. L. (1976). The potential use of freshwater tidal marshes in the management of water quality in the Delaware River. *In* "Biological Control of Water Pollution" (J. Tourbier and R. W. Pierson, eds.), pp. 173–186. University of Pennsylvania Press, Philadelphia, Pa.

Witter, J., and Croson, S. (1976). Insects and wetlands. *In* "Freshwater Wetlands and Sewage Effluent Disposal" (D. L. Tilton, R. H. Kadlec, and C. J. Richardson, eds.), pp. 269–295. The University of Michigan, Ann Arbor, Michigan.

Woodwell, G., Ballard, J., Clinton, J., and Pecan, E. (1976). "Nutrients, Toxins and Water in Terrestrial and Aquatic Ecosystems Treated with Sewage Plant Effluents." BNL50513, Brookhaven Nat'l. Lab. Assoc. Univ., Inc., Upton, N.Y.

ECOLOGY AND THE REGULATION OF FRESHWATER WETLANDS

Jack McCormick

WAPORA, Inc.
6900 Wisconsin Ave. N.W., Washington, D. C. 20015

Abstract Man's concept of the value of wetlands has changed dramatically during the past 3 decades as the inherent worth of wetlands has been recognized. The protection of our shrinking wetland resources has been bolstered in recent years with the enactment of several federal laws including the National Environmental Policy Act of 1969, the Federal Water Pollution Control Act of 1972, the Coastal Zone Management Act of 1972 and the Endangered Species Act of 1973. These laws together with the still powerful River and Harbor Act of 1899, several administrative directives and recently enacted state laws form the basis for the informed management of our wetlands. So that management decisions are truly based on "the best available data," scientists have a special responsibility to communicate their knowledge of wetlands in ways intelligible to decision makers who are most often laymen. The present state of wetland regulation is primitive, being derived more from common sense than from science. Future wetland regulation to be fully effective and rational must be conducted in conformity with the ecologically based, long-range planning with the focus on the wetland system as a dynamic resource. To achieve this goal, it is imperative that wetland ecologists, consultants, planners and regulators work toward standardization of methodology for obtaining, presenting and evaluating data thus permitting more direct comparison of data and reduction of variation in data produced by different groups.

Key words *Coastal Zone Management Act, communication, Endangered Species Act, Federal Water Pollution Control Act, management, Maryland, National Environmental Policy Act, planning, regulation, River and Harbor Act, standardization.*

INTRODUCTION

Science deals with sampling. Virtually all data are based on a minute fraction of the total universe they are intended to describe. Our data, therefore, have an inherent uncertainty and we express this by calculating standard errors and other statistics. Although our studies are conducted in the present, most seek to determine natural functions that will enable us to explain the past or, more commonly, predict future behavior. Much of the work of scientists, therefore, deals with predictions or forecasts. Our record of forecasts generally is good, but it varies from one facet of science to another.

The goal of science, and its end, is knowledge. This goal, in turn, is the beginning of another segment of man's efforts which we can call "appliance." Engineering, agriculture, forestry, range management, etc. are appliances. They sort among the facts of scientific knowledge to seek out those small bits and larger pieces that seem to be useful for application to some recognized human need.

Most of us fall at various points along a continuum that spans between the pure scientists, who may be driven only by a desire to know, and the applicator who is so totally engaged with day-to-day problems that he has little time to scan a newspaper, let alone to read a journal. To some degree, most of us are hybrids and our knowledge, interests, and priorities differ.

Scientists who seek to preserve the wetland system must strive to expedite the transfer of their expanding knowledge to the applicators who regulate wetlands and to the public who, through our political system, regulates the regulators. In turn, the applicators must recognize the gaps in the vital knowledge of the resources they control, must alert the scientist to their needs, and assist the scientist to obtain the time and money to collect and analyze new information.

Communication is the most important aspect of all transactions between human beings. It is the most difficult, but the most influential tool available in the politics of ecology. Several people who are not scientists, and several scientists who are not ecologists, have done much of the communicating about wetlands and about other sensitive environmental systems. Their efforts were useful and valuable initially, and they attracted the attention of the public. As long as they dealt with broad sweeping generalizations, those communicators benefited us. However, a few began to expound at lengths that quickly outdistanced the breadth of their absorbed facts, and they have laid a trail of exaggeration and misstatement that may be difficult to erase quickly.

Each of us should recognize the potential power we have if we learn to communicate accurate information and, especially, carefully formulated syntheses of knowledge in an understandable way. Jargon is the shorthand of science, but it is often unintelligible or misleading to the layman. Laymen constitute an overwhelming majority of the public; virtually all legislators are laymen, and many of the applicator–regulators are laymen. When scientists communicate with applicators, they should speak and write in simple, concise English. They should eliminate details, not list naked results, and drop any IFs, BUTs, or confusing discussions. Scientists should interpret data and use professional judgment to synthesize the factual pieces into an understandable whole.

The basic research of a scientist is conducted on 1 or more samples, usually discrete plots or individual objects collected at specific locations. However, data obtained from a specific site or a particular individual have limited value unless their relationship to the larger universes from which they were drawn is defined clearly. I recently reviewed a large number of published and unpublished reports on wetland areas of the Middle Atlantic States. I was pleased to learn of the intensity with which the wetlands have been investigated but I was dismayed to find that the value of many of those studies is substantially reduced by the fact that the investigators failed to recognize (or at least to explain) that they were dealing with particular community types. Largely, the studies are framed as investigations of particular marsh areas or mere geographical locations rather than as investigations of the areally related and functionally related stands of different wetland community types that, together, form the wetlands at those sites and are interactive components of a still larger system. No effort was made to stratify the wetlands by community types or to allocate sampling effort to determine the characteristics of these components. Consequently, the papers are descriptions of the particular wetlands at particular points in time. Unfortunately the information they contain has little potential for transferability, for synthesis, or for problem solving.

Wetlands can be classified in many ways. Cowardin et al. (1976) recently have developed a comprehensive classification for the United States Fish and Wildlife Service that is intended to apply nationwide. However, the level of discrimination is coarse, and this system presently is of relatively little use at the state scale, and does not discriminate between the components of most individual wetlands.

Most researchers have found that the community type is a reasonable approximation of the most definitive (finest-grain) unit of classification. In reality, the community type is usually identical to a vegetation type; however, there are some unvegetated community types so that name is appropriate. We intend to connotate "biotic communities," and our units are valid for this. However, few of the wetland animals have microranges that coincide precisely with the distribution of a particular type of vegetation. Most animals occur in (or range through) several to many types of vegetation and into the open waters of ponds, streams, bays, and/or the ocean. Nevertheless, current knowledge suggests that the animal complex of each community type is unique in terms of its total composition and the relative densities of the various taxa.

LEGISLATIVE BASIS FOR FEDERAL REGULATION OF WETLANDS

Man's concept of the value of wetlands has changed dramatically during the past 3 decades. Prior to 1950, the popular attitude was that wetlands were of little value except insofar as they could be diked, drained, or filled to render them more adaptable to farming, grazing, or real estate development for industrial, commercial, or residential purposes. Because wetlands bordered thousands of kilometres along the coasts and along major rivers, the sites they occupied were considered to be of particular value for port facilities; as locations for industries that require access to water for transportation, for process or cooling water, or as a medium into which to discharge wastes; as locations for marinas or residential developments with direct access to the water; or as economical areas for the placement of materials dredged during the construction or maintenance of channels, anchorages, and wharf facilities.

Federal laws that were addressed to wetlands generally were directed toward the conversion of wetlands to agricultural uses. Certain acts did authorize the establishment of specific wildlife refuges, parks, or similar reserves to include wetlands, but none sought to regulate wetlands to protect them for their inherent values. In part, at least, this probably was a reflection of the lingering frontier attitude that resources are limitless. According to this philosophy, if some wetlands are destroyed, it matters little, for there always are more wetlands.

During the 1950s and 1960s, a public awareness of the precipitous and accelerating loss of wetland hectarage through indiscriminate dredging, filling, and other work was developing. The United States Department of the Interior was influential in drawing attention to these losses and in explaining some of the broader consequences of such losses. Surveys of the wetlands of each state were compiled during the period from 1953 to 1955 and (for some states) reappraisals were made during the late 1950s and early 1960s.

The classifications were broad and attention was focused principally upon values to waterfowl. They were most important for their estimates of the rate of wetland destruction, and their appraisals of wetland values. They were products of the applicators and (in retrospect) it is amazing to discover how naive and incomplete they were in terms of a comprehensive understanding of the wetlands ecosystem. Nevertheless, they **communicated**. Although they contained only gross approximations, many of which are difficult to validate, the surveys represented the best available data. They allowed state officials and the public to convince legislators to enact laws to regulate the remaining wetlands and to slow drastically the rate of loss.

I do not intend to review all state wetland acts, but it is of interest to note that Massachusetts and Rhode Island enacted legislation to protect their coastal wetlands in 1963 and 1965, respectively. California established a commission with authority over the wetlands in San Francisco Bay during 1965. Massachusetts followed in 1968 with legislation to protect inland wetlands. All of the Middle Atlantic States now have laws to regulate activities in coastal wetlands, but only New York also regulates inland wetlands.

The oldest and one of the most powerful of the federal laws now used to protect wetlands by regulation is the River and Harbor Act of 1899 (Appendix I). Section 10 of this act prohibits the excavation of material from, or the deposition of material into, any navigable water of the United States without a permit or other authorization from the United States Army Corps of Engineers. It similarly restricts the accomplishment of any other type of work that would affect the location, course, capacity or condition of such navigable waters.

During the first 70 years, at least until about 1968, the U.S. Army Corps of Engineers emphasized or restricted its concern to issues of navigation in its administration of the 1899 Act (Committee on Government Operations, 1970). This narrow interpretation of regulatory responsibility was not inherent in the law, and as early as 1933 the United States Supreme Court declared that permits could be denied for proposed activities that would interfere with the public interest. This broad approach was amplified further by the U.S. Fish and Wildlife Coordination Act, as amended by the Act of 12 August 1968. The amendment declared as a national policy the recognition of "the vital contribution of our wildlife resources to the Nation," and specified that any proposal to impound, divert, deepen or otherwise modify any stream or other body of water under a federal permit or license requires consultation with the U.S. Fish and Wildlife Service to prevent loss or damage to wildlife resources.

At least in regard to the issuance of permits for nonfederal projects, conformance with the U.S. Fish and Wildlife Coordination Act was perfunctory until 1967. On 13 July 1967, the Secretary of the Interior and the Secretary of the Army endorsed a memorandum of understanding that established procedures by which the U.S. Army Corps of Engineers would obtain advice from the Department of the Interior in regard to the evaluation of all effects on fish and wildlife, recreation, pollution, natural resources and the environment that potentially could result from dredging, filling or other work for which a permit is required from the Corps under the authority of the 1899 Act.

On 7 December 1967, the U.S. Army Corps of Engineers revised its regulations to include a consideration of the effects of any proposal on the public interest, ". . . including effects upon water quality, recreation, fish and wildlife, pollution, our natural resources, as well as the effects on navigation. . . ." Effective 18 December 1968, these regulations again were revised. In the new wording, applications for permits were to be evaluated in regard to ". . .all relevant factors, including the effect of the proposed work on navigation, fish and wildlife, conservation, pollution, esthetics, ecology and the general public interest."

Regardless of the wording of the regulations, the regulatory practices of the District Offices of the Corps of Engineers left ". . .a substantial gap between promise [of the regulations] and performance [of the permits staff]" (Committee on Governmental Operations, 1970). In many cases, the Corps of Engineers allegedly approved, in a routine manner, all applications for new landfills, dredging and other work unless parties that opposed the issuance of the permit were able to demonstrate that substantial damage would result to the public interest. Following a review of these procedures, the Committee on Government Operations (1970) of the House of Representatives recommended that:

"The Corps of Engineers should permit no further landfills or other work in the Nation's estuaries, rivers and other waterways except in those cases where the applicant affirmatively proves that the proposed work is in accord with the public interest, including the need to avoid the piecemeal destruction of these water areas."

The significance of the 1899 Act in regard to the protection of wetlands was enhanced by the National Environmental Policy Act of 1969 (PL 91-190, Appendix I), or NEPA. Although NEPA assigns no new regulatory authority, Section 102(1) mandates that ". . .the policies, regulations, and public laws of the United States shall be interpreted and administered in accordance with the policies [of NEPA]." Section 102(2)(B) stipulates that all agencies of the Federal Government shall develop procedures to ". . . insure that presently unquantified environmental amenities and values may be given appropriate consideration in decision making along with economic and technical considerations." Furthermore Section 102(2)(C) of NEPA requires federal agencies to prepare a detailed statement, generally known as an environmental impact statement (EIS), on any federal action (including the issuance of a permit) ". . . significantly affecting the quality of the human environment." During the preparation of this EIS, the responsible federal official also must ". . .consult with and obtain the comments of any Federal agency which has jurisdiction by law or special expertise with respect to any environmental impact involved." In practice, the EIS is a public document and is subject to review and comment by any governmental official or agency and by any citizen or public interest group.

The principal effects of NEPA on regulatory procedures have been to force the coordination of all sources of expertise for the evaluation of applications; to apply simultaneously, or in close sequence, all relevant regulations and requirements; and to expose decision making to full public view. The National Environmental Policy Act also has effected a multidisciplinary approach to planning, in which environmental concerns are given a weight equal to economic, technical and other relevant concerns.

The geographic scope of the regulatory control of wetlands under the authority of the 1899 Act has been the subject of considerable legal challenge. The definition of "navigable waters," which had been interpreted rather narrowly by the Corps of Engineers, also was viewed narrowly by some, but much more broadly by others. In a sense, the question was rendered moot by the passage of the Federal Water Pollution Control Act Amendments of 1972 (PL 92-500), or FWPCA.

Section 404 of FWPCA authorizes the Secretary of the Army, through the

Chief of Engineers, to regulate the discharge of dredged or fill material into the waters of the United States. The selection and use of permitted disposal sites is conducted in accordance with guidelines developed by the United States Environmental Protection Agency (US-EPA) in collaboration with the Corps of Engineers. The Chief of Engineers can authorize the use of a disposal site on the basis of overriding economic impact on navigation. The Administrator of US-EPA, after consultation with the Secretary of the Army, can prohibit or restrict the use of any defined area if the proposed use would have an unacceptable, adverse effect on municipal water supplies, shellfish beds and fishery areas, wildlife or recreational areas.

The term "waters of the United States" includes all bodies of surface water and all areas of wetlands within the United States and its territories. Lakes and ponds ". . .created by excavating and/or diking dry land to collect and retain water for such purposes as stock watering, irrigation, settling basins, cooling or rice growing" are excluded by administrative decisions (Corps of Engineers, 1977).

Thus, Section 404 assigns to the Corps of Engineers regulatory control over all of the wetlands of the United States. In its current regulations, which became effective 19 July 1977, the Corps of Engineers (1977) included a general permit (Section 323.4-2) which authorizes without application the discharge of dredged or fill materials into the following waters of the United States:

(1) Nontidal streams, including impoundments and adjacent wetlands, located above a point at which the average annual flow is less than 5 cubic feet per second [= 0.14 cubic metres per second] (for intermittent streams, above the point at which such flow is equaled or exceeded 50 percent of the time);

(2) Natural lakes, including adjacent wetlands, which have a surface area of less than 10 acres [= 4.047 hectares] and are fed by or drained by a stream with an average annual flow of less than 5 cfs [=0.14 m^3/s], or which are isolated from any stream; and

(3) Other nontidal waters, other than isolated lakes larger than 10 acres [= 4.047 ha], that are not part of a system of surface drainage tributary to the interstate waters or navigable waters of the United States.

Discharges to such areas are conditioned to the extent that they must not destroy an endangered or threatened species, that they be free of toxic pollutants, that the fill be maintained to prevent erosion and that they not occur in a component of the National Wild and Scenic Rivers System or in a related state system.

The Corps of Engineers recognized that some discharges allowed by the general permit described above might produce significant adverse effects on the environment. Section 232.4-4 of the regulations, therefore, provides discretionary authority to the District Engineer to require individual applications for permits for proposed discharges to specified areas.

By administrative orders, US-EPA (Ruckelshaus, 1973) and US-DOT (Department of Transportation; Coleman, 1975) established broad policies to protect wetlands of all types. The policy of US-DOT applies to the planning, construction, and operation of transportation facilities and projects. The policy of US-EPA ultimately will apply to a geographically more extensive area, because it is relevant to the review of municipal water pollution control facilities that are funded under Section 201 of FWPCA, to the development of areawide management programs to control point and nonpoint sources of water pollution under Section 208 of FWPCA and to other programs that are under the jurisdiction of US-EPA. In practice, the agency requires that planning be conducted in accordance with its wetland policy, and that direct or induced encroachments into wetland areas be avoided or minimized.

The Coastal Zone Management Act of 1972 (PL 93-370, Appendix I) or CZM, is a law designed to coordinate planning and regulation, rather than to establish new regulatory jurisdiction. The act encourages the states to define the boundaries of their coastal zones, to inventory the resources within these zones, to develop comprehensive plans for the future uses of those resources and to coordinate existing authorities and develop any necessary new legislation to assure the implementation of, and adherence to, such plans. Following the federal approval of a comprehensive plan and a correlated plan for administrative management, the Federal Government will provide funds to assist the state in operating the management functions.

Insofar as wetlands are concerned, CZM will provide an opportunity to develop long-range planning and strategies. Actual regulation, however, will continue under state wetland acts (unless they are folded into more comprehensive state coastal legislation), Section 404 of the Federal Water Pollution Control Act Amendments, and Section 10 of the 1899 River and Harbor Act.

Section 307(c) of CZM requires federal agencies to comply (to the maximum extent possible) with the approved coastal zone management program of a state when the agencies conduct activities (including development projects) that directly affect the coastal zone of the state. It also requires any applicant, other than a federal agency, to submit a certification from the affected state that the project for which he seeks a federal license or permit will comply with the coastal zone management program of the state. Furthermore, Section 401 of FWPCA requires a nonfederal applicant also to submit a water quality certificate from the state in which the proposed project is located if the project may result in the discharge of a pollutant into the waters of the United States. These requirements for state certification prior to the issuance of a federal license or permit can serve as effective means to coordinate the 2 levels of regulatory jurisdiction.

Another federal law, the Endangered Species Act of 1973 (PL 93-205), and related state laws, could be important to the protection of at least some wetlands. Under the Endangered Species Act, species of animals and plants that are or soon may be in jeopardy of extermination can be designated as endangered or threatened, respectively. Critical habitats necessary to the survival of these imperiled species then can be established, and no federal agency can conduct a project, authorize funds for a project or issue a permit for a project that would affect the critical habitat adversely. Scientists will be particularly effective if they can provide the substantial information necessary to establish the imperiled status of wetland species and then to establish their critical habitats. For example, Stuckey and Roberts (1977) recently prepared a list of wetland vascular plants that they considered to be imperiled in Ohio, and R. J. Bartolotta and R. L. Stuckey (*personal communication*) compiled a list of wetland species that have been recommended by the Smithsonian Institution for designation as nationally endangered or threatened.

Two directives that were issued by the President on 24 May 1977 are relevant to the protection of wetlands. Executive Order 11990, which is titled "Protection of Wetlands," directs all federal agencies to " . . .take action to minimize the destruction, loss or degredation of wetlands, and to preserve and enhance the natural and beneficial values of wetlands. . ." (Carter, 1977*b*). This order does not apply to the issuance of federal permits, licenses or allocations to private parties for actions that are to be conducted in wetlands that are not federally owned. However, it does apply to the acquisition, management and disposal of federal lands and facilities, to construction or improvements undertaken, financed, or assisted by the Federal Government, and to the conduct of federal activities and programs which affect land use. Section 4 of the Executive Order requires that, when federally owned lands are leased, an easement is assigned, or they are disposed of to a nonfederal party, a reference be included in the conveyance to identify any wetlands and to indicate those uses which are restricted in such areas.

Executive Order 11988, titled "Floodplain Management," does apply to regulatory and licensing activities (Carter, 1977*a*). In regard to applicants for federal permits, authorizations, or funding, Section 2(c) of the Executive Order specifies that "Agencies shall also encourage and provide appropriate guidance to applicants to evaluate the effects of their proposals in floodplains prior to submitting applications. . ." Wetlands that are subject to flooding by streams or coastal water with a probability of 1% or greater in any given year (i.e., within the 100-year floodplain) are included under the authority of this Executive Order by virtue of Section 6(c). Particularly because it instructs agencies to discourage filling [Section 3(b)], this Executive Order adds weight to the regulatory power previously authorized by the 1899 Act and by Section 404 of FWPCA.

WETLAND REGULATION AND PLANNING

The regulation of wetlands is recent and the state-of-the-art is primitive. Most states in the Middle Atlantic region have prepared, or are preparing, maps to indicate the location of their coastal wetlands. New York is preparing maps also of inland wetlands, and other states are mapping inland wetlands in their coastal zone study areas. The quality of information contained in these maps varies considerably because there was no standardization of methods or definitions from one state to another.

To a large extent wetland regulation derives more from common sense than from science. This is a compliment to the individuals who regulate. Also, it is a recognition that scientific knowledge of wetlands is fragmentary, poorly organized, unstandardized and oriented more toward the interest of the investigator than toward the needs of society. In defense of scientists, society has not provided adequately for the funding of pertinent research, the regulators have not analyzed and expressed their needs clearly and there is no formal mechanism to expedite the essential communications between these interested groups.

The goals of men are multiple, and many goals are conflicting. The planner is a middleman. He must listen to and understand the resource scientists as well as people who express a desire for various future uses of the resource. After he has gained an understanding of the resource, he then must formulate a plan, and resultant recommendations which will be used to allocate the resource to those needs found to be valid. In the case of wetlands, some allocations will be for such nonconsumptive uses, absolute preservation, wildlife management, the harvest of natural products, hunting, etc. Others will be for such consumptive uses as filling for highways, dredging for port or marina development and similar water related activities. Ultimately, the recommendations of the planner are reviewed, often modified and finally approved by some body with authority over the lands.

Wetland regulation will not be fully effective and rational until it is conducted in conformity with an ecologically based, long-range resource plan. The fact that such planning still is embryonic and the fact that the state-of-the-art of regulation is still primitive represent opportunities to scientists to exercise great influence in the development of both aspects. Few regulators and almost no planners have extensive scientific backgrounds, and the people involved in these important professions are anxious to obtain information and professional opinions from scientists. The scientist, however, must learn to communicate in simple English or he will be frustrated by the lack of acceptance of his ideas. Scientific jargon is as undecipherable to most planners and regulators as ancient hieroglyphics are to most of us.

One further comment: one is more influential as a contributor to a developing plan than as a critic of a completed plan. In many situations a change in an adopted plan requires special action by an administrative body and may entail a public hearing. One must battle a certain bureaucratic inertia, and then swim against the tide of administrative red tape to change a plan that one believes to be wrong or defective. In contrast, a short visit or an informative letter may assure that your information will be reflected in a plan that still is being formulated.

What is Being Regulated?

Many state wetland acts were drawn up in haste by concerned citizens and legislators who had little or no scientific advice. The cut-and-paste origin of some of the acts is easily recognizable but unavoidable because the rush to the legislative floor was essential for passage. Unfortunately many acts contain restrictive statements of goals that actually could handicap the rational management and allocation of wetland resources.

In my opinion (and I think this point has not been explored adequately) our collective intent is to regulate and maintain the wetland **system**. With certain exceptions for unique areas, it is not necessarily our goal to preserve the wetlands on a particular **site**. If some nonwetland use is judged to be of more value to society on that site, can we maintain the system and continue to enjoy its many benefits by conditioning a permit to use the site through requiring the establishment elsewhere in the system of a new wetland that is equally or more productive of some resource? This would retain the integrity of the system, but would permit more flexible planning postures in regard to geographic allocations. The critical questions are: Can we identify functional units in the wetland systems? Within a functional unit, can we identify equivalent locations (one of which would be an existing wetland for which a priority nonwetland use is proposed, the other currently is not a wetland, but could be enhanced by transformation to a wetland)? Can we create, intentionally, new wetlands that will function as effectively as those that are of natural origin?

Standardization

Scientists continuously seek to improve their methodologies, but many of them view moves to standardize methods as an infringement upon their academic freedom. One conspicuous exception to this is the compilation of standard methods for the analysis of water (APHA, 1976). Although a research scientist does not necessarily feel bound to follow standardized methods, he often does in order to assure his colleagues that he has used repeatable techniques and to avoid the need to explain his methodology in detail.

Planners and regulators generally have little knowledge of ecological methods and they may not be aware that one's choice of method can affect the results of the study. Sampling methods can affect the type of information collected. For example, one person describing vegetation may report only rooted frequency while another may report only cover. These, of course, are not comparable and one cannot be transformed to the other. The sampling methods selected can also affect the quality of the data. The method used to select plot locations, for example, can bias the data positively or negatively and the use of some types of plotless methods can greatly underestimate such parameters as basal area or standing crop. Season of sampling can affect virtually any parameter.

Another member of the cast of characters who often becomes involved in the decision making process is the consultant. The consultant may act as an information collector and summarizer for the planner. He may recommend policy, draft regulations, formulate plans, assist regulatory personnel in evaluations of applications for permits or funding, be contracted by an applicant to prepare technical studies, advise on plans for a project and prepare an assessment of the environmental effects of the final plan that is submitted with an application for a permit.

The expertise of consultants varies nearly as widely as does that of planners and regulators. Full-time consultants are in business, and to remain employed continuously they must get jobs and must make at least a small profit. Therefore, most consultants will accept almost any job they are offered and will promote almost any job that is available. A common attitude is get the contract and (after you get the contract) buy the expertise you need. Nevertheless, at least some consultants who have engaged in work on important aspects of wetland identification, regulation, management, and/or planning have had no scientific training on, and little or no previous professional experience with, wetlands.

The quality of work produced by consultants is variable but most of the products satisfy the expressed needs of the client. Some products are exceptionally good and are of unanticipated value to the client.

I do not believe that individual consultants should be allowed (1) to choose the methods used in assessing a project for an applicant, (2) to evaluate an assessment for a regulatory agency or (3) to formulate resource allocation plans for an authority. Instead, at least for repetitive analyses, the scientists, regulators, planners, and consultants should convene and designate standard methods to obtain, present, and evaluate data. This agreement should be reflected in the resource inventory and analysis for planning, in assessments to accompany applications, and in reviews to arrive at regulatory decisions.

Standardization is essential if we are to have comparability between large scale resource inventories and the site-specific evaluations that are required in the regulatory process. Standardization also will serve to lessen the variations in data produced by different groups, particularly by consultants. And it also will avoid much of the misunderstanding and delay that occurs now in the environmental review process. Applicators are not expected to advance the state of the art of scientific methodology. It is not efficient to expect them, or to allow them, to introduce new methods or to select capriciously from among the dozens or hundreds of methods available in the literature to evaluate most parameters.

Those of us who are scientists and consultants generally have little difficulty in remembering which hat we are wearing at a particular time. When we are consultants, we expect to work under more or less specific limitations. When we are scientists, each of us can expect the same degree, or more, of academic freedom as our full-time scientist colleagues.

Standardization also need not be eternal, stagnating or intellectually stifling. As better methods evolve, they can be included as standards. These revisions, of course, should be orderly and carefully considered. Proper correlations of the old and new methods should be developed, so that they are relevant to the resource base.

A Dynamic Resource

The various Middle Atlantic States have mapped the location, extent, and nature of their coastal wetlands since 1970. The mapping was done by analyses of aerial photographs and was verified to some degree by fieldwork. Basically, however, the mapping represents conditions at the instant that the aerial cameras exposed the film.

Based on our experience in Maryland, some wetlands are eroding rapidly and were reduced in size by 25% to 50% between the autumn of 1972 and the autumn of 1976. Most of these were small, peninsular wetlands along the shores of Chesapeake Bay. In other places the trees in wooded wetlands have died since 1972, and the areas are transforming to marshes. There also are areas in which one marsh type is recorded on the 1972 photographs, but another marsh type occupied the site in 1976. These changes do not reflect seasonal variations. Rather, they appear to be valid displacements of one community type by another. This emphasizes that we are dealing with a very dynamic resource that can change rapidly in nature, size, and location. The upper inland boundary of a particular wetland also may shift several metres horizontally. The shift is landward if it reflects rising sea level and seaward if it reflects rapid siltation, organic accumulation, or uplift.

Planners and regulators need to recognize the dynamic nature of the resource and incorporate provisions for frequent revision of their base-line inventories, as well as their plans and regulatory postures. Legislation should include (or be amended to include) full authorization for these revisions by the administrators, subject to public review. Scientists should investigate the changes to develop predictive tools. In particular, we should seek methods to anticipate human actions which lead to the degradation of wetlands not directly affected by those actions. We also should be able to anticipate whether or not a site that currently is of minimal value as a wetland may, within a reasonable period of time and with reasonable certainty, become a high value wetland site.

The administrative decision adopted by the State of Maryland is to map wetland features as they appear on the 1972 aerial photographs. We currently are completing type mapping of the wetlands and are aware of substantial changes that have occurred during the intervening 4 years. Our field examinations, however, now have extended over a period of more than a year, and before the job is completed, they will span at least 2 years. Because wetlands can change significantly during a single storm, there would have been a great variation in the chronology of map conditions from place to place in the state had the decision been to map current conditions. The decision also is of great scientific value, if the next resource inventory is as detailed and accurate, and follows the same strategy. This next inventory will give us a second point in time, and will allow us to determine, in detail, area, locational and community type changes. Insofar as I know, no other Middle Atlantic State has a resource base in equally fine detail, and none will be able to track changes so accurately as will Maryland.

Ecological Assessment of Wetland Values

In our modern society it is unrealistic to believe that every hectare of wetland can or should be declared inviolate and, thus, preserved in perpetuity. The accurate delineation of wetlands, therefore, may be the easiest task we face. Much more difficult is the task to sort among this inventory of valuable, sensitive environmental features and determine which are of greatest value and which are of least value.

During 1972, one of our jobs required me to question many colleagues about the values of coastal wetlands, including the freshwater segments. One question was: *Can we estimate the proportion of wetlands that can be removed without significant impairment of the ability of the estuarine system to function and to produce the renewable resources harvested by man?*

Most colleagues pleaded that our knowledge is too limited to allow that question to be answered. Especially when I emphasized that if we as scientists, do not frame an answer based on our experience and training, the administrators and legislators will develop a "best available data" answer. In response, my colleagues uniformly answered: *The loss of wetland hectarage will be reflected by a proportionately similar loss in estuarine resource production, but it is not possible to predict the magnitude of the loss that will result in the failure of the system. It might be 10% of the wetland hectarage or 80%.*

There is no standard method for the assessment of the ecological value of wetlands. This is a critical point at which scientists can aid and influence the value system of society. If we rest on our laurels, other sectors of society will proceed beyond us. Their decisions will be based on the best data they have available, and probably will not reflect the information reported here.

LITERATURE CITED

American Public Health Association. (1976). Standard methods for the examination of water and wastewater. Fourteenth edition. Prepared and published jointly by the American Public Health Association, American Water Works Association, and Water Pollution Control Federation, Washington, D. C. 1193 pp.

Carter, J. (1977a). Executive Order 11988. Floodplain management. *Federal Register* 42(101), 26951–26957.

Carter, J. (1977b). Executive Order 11990. Protection of wetlands. *Federal Register* 42(101), 26961–26965.

Coleman, W. T. (1975). Preservation of the Nation's wetlands. Office of the Secretary, Department of Transportation, Washington, D. C., Order DOT 5660.1. 3 pp.

Committee on Government Operations. (1970). Our waters and wetlands: How the Corps of Engineers can help prevent their destruction and pollution. Twenty-first Report by the Committee. 91st Congress, 2d Session, House Report 91-917. 18 pp.

Corps of Engineers. (1977). Regulatory programs of the Corps of Engineers. Final rules. [Amendments to Chapter II of 33 CFR.] *Federal Register* 42(138) 37121–37158.

Cowardin, L. M., Carter, V., Golet, F. C., and Latroe, E. T. (1976). Interim classification of wetlands and aquatic habitats of the United States. Office of Biological Services, U.S. Fish and Wildlife Service, Washington, D. C. 109 pp.

Ruckelshaus, W. D. (1973). EPA policy to protect the Nation's wetlands. Environmental Protection Agency, Washington, D. C., Administrator's Decision Statement 4(Revised). 3 pp.

Stuckey, R. L., and Roberts, M. L. (1977). Rare and endangered aquatic vascular plants of Ohio: an annotated list of the imperilled species. *Sida* 7, 24–41.

APPENDIX I

Selected Public Laws (Federal) Related to Freshwater Wetlands

River and Harbor Act of 1899 (33 USC 401 et seq.; approved 3 March 1899; as ammended and supplemented).

PL 91-190. National Environmental Policy Act of 1969 (42 USC 4321-4347; approved 1 January 1970; as amended).

PL 92-500. Federal Water Pollution Control Act Amendments of 1972 (33 USC 1251, et seq.; enacted by Congress 18 October 1972, overriding the President's veto of 17 October 1972; as amended).

PL 93-205. Endangered Species Act of 1973 (16 USC 1531-1543; approved 28 December 1973; as amended).

PL 93-370. Coastal Zone Management Act of 1972 (16 USC 1451-1464; approved 27 October 1972; as amended).

MANAGEMENT POTENTIAL:
SUMMARY AND RECOMMENDATIONS

Forest Stearns

Department of Botany
University of Wisconsin–Milwaukee
Milwaukee, Wisconsin 53201

Freshwater wetlands comprise a valuable national resource. Once classed only as wastelands, the vastly increased pressures now focused on the land have stimulated many suggestions for wetland use. For example, the potential of wetlands for processing human waste, in contrast to merely using them as dumps, has attracted the attention of engineers, administrators and the general public (Tourbier and Pierson, 1976). As is often the case with such intriguing ideas, there is a risk that operational systems of waste disposal will be implemented on a large scale before the effects of such utilization are understood. The preceding sections of this volume have demonstrated that our knowledge of wetlands, although greatly increased in the last 10 or 20 years, is still miniscule in relation to the complexity of these systems.

Wetlands were managed long before there was any knowledge of the physical and chemical interactions or of the relationships of plants, animals, soil, and water. As in the past, wetland management today is concerned with the preservation, use or establishment as dictated by human needs. The 5 papers in this section obviously could not cover the entire span of wetland management; rather thay represent a sample, progressing from the best known, i.e., management for wildlife, to new developments in wetland management for nutrient assimilation and the creation of wetlands.

In North America, wetland management has been directed primarily toward providing habitat for wildlife. Wildlife managers have had the most experience with wetland manipulation and, as Weller (1978) indicated, there is a wealth of information on the results of such manipulations. Stone et al. (1978) remind us that, along the Gulf of Mexico as elsewhere, land manipulation for agricultural and industrial purposes has produced many changes, some foreseen and some counterintuitive. In Louisiana, the hydrologic regime provides a considerable energy subsidy, and human activity, including diking, fertilization, drainage and widening of canals has resulted in considerable wetland loss.

The creation of marshes *de novo*, and the management of marshes for nutrient assimilation discussed by Garbisch and Coleman (1978) and by Sloey et al. (1978) respectively, are techniques in which new knowledge is developing rapidly. It is increasingly important also that problems in the regulation of wetlands, elucidated by McCormick (1978) are understood and appreciated.

Among the many generalizations pertinent to management is the concept that wetlands are constantly undergoing changes in water level, productivity, and composition of the plant and animal populations (Weller, 1978). Most wildlife species have developed a wide tolerance to these changes or (by virtue of mobility) can disregard them. Weller suggests recognition of these short-term changes in management strategy. Although well defined in the upper Midwest, where they result largely from shifts in precipitation and muskrat activity, cycles of change occur in all wetlands depending upon the hydrologic cycle and other factors. In contrast, human interference, such as the digging of a canal may have major and frequently irreversible effects upon wetland composition and survival (Stone et al., 1978). When channeling occurs not only is the wetland altered or destroyed but it no longer serves its various roles thus passing the impact along to adjoining systems. Sloey et al. (1978) indicate that seasonal changes also are of major importance in nutrient retention.

Purposeful, in contrast to inadvertant, creation of wetlands is a recent approach, largely stimulated by the need to utilize dredge spoil deposits and by belated recognition that marshes serve a variety of functions from shoreline protection to energy sources. Garbisch and Coleman (1978) report on one of these developments in which problems of site, inelastic contract stipulations and unanticipated wildlife depredations were involved. We may anticipate a considerable increase in the energy and funding devoted to wetland establishment both in the brackish tidal zone and on the shorelines of the Great Lakes.

Intentional wetland development or modification will provide processing areas for disposal of domestic sewage, a use that may be expected to spread rapidly. To date there have been successful developments on a small scale, i.e., campground size, and some progress on a larger scale. However, as Sloey et al. (1978) and others suggest, wetlands may serve only as a temporary trap for nutrients releasing them in the spring flush as the winter snow melts. Although considerable work is in progress on a variety of wetland types (Tilton et al., 1976) much more is needed.

Wetland controls have been expanded greatly in the last decade. Yet in most areas, ". . . wetland regulation derives more from 'common sense' than from science" (McCormick, 1978). Much of the recent legislation will probably undergo revision; however, it has greatly retarded the loss of coastal wetlands. In the agricultural Midwest, wetland regulation suffers from some lack of understanding of wetland values. It will be some time before the remaining wetlands may be considered secure.

Interactions among wetlands, agriculture, industry, and transportation, as seen in Louisiana, are evident across the continent. The need to analyze plans for deleterious effects before work begins using computer simulation techniques (Stone et al., 1978), applies throughout. Finally, we are all aware that private wildlife managers, state departments of natural resources and several United States government agencies have established large numbers of undocumented experiments in wetland management. Few of these actions have been adequately evaluated, so that an intensive evaluation effort should be profitable not only for wildlife management but for other goals as well.

At our current level of knowledge, wetland management remains an art with diverse objectives. However, to be effective management requires direction or purpose solidly based in values. Among other things, the basic values provide some limits to the degree of alteration permissable. Marsh exploitation, by peat mining, for example, cannot be called management.

Management alternatives depend upon system attributes discussed in detail elsewhere in this volume. The role of the wetland system in influencing offshore or downstream systems, circulating minerals and energy, acting as a sink and influencing water levels and stream flow are drawn upon to achieve management goals. To be attainable, the goal must be in step with system attributes.

Management requires that decisions be made and plans implemented; these decisions involve the maintenance of a functioning wetland system for a specific goal or goals. A functioning system implies a biotic community; an ecosystem functioning within certain limits and including both biotic and abiotic components.

For what reasons are wetlands managed? There are usually several goals, and they are better considered together than separately. Management goals fall in 3 areas: (1) environmental protection, (2) recreation and aesthetics, and (3) production of renewable resources. Specifically, wetlands may be managed to:

1) Maintain water quality. Impound nutrients temporarily or permanently, retain sediment and particulate materials, process organic substances and serve other functions in improving water quality

2) Reduce erosion. Protect areas affected by high energy impact from waves, motorboats, etc.

3) Protect from floods. Serve as buffers to retain or detain runoff, and provide resilient vegetation to sustain periods of temporary floodwater storage

4) Provide an unsubsidized system which can process airborne pollutants

5) Provide a buffer between urban residential and industrial segments by ameliorating extremes of climate and physical impact such as noise

6) Maintain the gene pool of marsh plants and to provide examples of communities complete and capable of functioning as unsubsidized support systems. (This goal requires a positive management decision)

7) Provide aesthetic and psychological support for human beings

8) Produce wildlife. Provide habitat for waterfowl and other wildlife which may in turn serve both aesthetic and practical consumptive functions, i.e., hunting and trapping

9) Control insect populations

10) Provide needs of fish for spawning and the production of food organisms

11) Produce food, fiber and fodder. Production may take the form of wetland forestry, management which has been practiced in North America for several hundred years, or food or fiber crops, ranging from cranberries to cattails

12) Expedite scientific inquiry. Management may assist in furthering knowledge of marshes.

Most wetland management is oriented to several goals and most wetlands serve multiple uses. It should be clear that management requires more than scientific understanding. It involves decisions by laymen, planners, administrators and managers. In wetlands, as in other ecosystems, successful management requires coordination and cooperation.

In management one must look beyond the wetland. The effect of inescapable regional pollution (e.g., the acid rain from industrial concentrations) or the uncontrolled contributions of N, heavy metals and organic toxins must be considered. What will be the effect of dustfall from agricultural fields with addition of P and Ca, or of agricultural runoff carrying heavy nutrient loads, especially of N and K. The overall regional hydrologic regimes, including drainage and flooding, the impacts of traffic density and noise, the economic pressures subsidizing drainage and those encouraging wetland retention are all factors in management decisions.

At this time, those responsible for managing wetlands see numerous gaps in our understanding of wetland ecology including (1) lack of knowledge of substrate—nutrient interactions and of decomposition processes, (2) lack of knowledge of the behavior of invertebrates as well as relatively limited knowledge of secondary consumers, (3) lack of information on the life cycles, germination, growth, and physiology of wetland vascular plants including internal processes of translocation, mineral nutrient mobilization and senescence and their competitive strategies, (4) lack of adequate knowledge of algal and bacterial functions in wetlands and (5) inadequate understanding of the role of freshwater wetland systems in the regional physical and biotic landscape. Additionally, questions of wetland size, heterogeneity, colonization and species extinction await exploration.

Despite this missing knowledge, we know enough to state that certain approaches to management may be unprofitable or perhaps dangerous to implement or, conversely, that other approaches have been proven safe by long experience. However, for intelligent management of wetlands in the future, several specific lines of research need urgent attention. In the area of nutrient dynamics they include studies of (1) nutrient and heavy metal retention capacities of various wetland types and substrates including rates and timing of nutrient flushing under natural conditions or as modified by man, and (2) the relative importance, rates and limits of various chemical processes involved in nutrient transformations (e.g., denitrification, nitrification, and absorption, etc.) as they relate to different substrates. In the area of wastewater treatment they include studies to (1) identify diagnostic symptoms in wetland communities which indicate heavy metal and nutrient concentrations and their impact on species diversity and productivity and (2) determine the potential inherent in waste disposal for simplification and eventual destruction of wetlands including the response of indicator species to wastewater application. In the area of wetland stability and secondary production they include studies to elucidate the trophic structure of wetlands and the energetics of consumer levels in various wetland systems with special attention focused on the invertebrates. Finally research is needed to identify signs of wetland disruption and change, especially identification of species indicative of wetland dysfunction and potential degradation, useful of wetland managers.

The above statements apply not only to wetlands but to all other ecosystems. We must ask the question "what is good management?" Assume that we know how to manipulate a marsh to cause rapid organic decomposition. Is this advisable? What will it do to wetland structure and stability? Assume we can convert wetland type A to wetland type B. How do we determine which has the greater value: should conversion be permitted or encouraged? Is rapid export of nutrients desired or will delaying outflow of nutrients best achieve our objectives? How valuable is a diversity of species and what is the value of wetland production? Above all, can we identify a set of wetland values?

We already allocate lands for various uses. Increasingly, we must justify these allocations, and to do so it is necessary to have a reasonable knowledge of system values and of system response to perturbations. Wetlands have too long been considered wasted landscape by many decision-makers because only their negative aspects were considered. Preferably, these values would not be related to economics but to survival. They would be energy and food-chain oriented and the aesthetic and psychological support values would be integrated into the more readily documented concepts.

For the manager and decision-maker there is an equally urgent need to develop techniques for the synthesis of information and communication of that knowledge to managers, users of wetlands and administrative and legislative decision-makers. Specifically continuous extraction of research information and transmission of this information (in suitable form) to persons who manage and make decisions (i.e., congressional committees, federal, state and local officials, etc.) is essential. Action by management may be complex and fast moving. A series of workshops on the problems and potentials of wetland management should begin shortly. These workshops should be staffed by experienced managers and researchers with the intent of assisting in establishment of goals and values and to warn of problems. The public needs more information on wetland management, the potentials, drawbacks and values involved. A short factual treatise on wetlands should be prepared to provide ideas and serve to bridge the gap between manager, administrator and researcher.

Much may be learned by evaluation of past wetland management. Energy and cost accounting techniques should be used and both social and biological values examined. With the demonstrated complexity of wetlands, the need for greater long-term support for research is clear. Research has been hampered by the lack of permanent study areas. Certain wetlands should be set aside for long-term studies permitting establishment of applied and manipulative studies adjacent to control areas. As discussed elsewhere in this volume, much remains to be done in the area of sampling methodology. Sampling vegetation and environments to provide data for management decisions can be greatly improved especially with the introduction of system modeling and the use of conceptual models in communication of research results.

While essential basic information is being obtained and values are being developed, present and additional legislation must be implemented to slow the rate of wetland loss. If wetlands indeed make valuable contributions to the landscape, they should not be destroyed before their contributions are understood.

The papers presented in this section and data gained from experience of managers, researchers, and scientists led to several general conclusions:

1) Management decisions whenever possible should complement natural functions and allow natural processes to accomplish the desired results.
2) Wastewater should not be applied to natural freshwater wetlands, other than experimentally, until more is known about long-term effects. Present information suggests that the risk of damage may not be worth the gain. Local tests are always necessary to determine suitable loading rates.

3) The rate of wetland conversion should be slowed until more is known about the functions of wetlands in regional systems.

4) Management should hold to a minimum, factors which tend to degrade marsh structure and function. Biological as well as nonbiological approaches must be included in the evaluation of the health and future of wetlands.

5) In creating wetlands, the manager should remember that plants of different species vary greatly in vulnerability to physical stress and to animal damage as well as in adaptation to water depth and other factors.

6) In modifying wetlands (where this is essential), attempt to avoid disturbance; conduct the physical operation rapidly, reduce height of spoil banks, limit impoundments and maintain normal water circulation.

7) Natural perturbations may occur and management techniques successful at one point in the climatic cycle may not be applicable at others.

8) Informed and conservative management is essential in all wetlands; so large an area has already been lost that the remaining wetlands must be protected. This concept has been embodied in much recent legislation.

ACKNOWLEDGMENTS

The ideas expressed in the preceding pages originated and were discussed in several evening sessions during the symposium. The contributions of the authors of management papers and of others present at these discussions are greatly appreciated. Special thanks are extended to James Stone, William Sloey, Jack McCormick, Milton Weller, Vernon Laurie and others who commented on a draft of this paper.

LITERATURE CITED

Garbisch, E. W., and Coleman, L. B. (1978). Tidal freshwater marsh establishment of Upper Chesapeake Bay: *Pontederia cordata* and *Peltandra virginica*. *In* "Freshwater Wetlands: Ecological Processes and Management Potential" (R. E. Good, D. F. Whigham and R. L. Simpson, eds.), pp. 285–298. Academic Press, New York.

McCormick, J. (1978). Ecology and regulation of freshwater wetlands. *In* "Freshwater Wetlands: Ecological Processes and Management Potential" (R. E. Good, D. F. Whigham and R. L. Simpson, eds.), pp. 341–355. Academic Press, New York.

Sloey, W., Spangler, F., and Fetter, C. (1978). Management of freshwater wetlands for nutrient assimilation. *In* "Freshwater Wetlands: Ecological Processes and Management Potential" (R. E. Good, D. F. Whigham and R. L. Simpson, eds.), pp. 321–340. Academic Press, New York.

Stone, J., Bahr, L., and Day, J. (1978). Effects of canals on freshwater marshes in coastal Louisiana and implications for management. *In* "Freshwater Wetlands: Ecological Processes and Management Potential" (R. E. Good, D. F. Whigham and R. L. Simpson, eds.), pp. 299–320. Academic Press, New York.

Tilton, D. L., Kadlec, R. H., and Richardson, C. J., eds. (1976). "Freshwater Wetlands and Sewage Effluent Disposal." University of Michigan, Ann Arbor, Michigan.

Toubier, J., and Pierson, R. W., eds. (1976). "Biological Control of Water Pollution." University of Pennsylvania Press, Philadelphia, Pa.

Weller, M. (1978). Management of freshwater marshes for wildlife. *In* "Freshwater Wetlands: Ecological Processes and Management Potential" (R. E. Good, D. F. Whigham and R. L. Simpson, eds.), pp. 267–284. Academic Press, New York.

INDEX

A

Aboveground standing crop, 285, 287, 296
Acer, 234
Acnida, 5, 13
    altissima, 211
    cannabina, 8, 17, 334
Acorus, 5, 11, 13, 14, 16, 84
    calamus, 6, 13, 15, 17, 174, 288
Actinomycetes, 117
Aeration, effect on
    decomposition, 133, 328
    nutrients, 58, 224, 228, 230
    roots, 44
Aerial photography, 353
Afterripening, 290
Agassiz National Wildlife Refuge, Minnesota, 190
Agelaius phoeniceus, 270, *see also* Redwinged blackbird
Agricultural runoff, 100, 101, 305, 314, 315, 317, 329
Agriculture, 305, 307
Aix sponsa, 279
Alcohol, 117
Alkalinity, 197, 199
Alkaloid, 164
Alligator, 304
Alnus rugosa, 218
Alternanthera philoxeroides, 303
Aluminum, 99, 105, 106, 223, 224
Ambrosia, 5
    trifida, 8, 244
Amine, 117
Amino acid, 164, 246
Ammonia, 178, 187, 197, 198, 201, 209, 217, 223, 227–229, 233, 243, 246, 248, 249, 252–254, 261, 329, 330
    excretion of, 261
Amphibian, 303, 335
Amphipod, 96, 261

Anaerobic conditions, effect on
    nutrients, 69, 201, 223, 227, 228, 333
    root, 44, 70, 224
Anas acuta, 275
Andromeda polifolia, 57
Animal
    depredation, 285, 291, 293, 294
    role in nutrient cycling, 261
Anthoxanthum odoratum, 47
Arctic plant growth, 170
Arctophila fulva, 174, 175
Arrow arum, 90, *see also Peltandra virginica*
Arrowhead, 90, *see also Sagittaria latifolia*
Ash content, 131, 133, 156, 158–162, 165, 203, 205, 206, 209, 221
Aspen, 234, 236
Aster, 334
Atchafalaya Bay, Louisiana, 305, 312
ATP, 131, 139, 140
Aulacomnium palustre, 57
Avifauna preservation, 271
Azolla caroliniana, 334

B

Baccharis, 8
Bacteria, 95, 96, 117, 125, 199, 204, 261
    coliform, 336, 337
    decomposer, 92
Barataria estuary, Louisiana, 314, 315
Base exchange site, 212
Belowground biomass, 29–31, 184, 231
Belowground phosphorus deficit, 183
Belowground production
    freshwater tidal wetland, 4, 13, 14, 17, 79, 80
    prairie glacial marsh, 21, 29–32, 80
    sedge wetland, 48, 49, 80
Belowground productivity
    bog marsh, 56–58, 80

estimation of, 55, 56, 83
freshwater wetland, 79
sampling, 83–85
Betula, 221, 234, 329
  nana, 57
  pumila, 57, 60, 61, 115, 118, 218, 225, *see also* Bog birch
Bicarbonate, 217, 227
Bidens, 5, 8, 13, 14, 16, *see also* Bur marigold
  cernua, 22, 25, 29, 32, 211
  laevis, 17, 89, 90, 94, 244, 334
Big cordgrass, 4, *see also Spartina cynosuroides*
Bird, 303
  breeding, 275
  evolutionary history, 274
  migration, 275
  nest number, 269, 270
  nest sites, 270, 274, 279
  nesting chronology, 275
  populations, 267, 273, 280
Bison bison, 277
Black spruce, 64, 67
Blue-green algae, 95, 261
Boehmeria cylindrica, 17
Bog, 56, 61, 64, 217, 218, 221, 236
  birch, 120–127, 226, 227, 230–232, 329, *see also Betula pumila*
  black spruce, 221
  blanket, 63, 66, 73, 218, 221–223, 231, 233–236
  depression, 70
  lacustrine, 73
  minerotrophic, 71
  ombrotrophic, 68, 71–73, 217, 218, 221, 222, 224, 227, 228
  raised, 63, 218, 221, 231
  sphagnum, 67, 75, 223
  string, 67
  succession, 72, 73
Boron, 160–162
Brackish wetland, 4, 10, 255, 303, 304, 309
Brasenia schreberi, 161, 162
Bryophyte, 53
Bulrush, 278, *see also Scirpus*
Bur marigold, 90, *see also Bidens*

C

Calamagrostis canadensis, 57
Calcium
  accumulation in sediment, 211

availability, 259, 325
budget, 212, 233–235
carbonate, 100, 139, 199, 202, 325
changes during decomposition, 100, 108, 110, 111, 115, 124, 127
dry fall, 360
enrichment, 59
export from wetland, 233, 236
flux, 207, 208, 230, 232
in interstitial water, 228, 229, 231
leaching from litter, 99, 179
leaching from plant, 178, 199, 260
in litter, 105, 106, 232
in peat, 217, 221–224
in plant, 156–163, 171–175, 177, 195, 203, 205, 206, 208, 210, 224, 233
phosphate, 325
plant accumulation, 177, 205, 209
in soil, 159, 195, 200, 202, 217, 223, 233
in water, 100, 159, 197–199, 217, 227–229
wet fall, 232
Calluna, 221
  vulgaris, 58, 224
Canal, 358
  alignment, 311
  effect on nutrients, 314, 315, 317
  management implications, 299–318
Capillarity, 74
Carbohydrate, 131, 133, 137, 179, 181, 182, 185
  storage, 30, 184, 186
  total available, 164
  total nonstructural, 137
Carbon, 106, 117, 131–133, 142, 156–160, 163, 178, 186, 229, 314, 315
  dioxide, 134, 199, 201, 202, 243, 246, 247, 250, 251, 253, 254
  flow, 178
  nitrogen ratio, 95, 96, 131, 140, 141, 147, 179
Carex, 23, 25, 27, 34, 47, 60, 61, 82, 115, 179, 218, 221, 225, 269, 287, 333
  aquatilis, 42, 47, 48, 57, 84, 119, 174–179, 187, 189, 225
  atherodes, 24, 25, 28, 31, 33, 35, 47, 99, 101–103, 105, 107
  comosa, 225
  lacustris, 25, 30, 31, 39, 41, 42, 44–49, 158, 178, 190, 202, 204, 206, 209, 210, 225
  lanuginosa, 47
  lasiocarpa, 119, 225
  life history, 39–50

lyngbyei, 40
nigra, 40, 58
oligosperma, 119, 225
rostrata, 25, 28, 31, 34–36, 39, 41–49, 57, 80, 105, 119, 225
stricta, 42, 43, 48
Carnivore, 304
Carp, 277
Carr, 73
Cation uptake, 212
Cattail, 4, 13, 68, 90, 270, 272, 278, 328, *see also Typha*
Cattle, 104, 277, 304
Cedar, 67
Cellulose, 96, 121, 131, 133, 135, 136, 163, 164
Cell-wall material, 163, 164
Ceratophyllum demersum, 23, 25, 161, 162
Chamaedaphne, 221, 329, *see also* Leatherleaf
calyculata, 57, 60, 61, 115, 118, 218, 225, 226
Chara, 23, 158, 162, 163
Chemical composition
interspecific variation, 160–164
intraspecific variation, 156–160
wetland plant, 155–165
Chesapeake Bay, 285, 287, 296, 353
Chloride, 197, 199, 228, 229
Chloris acicularis, 47
Chlorophyll, 164
Cicuta maculata, 17
Cistothorus
palustris, 270
platensis, 270, *see also* Short-billed marsh wren
Cladium
jamaicense, 159, 321, 330, 333, *see also* Sawgrass
mariscus, 44
Cladophora, 162, 163
Climate, 64, 112, 271, 327
effect on productivity, 61
Coastal Zone Management Act, 341, 348
Cohort, 42
Common reed, 90, *see also Phragmites communis*
Community composition, 81, 300
Competition, 270, 274, 276
Conductivity, 197
Consumer, 142, 299
activity, effect on productivity estimate, 55

migration, 186, 190, 304
Coot, 275
Copper, 160–162, 222
Cornus, 323
Cover-water interspersion, 267, 269, 270, 278
Crayfish, 304
Crustaceans, 304
Cuscuta, 17
Cyperus, 22
papyrus, 156–158
Cypress, 64, 75
dome, 75, 321, 330, 334, 335
swamp, productivity of, 75

D

Danthonia caespitosa, 47
Debris deposition, 285, 291, 293, 295
Deciduous forest, 223
Decomposer activity, 179
Decomposition, 201, 204, 208, 260–261, 328, 360, 361
aerobic, 115, 116, 119, 131, 133–137, 139–141, 146
anaerobic, 97, 115, 116, 119, 131, 133, 134, 140, 146
fen, 232
freshwater wetland, 145–150, 325, 326
intertidal freshwater marsh plant, 89–97, 254
laboratory studies, 92, 93, 131, 133–139
littoral zone of lake, 131–142
microenvironment, 149
model, 147–150
northern wetland, 115–128
prairie glacial marsh, 99–112
rate, 91–94, 116, 121–123, 131, 135–137, 140, 141, 145–147
weight loss, 135, *see also* Litter, weight loss
Deer, 104
Dehydrogenase activity, 131, 133, 140
Delaware River, 5, 12, 244
Delta formation, 305, 312
Denitrification, 117, 170, 187, 188, 191, 197, 198, 201, 228–230, 261, 314, 321, 324, 326, 328, 330, 332–334, 337, 361
Detrital system, 304
Detritivore, 92, 96, 97, 304
Detritus, 96, 97, 142, 208, 260, 287, 302, 315, 317
food web, 97

Dewatering, 278
Diking, 358
Dissolved
  organic carbon, 137–139
  organic matter, 91, 93, 96, 97, 131–134,
    148, 300
  solids, 71
Distichlis, 82
  spicata, 8, 11, 303
Diversity, 74, 196
Drainage, 304, 305, 358, 360
Drawdown, 23, 32, 112, 195, 197, 211, 212,
  262, 272, 273, 276, 277, 279, 280
Dredging, 305, 307, 312, 314, 317, 345, 347,
  350
Driftwood, 291, 293, 295
Drought, 22, 23, 275, 277
Dry fall, 186, 187, 219, 224, 260
Duck, 270, 274, 275, 277–280
  food, 287
  nesting population, 269
Dupontia fischeri, 57

E

Eagle Lake, Iowa, 23–36, 99–112
Earthworms, 117
Ecosystem theory, 73, 74
Eichhornia crassipes, 156–159, 164
Elemental composition, 91
Eleocharis
  acicularis, 161, 162
  quadrangulata, 161, 162, 173
Empetrum nigrum, 57
Endangered Species Act, 341, 348, 349
Energy
  content, 164, 165
  cost accounting, 299
  flow, 96
  subsidy, 300, 358
Environmental impact statement, 346
Epiphytic diatoms, 162
Equisetum fluviatile, 57
Erechtites hieracifolia, 334
Erica tetralix, 58, 224
Eriophorum, 60
  angustifolium, 58, 59
  scheuchzeri, 57
  vaginatum, 57–59, 61
Erosion, 233, 305, 312, 314, 347, 353, 360
Ether extract, 163, 164
Eupatorium
  compositifolium, 334
  perfoliatum, 288
Eutrophication, 187, 299, 313, 315, 317

Evapotranspiration, 70, 74, 197
Everglades, 63, 66, 73
Evergreenness, 227

F

Fagus, 234
Fallen litter compartment, 106–108
Federal Water Pollution Control Act, 341,
  346, 348, 349
Feedback loop, 72, 73
Feedlot runoff, 337
Fen, 53, 54, 64, 73, 217–219, 221–224, 226,
  227, 229–231, 236
  minerotrophic, 217, 227
Fermentation, 117
Fertilization, 285, 287, 289, 291–296, 305,
  358, *see also* Nutrient experimental en-
  richment
Fetch, 287, 312
Fiber, 131, 136, 137
  production, 360
Filamentous algae, 243, 244, 246
Finfish, 304
Fire, 104, 106
Fish, 277, 293, 360
  fishponds, 175, 187
  harvest, 299
  processing waste, 330, 333, 337
Flood
  control, 306, 312, 360
  transport, 186
Flooding, 41, 43, 63–66, 69, 70, 72–75, 117,
  189, 197, 211, 243, 251, 252, 254, 276, 303,
  328, 360
Flower, 156
  stalk, 156
Flowering, 40–43, 291, 293–295
Flushing, 70, 243, 254, 325, 328, 329, 331–
  333, 358, 361
Fodder production, 360
Food
  plants, 270
  production, 360
  resource, 267, 280
Forest, 234, 236
Forestry, 307
Fragmentation, 99, 103, 104, 109, 110, 116,
  119, 148
Freezing, 199
Freshwater
  coastal marsh, Louisiana, 299–318
  marsh, management for wildlife, 267–281
  riverine marsh, nutrient dynamics, 195–
    213

tidal marsh, 63, 89, 175, 321
  decomposition, 89–97
  establishment, 285–296
  seasonal nutrient pattern, 243–256
tidal wetland
  biomass, 3–18
  Middle Atlantic Coast, 3–18
  primary production, 3–18
wetland
  hydrology, 63–76
  management, nutrient assimilation, 321–337
  management potential, 357–363
  regulation, 341–354
Fringe marsh, 63
Frost, 102, 103
Fulvic acid, 230
Functional unit, 351, 354
Fungi, 92, 117, 204
Furbearer, 304

G

Gaseous exchange, 186, 187, 197
Gastropod, 304
Geese, 270
Geomorphology, 142
Germination, 22, 272, 273, 276, 280, 290–294
Glyceria, 47
  maxima, 47, 174
Goose Lake, Iowa, 23–36, 99–112
Grazer, 83, 262
Grazing, 4, 177, 277, 304
Groundwater, 186, 188, 189, 260
Growing season length, 35, 53, 56
Growth, 292
  seasonal pattern, 13
Gull, 190
Guttation, 178

H

Habitat, 268, 281, 349, 357
  islands, 271
  loss, 313
  quality, 267, 280
  utilization, marsh birds, 269
Hamilton Marshes, New Jersey, 12, 14–18, 244, 245
Heavy metals, 360, 361
Hemicellulose, 121, 131, 133, 135, 136
Hemimarsh stage, 269, 272, 273, 278
Herbicide, 70, 278

Herbivore, 111, 272, 273, 304
  influence, 268
Herbivory, *see* Grazing
Hibiscus, 4, 5
  palustris, 7, 9
Highmoor, 218
Houghton Lake fen, Michigan, 218, 219, 222, 226, 227, 229–231
Houghton Lake, Michigan, 118, 223, 225
Humic compounds, 137, 139, 230
Hunting, 101, 350, 360
Hurricane protection, 306
Hydrilla verticillata, 164
Hydrocotyle umbellata, 161, 162
Hydrodictyon reticulatum, 163
Hydrogen
  gas, 117
  ion, 217, 227
  sulfide, 65, 66, 117, 201
Hydrologic
  recharge, 196
  regime, 63, 67, 299, 300, 302, 307, 321, 358, 360
Hydrology, 22, 217, 227, 260, 262, 304, 305, 314, 358
  effect on freshwater wetland, 63–76, 81, 85
  effect on nutrient cycling, 63, 305
  model, 65
Hypnum pratense, 57

I

Immobilization, 197
  limit, 179
Impatiens, 13, 25
  capensis, 17, 244, 334
Impoundment, 278, 304, 309–311
Insect, 304, 360
  control, 360
  damage, 291
  larval damage, 32
Insolation, 304
Inundation, *see* Flooding
Invertebrate population, 276
Iris, 335
  prismatica, 288
  pseudacorus, 288
Iron, 99, 105, 106, 160, 162, 201, 222–224, 227–229, 325
  enrichment, 60
Isopod, 96
Isotopic tracer, 177
Iva, 8

J

Juncus, 82, 84, 127
  effusus, 157, 161, 165
  roemerianus, 9, 11, 80, 93
  squarrosus, 122
  tracyi, 121
Jussiaea repens, 156
Justicia, 158
  americana, 156, 157, 177, 288
Juvenile
  organs, 183
  rhizome, 182

L

Lacustrine wetland, 323, 331
Lagg, 54
Lake Mendota, Wisconsin, 169, 171, 172, 176, 183
Lakeshore marsh, nutrient movement, 169–191
Landfill, 345, 346
Land loss, 299, 305, 311–314
Larch, 67, *see also Larix laricina*
Larix laricina, 224, 226
Latitude, effect on productivity, 47, 48, 81
Lawrence Lake, Michigan, 132
Leaching
  from litter, 99, 103, 106, 107, 109–111, 116, 122, 124, 127, 137, 146, 148, 179, 261
  from plants, 92, 108, 124, 127, 148, 177, 178, 180, 189, 199, 205, 208, 226, 231, 232, 243, 260, 261
  from sediment, 202, 211, 260, 278
Leaf
  area, 36
  blade, 156
Leatherleaf, 120–127, 227, 230–232, *see also Chamaedaphne calyculata*
Ledum palustre, 57
Leersia, 6
  oryzoides, 17
Lemna, 32, 158
  perpusilla, 334
Lentic marsh, 73–75
Levee, 309–311
Life
  history, 80, 81, 83, 181
    Carex, 39–50
  span, 176
  strategy, 42
Light, 159, 262, 276

Lignin, 96, 121, 131, 133, 135, 136
Litter
  decomposition, *see* Decomposition
  nutrients, 95, 99, 105–111, 116, 146, 179, 231, 255
  removal, 35
  weight loss, 89, 91–95, 103–107, 115, 116, 120–122, 145, 146
Litterbag, 90–94, 101, 104, 115, 118, 119, 133, 145, 179, 256
Litter fall, 103, 123, 231, 233, 314
Littoral zone of lakes, decomposition, 131–142
Loading limit, eutrophication, 299
Lobelia cardinalis, 288
Long-range planning, 348, 350
Lotic marsh, 73–75
Louisiana coastal freshwater marshes, 299–318
Luxury consumption, 205
Lyngbya, 162, 163
Lyonia
  lucida, 334
  mariana, 334
Lythrum, 5, 11, 13
  salicaria, 8, 244

M

Macronutrients, *see* specific elements
Macrophytic algae, 160, 163
Magnesium
  carbonate, 202
  changes during decomposition, 115, 127
  leaching from litter, 124, 179
  leaching from plant, 205, 260
  as a limiting factor, 172
  in litter, 99, 105, 106
  in plant, 156, 158, 161, 171, 173, 174, 176, 177, 195, 204–206, 209, 210, 217, 224
  in soil, 159, 200, 202, 221–223, 229, 231
  in water, 100, 159, 197–199, 211, 229
Mammal, 303
Management
  freshwater wetland, 321–337, 341
    potential, freshwater wetland, 357–363
    public needs, 267
    recommendations, freshwater marsh, Louisiana, 316–318
    strategy, 267, 274
  for wildlife, freshwater marsh, 267–281
Manganese, 158, 160–162, 201
Marl, 162

Marsh
artificial, 321, 322, 328, 335–337
basin shape, 267, 278
classification, 22
drainage, 100
establishment, 285–296, 358
function, model, 302, 304–305
habitat cycle, 275
mallow, 4, *see also Hibiscus*
management evaluation, 280
management theory, 280
minimal functional size, 271
modification, 278–279
sediment, 70
Maturity, 74, 294
effect on chemical composition, 157, 158
Maximum standing crop, *see* Peak standing
crop
Meadow marsh, 73, 74
Mercaptan, 117
Methane, 117, 201
Method
field, 351–353
harvest, 5, 16, 18, 55, 82–84, 181
Mice, 261
Microbial activity, 104, 109, 117, 118, 124,
131, 133, 139–142, 146, 148, 149, 205, 210,
223, 229, 232, 256, 259, 262, 315
decomposition, 104, 117, 132, 139–141
electron transport system, 140
enrichment, 209
nutrient uptake, 109, 110, 179
processes, 230
storage module, 304, 321, 326
Micronutrients, 70, 222, *see also* specific
elements
Microorganisms, 116, 125
Midge, 261
Mineralization, 179, 197, 201, 209, 224, 227
Minerotrophic wetland, 73, 218
Mire, 54, 64
Mobilization, 169
Model
computer simulation, 299, 359
decomposition, 147–150
freshwater marsh function, 302, 304, 305
hydrodynamic, 307–311
input-output, 169, 170, 186
nutrient, 205, 207, 208
nutrient flow, 107–111
nutrient flux, 230
nutrients, wastewater, 326, 327
wetland ecosystem hydrology, 65, 72, 73

Molinia caerulea, 58, 59, 61, 224
Morphology, plant, effect on
chemical composition, 163
decomposition, 131, 132
Mortality, 4, 5, 40, 79, 80, 83, 208, 243, 246,
254, 256, 261, 277, 291–293, 295, 296
leaf, 16, 18, 103, 111, 177, 178, 180
shoot, 39, 41–43, 45–47, 49, 99, 102, 103,
108, 109, 111, 177, 178, 180
Muskrat, 23, 104, 106, 261, 268, 270, 273,
275, 277, 278, 280, 358
reproductive
rate, 275, 277
success, 275
Myocaster coypus, 273
Myrica gale, 58
Myriophyllum, 139, 141
exalbescens, 23
heterophyllum, 135–138, 140, 161
spicatum, 158

N

Najas, 139, 141
flexilis, 23, 25, 135–137, 140
guadulapensis, 161, 162
Narthecium ossifragum, 58, 224
National Environmental Policy Act, 341,
346
National Wild and Scenic Rivers System,
347
Navigation, 306, 345
Nelumbo lutea, 161, 162
Nematode, 117
Niche, 66, 74, 203, 270, 274, 276
Nitella, 23, 25
Nitrate, 178, 188, 196, 198, 201, 211, 217,
223, 227–229, 233, 235, 243, 246, 248, 251,
254, 255, 261, 329–331, 333
Nitrification, 117, 197, 201, 229, 321, 328,
332, 361
Nitrite, 197, 250
Nitrogen
availability, 53, 58, 117, 212, 217, 222, 259
budget, 211, 212, 233, 234
change during decomposition, 95, 96, 100,
107–110, 115, 124, 125, 140, 141
dissolved organic, 178, 261
dry fall, 187
enrichment, 59–61, 229, 232, 235
exchangeable, 232
export from wetland, 189, 233, 261, 314,
315

fixation, 187, 261, 326
flux, 76, 99, 187, 189, 190, 207, 208, 211, 212, 230–232, 235, 236, 256, 262
gas, 324
in groundwater, 189, 330
immobilization, 179, 197
in interstitial water, 189, 228
leaching from litter, 106, 127, 256, 260
leaching from plant, 178, 260
as a limiting factor, 172, 209, 220, 222, 262
in litter, 89, 96, 99, 105, 131, 132, 179, 232
loss from sediment, 211
mineralization, 197, 201, 209, 227, 259
in peat, 217, 221, 223
in plant, 89, 91, 156–161, 163, 170–177, 180, 187, 189, 195, 203, 204, 206, 208, 210, 217, 224–227, 232, 255, 259, 321, 329
plant accumulation, 175–177, 205, 206, 209, 210, 217, 232, 243, 259, 324
sink, 191, 243
in soil, 188, 200, 209, 210, 212, 217, 321
turnover, 217, 231, 255
in wastewater, 324, 334
in water, 35, 188, 189, 197, 198, 211, 228, 243, 246, 248–256, 260, 314, 331, 360
in wet fall, 187, 233, 235
wetland retention, 188, 260, 324
wetland uptake, 205, 212, 329, 330, 337
Northeast River, Maryland, 287
Northern wetland
decomposition, 115–128
nutrient dynamics, 217–237
primary production, 53–61
Nuphar, 5, 11, 13, 82, 139, 141, 256
advena, 4, 6, 90, 162, 244
luteum, 89, 90, 94, 161
variegatum, 135–137, 140
Nursery species migration, 304
Nutrient, 155–263, *see also* specific nutrients
accumulation, 175, 177, 178, 180, 184, 205, 243, 324, 330
analysis, 102, 181, 182, 219, 220, 245
assimilation, 197, 321–337, 358
availability, 65, 74, 80, 116, 117, 202, 205, 210, 217, 221–223, 232
effect on standing crop, 53, 56, 58, 68, 81, 172, 217, 232, 259, 260, 333
budget, 76, 169, 170, 184, 187, 205, 211, 218, 233, 234, 261
burning, effect of, 277
compartment, 207, 208, 218, 231, 232, 321, 326–330

concentration, 146, 259, 260
plant, 91, 203, 224, 327, 333
water, 197
conservation, 180, 190
cycling, 49, 71–75, 90, 100, 108, 112, 170, 178, 186, 195, 196, 199, 208, 217, 218, 231, 259–262, 276, 277, 279, 314, 324, 325, 327
experimental enrichment, 58–61, 230, 232, 235, 236, 262, 278, 279, *see also* Fertilization
exchange, 261, 262, 325, 326
flow, 107–111, 178, 187, 189
flux, 64, 99, 100, 104, 108, 196, 199, 202, 204, 210, 230, 231, 233, 236, 256
input, 76, 186–191, 199, 218, 222, 224, 232, 234, 235, 260, 304, 326–328
load, 71, 188, 244, 329, 334, 335
loss, 187, 190, 231, 232
movement
freshwater tidal marsh, 243–256
lakeshore marsh, 169–191
output, 199, 211, 218, 234–236, 256, 260
pool, 304, 324
plant, 170, 175, 182, 183, 230, 232, 261
sediment, 259, 324
pump, 177, 186, 208, 212, 260
reallocation, 177, 178, 180
release from
ice, 198
litter, 99, 100, 106, 108, 109, 111, 112, 124, 146, 261
plant, 176–179, 199, 261
sediment, 201
reserve, 170, 171, 184
seasonal pattern, 111, 158, 169–171, 176, 177, 182–186, 195, 197–200, 202–204, 209, 212, 222–225, 227, 228, 243–256, 326–329
sink, 100, 170, 179, 180, 191, 196, 199, 201, 231, 243, 244, 255, 260, 322, 325, 328, 330, 358, 359, 361
standing stock, 172–176, 178, 182–186, 191, 255
storage, 176, 199, 236, 321, 326, 327–329, 331
transformation, 324–326, 335, 361
translocation, *see* Translocation
uptake
litter, 99, 100, 111, 146
microorganism, 199, 261
plant, 61, 106, 159, 169, 170, 172, 176–178, 181, 185, 186, 195, 205, 217, 228, 230–232, 243, 259, 261, 321, 324, 331

Nymphaea odorata, 161, 162, 334
Nymphaeaceae, 156

O

Odocoileus virginianus, 277
Oedogonium, 163
Olney three-square, 4
Ombrogeneous, 218
Ombrophilous, 218
Ondatra zibethicus, 267, *see also* Muskrat
Organic
  accumulation, 72, 75
  acid, 117
  carbon export, 304
  colloid, 325
  concentration, sediment, 70, 71
  constituent, 163, 164
  deposition, 63, 64, 70, 72
  export, 70, 72, 75
  flux, 63, 70
  matter, 200–202, 211, 221
    export, 302
Orontium aquaticum, 288
Osmunda, 287
Outflow regulation, 331, 332
Overwintering green shoot, 41, 45, 46, 49,
  80, 170, 182, 190
Oxidation reduction, 117
Oxygen, 64, 243, 246, 247, 250, 251, 253, 254
  availability, 65, 117, 131, 132, 134, 135,
    148, 325, 326
  biological demand, 314, 331, 336, 337
  depletion, 70
  saturation, 64

P

Palustrine wetland, 323, 324
Panicum
  hemitomon, 161, 162, 303
  virgatum, 9
Papyrus, 187, 188
Par Pond, South Carolina, 161, 162
Parasite, 83, 111
Particulate matter, 64, 131, 132, 137, 261,
  300
Pathogen, 83, 111
Peak
  nutrient standing stock, 172–176
  standing crop
    fen, 226, 231, 232
    freshwater tidal wetland, 3, 5–11, 13,
      16, 17

prairie glacial marsh, 21, 24, 29, 33, 34,
  103, 106, 108
  temperature, effect on, 56
Peat, 53, 71, 81, 116, 217, 220–223, 228–232,
  234, 329
  accumulation, 70, 72, 73, 75, 146
  filter, 321, 336
  turnover, 115
Peatland, 321, 325, 329
  minerotrophic, 53, 54, 73
Peltandra, 5, 6, 10, 11, 13, 14, 16, 17, 82,
  84, 256
  virginica, 17, 89–96, 244, 285–291, 294
Perched bog, 74, 76, 234
Percolation, 72, 74
Pest control, 305
Pesticide, 65, 70
Petiole, 156
pH, 64, 71, 117, 159, 200, 221–229, 325, 326
  nutrient availability, 58, 61
Phalaris, 25
  arundinacea, 8, 202, 206, 211
Phenology, 83
Phenophase, 205
Phenotypic variation, 203
Phosphate, 178, 217, 230, 235, 243, 249–254,
  314, 325
Phosphorus
  allocation, 169, 170, 180–186
  availability 53, 117, 195, 202, 212, 220–
    222, 259
  budget, 184, 205, 207, 208, 233
  change during decomposition, 100, 108,
    110, 111, 115, 125, 126
  cycle, 325
  enrichment, 59–61, 232, 235
  export from wetland, 188, 189, 211, 232,
    233, 260, 314, 315, 331
  flux, 187, 190, 198, 199, 207, 212, 230, 231,
    235, 236, 243
  in groundwater, 189
  in interstitial water, 189, 229
  leaching from litter, 99, 124, 179
  leaching from plant, 127, 178, 189, 232
  as a limiting factor, 58, 217, 220, 260, 262
  in litter, 105, 232, 329
  loss from sediment, 187, 211
  in peat, 217, 221, 223
  in plant, 156–161, 163, 169, 170–174, 176,
    178, 180, 182, 183, 187, 195, 203, 204,
    206, 209, 224–227, 232, 321, 328, 329
  plant accumulation, 106, 169, 170, 175,
    177, 181, 184–186, 205, 210, 217, 232,
    243, 259

reallocation, 184–186
seasonal stock, 176
sink, 199, 231, 243, 325
in soil, 159, 200, 201, 210, 212, 220, 224, 231, 260, 333
storage compartment, 328–330
total dissolved, 228, 229
translocation, 180–186, see also Translocation
in wastewater, 324, 325, 330, 331, 333, 334
in water, 35, 159, 181, 188, 189, 197–199, 228, 229, 249–254, 260, 325
wet fall, 187, 233, 235
wetland retention, 218, 330
wetland uptake, 336, 337
Photosynthesis, 199
Phragmites, 5, 25, 47, 82, 84, 124, 175–178, 180, 269, 270, 277, 331, 335
  communis, 7, 24, 25, 30, 31, 34, 66, 90, 122, 158, 171–173, 189, 202, see also Common reed
Phytin, 210
Phytoplankton, 199
Picea, 221, 234, 323
Pickerelweed, 90, see also *Pontederia cordata*
Pioneering, 274
Pipeline ditch, 314
Pistia stratioides, 164
Pithophora kewensis, 158–160, 162, 163
Plant harvesting, 321, 330, 333, 334
Plant part, effect on chemical composition, 156
Plant-soil-water interaction, 208–210
Planting, 278
Poa arctica, 57
Pocket marsh, 63
Podilymbus podiceps, 270
Pollutant, 305, 347, 348, 360
Pollution, 35, 345
  control, 198, 208
Polygonum, 6, 22, 25, 206, 287
  arifolium, 13, 16, 17, 244, 334
  lapathifolium, 206, 211
  punctatum, 13, 17
  sagittatum, 244
Pontederia, 5, 6, 10, 13, 84
  cordata, 89, 90, 94, 161, 162, 285–290, 293–296
Population shift, 275
Populus, 234, see also Aspen
  tremuloides, 218
Potamogeton, 23, 323

diversifolius, 161, 162
  pectinatus, 25
Potassium
  availability, 53, 195, 202
  budget, 234
  change during decomposition, 100, 108, 109, 111, 115, 126, 127
  enrichment, 59, 235
  flux, 207, 235, 236
  in interstitial water, 228, 229
  leaching from litter, 99, 106, 124, 233
  leaching from plant, 178, 260
  as a limiting factor, 58, 220
  in litter, 105, 179
  in peat, 221–224, 231
  in plant, 156–160, 171–174, 180, 195, 199, 203–210, 224
  plant accumulation, 175–177
  in soil, 159, 200, 202, 210, 212, 223
  in water, 159, 197–199, 212, 360
  wet fall, 233
  wetland retention, 218
Pothole, 277
Prairie glacial marsh
  litter decomposition, 99–112
  production, 21–36
  vegetation cycle, 22, 23, 32, 33
Prairie pothole marsh, 269, 275
Precipitation, 101, 107, 109, 358, see also Wet fall
Primary production, 79–85, 170, 299, 311, 358
  cypress swamp, 75
  estimation of, 4, 12, 33, 34, 40, 41, 49, 55, 79, 80, 83–85, 178, 181
  freshwater wetland, 81
  Louisiana, coastal marsh, 303
  northern bog marshes, 53–61, 80
  prairie glacial marsh, 21–36, 80
  sedge wetland, 39–50, 80
  tidal freshwater wetland, 3–18, 80, 90, 255
Productivity
  hydrology, effect on, 63, 68–70, 80
  limiting factors, 53, 56, 58, 65, 66, 68, see also Nutrient availability, effect on standing crop
  management, 268, 271
Protein, crude, 163, 164
Protoplasmic material, 163
Proximate analysis, 163, 164

Q

Quercus, 234

R

Radiophosphorus uptake, 181, 208
Rainfall, 222, 260, 304
Raised-convex marsh, 73, 74
Ranunculus, 47
Recharge, 188
Redox potential, 117, 149, 201, 229, 325, 326
Red-winged blackbird, 274, *see also Agelaius phoeniceus*
Reference quadrat technique, 40
Reflooding, 273, 280
Regulation, freshwater wetland, 341–354, 358, 359
Remote sensing, 276
Reptile, 303, 304
Respiration, 116, 117, 146, 148
Revegetation, 272, 279
Rheophilous, 218
Rhizoclonium, 246, 253, 330
Rhizome, 156, 171, 172, 176, 181–184, 226
    decomposition, 111, 147, 208
    development, 43
    production, 14
Rhizosphere, 149, 210
Rhynchospora, 221
Riccia fluitans, 32
River and Harbor Act, 341, 344, 348
Riverine wetland, 323
Root, 156, 171, 176, 177, 181, 184, 208, 226
    decomposition, 97, 111, 146, 147
    development, 43
    growth, 61, 224
    morphology, 43, 44
    production, 10, 14
Root shoot ratio, 29, 44
Rubus chamaemorus, 57
Rumex, 22
Runoff, 53, 74, 181, 186, 188, 197, 199, 222, 251, 254, 260, 299, 305, 313, 360

S

Sagittaria, 5, 13, 14, 16, 84
    falcata, 303
    latifolia, 5, 8, 15, 17, 22, 23, 25, 26, 57, 89, 90, 94, 101, 161, 162, 244
Saline
    marsh, 69, 89, 303, 304
    wetland, productivity, 4, 10, 17, 80
Salinity, 4, 96, 277, 288, 300, 303, 312, 313, 315
Salix, 115, 121, 208, 218, 270, 329, *see also* Willow

bebbiana, 57
discolor, 119, 225
herbacea, 58
lapponum, 58
lucida, 225
interior, 195, 208
pedicellaris, 119, 225
pellita, 119
serissima, 57
subserica, 225
Salt
    dissolved, 199
    marsh, 70–71
    marsh cordgrass, 4, *see also Spartina alterniflora*
Saltmeadow cordgrass, 4, *see also Spartina patens*
Salts, 69, 70
Saururus cernuus, 158, 159
Sawgrass, *see Cladium jamaicense*
Scanning electron photomicrograph, 148
Schoenus nigricans, 58
Schoeonoplectus lacustris, 174
Scirpus, 5, 9, 58, 82, 84, 204, 221, 269, 287, 333, 334, 336
    acutus, 24, 25, 30, 31, 135, 136, 139–141, 270, 335
    americanus, 161, 172, 174, 288
    fluviatilis, 22, 24–28, 30, 31, 33–36, 101–111, 180, 195, 202, 205–210, 335
    lacustris, 335
    olneyi, 4
    subterminalis, 135–137, 139–141
    validus, 17, 22–29, 31–35, 99, 101, 102, 105–107, 288, 328, 335
Season, 91
Seasonal biomass pattern, 4, 11–15, 17, 26–31
Secondary production, 90, 97
Secretion, 148, 327, 333
Sedge, 120–126, 227, 232, 270, *see also Carex*
    meadow, 53, 56, 63
    shrub, 68
    swale, 68
    wetland, 39–50, 54, 61
Sediment, 71, 72, 91, 199, 201, 233, 259–261, 288, 296, 304, 314, 324, 325, 330
Sedimentation, 72, 90, 142, 149, 170, 305
Seeding, 285, 287, 293, 296
Senescence, 13, 14, 23, 55, 102, 140, 148, 169, 171, 180, 182, 183, 185, 260, 261
Sewage, 61, 199, 244, 246, 250, 254–256, 260, 315, 321, 329, 331, 333, 335

Shoot base, 172, 176, 181–184
  development, 41, 43
  emergence, 40, 41, 45, 46
  growth, 171, 184
  population dynamics, 45–47, 49
Shorebird, 335
Shore erosion, 286
Short-billed marsh wren, 275, 278, *see also*
  Cistothorus platensis
Short-term marsh change, 272
Silicon, 199, 260, 314
Silt contamination, 111
  input, 70
  trapping, 72, 73
Sloughing, 177, 184
Smooth cordgrass, 68, 69, *see also Spartina
  alterniflora*
Snail, 261
Sodium, 99, 105, 106, 124, 156, 158–161,
  163, 171, 173, 174, 178, 179, 199, 202, 221–
  224, 260
Sodium enrichment, 60
Soil, 201–202
  alfisol, 201
  bulk density, 223
  cation-exchange capacity, 202, 220–222
  cations, 209
  chemistry, 220–223, 276
  density, 69
  fauna, 117
  forest, 209
  histosol, 118, 220
  mollisol, 201
  nutrients, 210, 217
Soligeneous, 218
Solubilization, 117
Sorption, 229, 230, 232, 236, 321, 330, 361
  potential, 223
South Carolinian Lakes, 175
Sparganium, 84, 178
  eurycarpum, 22–28, 30, 31, 33–36, 99,
    101–103, 105–107, 202, 204, 206, 335
Spartina, 25, 84, 122
  alterniflora, 4, 9, 11, 68, 69, 79, 80, 92–
    96, 303
  cynosuroides, 4, 5, 7, 9, 82, 96
  patens, 4, 8, 11, 82, 303
Spatial heterogeneity, 66
Spatterdock, 4, 90, *see also Nuphar*
Species
  composition, 63, 66, 112, 276, 299, 312,
    334, 335, 358
  diversity, 4, 196, 267, 303
  richness, 63, 64, 66, 67, 68, 267, 270, 271

Sphagnum, 58, 64, 67, 75, 221–224, 323
  fuscum, 224
  heath, 67
  papillosum, 58
Spirodela
  oligorhiza, 334
  polyrhiza, 32
Spirogyra, 163
Spoil, 314, 358
  bank, 309–312
  disposal, 305
Spruce, 223
  forest, 223
Stability, 271
Stachys hyssopifolia, 288
Stand
  history, 81
  location, effect on chemical composition,
    156, 157
Standardization, 341, 350–353
Standing litter compartment, prairie glacial
  marshes, 102–106, 109
Starch, 184
Streamflow, 260, 359
Stress, 70, 294, 295
Submergents, 272, 277
Subsidence, 305, 312
Substrate, 65, 66, 91, 159, 285, 287, 288, 294,
  303, 321
  effect on productivity, 61
Subsurface flow, 197
Succession, 72, 73, 90, 267, 271, 276, 300
  seasonal, 13
Sulfate, 100, 197, 199, 201, 217
Sulfur, 158, 160–162, 170, 172–175, 211, 261,
  262
Sulfuric acid, 314
Sunken-convex marsh, 74
Surface-groundwater interactions, 323, 324
Survival, *see* Mortality
Suspended matter, 199
Swamp, 217, 236
  conifer, 224
  forest, 67, 304
Sweet flag, 13, *see also Acorus*
System management, 275

T

Tannin, 164
Taxodium, 330
Temperature, 53, 56, 96, 101, 109, 131–134,
  137, 139–141, 146, 159, 184, 185, 228, 229,
  328
  effect on standing crop, 47, 81

Theresa Marsh, Wisconsin, 196–201, 203, 204, 206
Throughflow water, 170, 186, 188
Tidal
    action, 243
    amplitude, 288, 289, 296
    current, 312
    cycle, 245, 250–255, 307
    elevation, 70, 285, 287–296
    marsh, 73–75
    subsidy, 11
Tinicum Marsh, Pennsylvania, 255
Tissue nutrient level, 170–174, 182–184
Toppling, 103, 148
Toxin, 63–65, 69, 83, 304, 360
Translocation
    carbohydrate, 55, 185
    nitrogen, 108, 180, 243, 256
    nutrients, 112, 171, 176, 177, 179, 180, 187, 189, 190, 208, 217, 225, 261, 328
    phosphorus, 110, 169, 177, 178, 180, 182–186, 226, 243
    potassium 111, 180, 205
Transportation, 306
Trapping, 306, 360
Trichophorum, 221
    caespitosum, 58
Trophic structure, 304
Tropical swamp nutrients, 187, 188
Transplanting, 285, 287, 290, 293, 296
Tundra, 53
    meadow, 54, 226
    pond marsh, 176, 178, 179, 187, 189
    soil, 175
Tupelo, 64
Turbulence, 64
Turnover
    leaf, 5, 123, 255
    nitrogen, 217, 231
    nutrient, 112, 177, 208
    plant, 24–28, 34, 35, 47, 83, 116
    stem, 123
Typha, 4, 5, 7, 9, 11–14, 22, 23, 47, 80, 82, 84, 124, 160, 181, 203, 204, 221, 256, 260, 269, 287, 323, 329, 333, 335, *see also* Cattail
    angustifolia, 34, 122, 158, 173, 189, 244
    glauca, 23–28, 30, 31, 33–36, 99, 101
    latifolia, 17, 90, 157–162, 165, 169–172, 175–182, 184, 186, 197, 202, 206, 209, 210, 218, 244

U

Umbel, 156
Urban
    development, 307
    runoff, 314, 315, 317
U.S. Fish and Wildlife Coordination Act, 345
U.S. Fish and Wildlife Service, 345
Utricularia, 334
    inflata, 161, 162

V

Vaccinium uliginosum, 57
Vallisneria, 323
Vegetation
    cycle, 21, 22, 32, 33
    macrofauna association, 274
    pattern, 267
    zonation, 66, 269, 270
Vitamin, 164
Volatilization, 117

W

Warbler, 270
Washout, 321
Waste treatment design, 336, 337
Wastewater, 321–323, 324–326, 328, 331, 335, 337, 361, 362
    application method, 331–333
    biological treatment, 286
Water
    capillary flow, 74
    chemistry, 64, 100, 223, 227–230, 276
        seasonal, 218, 227–230, 246–256
    circulation, 307
    depth, 159, 269
    flow, 72, 74, 80, 199, 232, 254, 260, 299, 300, 302, 305, 307–309
    level, 272, 274–277, 300, 307, 308, 324, 335, 358
    permanence, 271
    purification, 322
    quality, 299, 305, 326, 345, 348, 359
    regime, 267
    renewal rate, 63, 64, 69, 70
    source, 63, 64, 71, 74
    stress, 68
    table, 53
    velocity, 63–66, 69, 70, 74, 196, 307, 308

Waterfowl, 100, 268, 269, 304, 335, 344, 360
  habitat, 211
  hunting, 268
Water Resources Development Act, 286
Waterlogging, 44, 53, 58, 61, 81, 117, 220,
  223, 224
Wave
  action, 291, 299, 312
  stress, 285, 287, 293
Wet
  fall, 76, 186, 187, 260
  tundra macrophytes, 180
Wetland
  classification, 73–76
    wildlife, 269–271
  destruction, 344
  elevation, 47, 66, 72, 73
  filling, 350
  heterogeneity, 269, 271, 276
  legislation, 343–349
  loss, 358
  maintenance, 306
  management, 150
    canal effect, 299–318
  mapping, 350, 353
  plant, chemical composition, 155–165
  preservation, 268, 278, 350
  protection, 268, 279, 344, 346
  restoration, 286
  terms, 218
  utilization
    energy extraction, 306
    environmental impacts, 305
    fishing, 279, 306
    mineral extraction, 306

  recreation, 306, 345
  tourism, 306
  use-issue categories, 306
  value, ecological assessment, 354
Wildlife, 281, 335, 345, 357, 358
  behavioral response, 268
  habitat, 286, 360, see also Habitat
  management, 267–281, 350, 359
  needs, seasonal, 270
  population response, 268, 280
  reproductive cycle, 270
Wild rice, 4, see also Zizania aquatica
Willow, 120–127, 337, see also Salix
Wind, 103, 104, 287, 307
Wood, 208

X

Xanthocephalus xanthocephalus, 270, see
  also Yellow-headed blackbird
Xanthophyll, 164

Y

Yellow pondlily, 90, see also Nuphar
Yellow-headed blackbird, 274, see also
  Xanthocephalus xanthocephalus
York River, Virginia, 90

Z

Zinc, 158, 160–162
Zizania, 5, 10, 11, 13, 16, see also Wild rice
  aquatica, 4, 7, 12, 17, 75, 89, 90, 93, 94,
  244, 334
Zooplankton, 199